137
The Riddle of Creation

The Torah and Science Series

137

THE RIDDLE OF CREATION

Rabbi Yitzchak Ginsburgh

ג	ל	ע	י	נ	י
ו	א	ב	י	ט	ה
נ	פ	ל	א	ו	ת
מ	ת	ו	ר	ת	ך

Gal Einai

137: The Riddle of Creation
Rabbi Yitzchak Ginsburgh

Research and editing by Rachel Gordon

Printed in the United States of America
First Edition

For information:

Israel: Gal Einai
PO Box 1015
Kfar Chabad 6084000
tel. (in Israel): 1-700-700-966
tel. (from abroad): 972-3-9608008
email: books@inner.org

Web: www.inner.org
Twitter: @RabbiGinsburgh

Cover design: Ben Gasner

Design and Layout: David Hillel

ISBN: 978-965-532-042-8

TABLE OF CONTENTS

INTRODUCTION

Viktor Weisskopf was former assistant to quantum physicist Wolfgang Pauli. Gershom Scholem was a professor of Jewish mysticism at the Hebrew University in Jerusalem. When they met, Scholem asked Weisskopf to name one of the deepest unsolved problems of physics. Weisskopf replied, "Well, there's this number 137…" In a flash of wisdom, Scholem responded, "137 is the numerical value of Kabbalah (קַבָּלָה)!"

The Enigma

Imagine you have all the pieces of a gigantic jigsaw puzzle. You have succeeded in putting some of the pieces together to form part of the picture. It depicts a landscape. Some other pieces form another part of the puzzle. It depicts a still life. It seems to you that these two parts must belong to different puzzles, yet you know that this cannot be so. How can you ever complete it?

In both parts of the puzzle (the landscape and the still life) there are some pieces that cannot be moved. These are the pieces that will guide you to the eventual solution.

As you continue to piece the puzzle together, you come across another enigma, which completely baffles you. In the center of the puzzle there are pieces that apparently depict a portrait of a human face. This portrait is a mirror image of you.

Such a puzzle is a metaphor for the quandary of modern physics. The ultimate goal of physics is to discover an elegant "theory of everything" that will unite all four forces of nature. The particles necessary to unite three of these forces are accurately explained by the standard model of quantum mechanics. Gravitation, the fourth force, is adequately described by the general theory of relativity. Just as no apparent connection exists between the landscape and the still life in your jigsaw puzzle, no solution has been verified that makes the standard model and general relativity mathematically compatible.

> *…the standard model cannot be a complete or final theory because it does not include gravity. Moreover, attempts to incorporate gravity into its quantum-mechanical framework have failed due to the violent fluctuations in the spatial fabric that appear at ultramicroscopic distances — that is distances shorter than the Planck length. The unresolved conflict has impelled a search for an even deeper understanding of nature.[1]*

A century of rigorous research has almost unequivocally proven quantum mechanics at the microscopic level. However, the results of the research imply that the consciousness of the human observer is also a factor in obtaining the results. This is the self-portrait that is becoming apparent in the puzzle you are trying to piece together.

The Scientific Method

The general procedure for discovering how the world works is first to gather data through observation. When a query arises about the facts that have been accumulated, a preliminary hypothesis is proposed that offers a reasonable solution to the query. Predictions (or logical consequences) are then derived from the hypothesis. The next stage involves testing the hypothesis by conducting experiments based on those predictions. A scientific hypothesis must be refutable; otherwise it cannot be meaningfully tested. A recurrent back-and-forth interplay between observable facts and experiment follows. In order to validate a theory as a law of nature, it must fit with all facts all the time.

Relativity and quantum theory predict undeniably counter-intuitive phenomena. Since their discovery, it has become necessary to complete the scientific method by including an additional level. Once a hypothesis has been formulated to adequately explain the facts, a metaphysical interpretation of reality must be presented that offers a rational explanation for the phenomenon. This level has always been

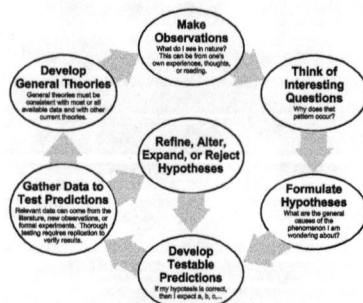

The Scientific Method as an Ongoing Process

2

present in the scientific method but, until the advent of quantum mechanics, it was not always a necessary part of the total structure. Scientists in our era are thus forced to become not just physicists, but metaphysicists.

Generally speaking, science believes in a concrete, phsyical world. Upon investigation, however, scientists have discovered—to their chagrin—that the wave equations of quantum mechanics may not define a tangible reality. This question of existence is the ultimate philosophical query. As new theories are being hypothesized, the scientific community finds itself increasingly troubled by such meta-physical questions.

In contrast, one essential philosophical query of Judaism is whether or not *we* exist in the world that God created.[2]

Torah and Science: Induction vs. Deduction

The early roots of the scientific method lie in deductive reasoning. Deductive reasoning derives conclusions from first principles that are accepted as truth. For example, "Whenever it rains, it is cloudy; it is raining, therefore, it must be cloudy." If the general premise is true and the inference drawn from that premise follows the rules of logic, deduction holds that the conclusion must also be true. Ancient natural philosophers began with such "absolute" axioms and deduced new statements implied by them.

In order to disprove such conclusions, the opposition must disprove the original premise (in our case, we would need to prove that it sometimes rains, even when there are no clouds visible). Thus, it is very difficult to oppose cases of deductive reasoning.

Modern science has refined its desire to always be correct. Nowa-days, the scientific method depends exclusively on inductive logic to come to conclusions. Induction attempts to abstract general principles from specific observations. However, the conclusions of such logic are *uncertain*. For example, "All swans we have observed are white;

therefore, all swans must be white." Although the conclusion is statistically reasonable, it is not certain. The next swan we observe may be black.

Basing our understanding upon induction requires humbling ourselves with the knowledge that our theories are subject to revision. Discovery of new facts or phenomena may completely overthrow our earlier theoretical model.

Nonetheless, theorists continue to devise new hypotheses that suggest that there is an absolute axiom from which all natural phenomena can be deduced—the elusive Theory of Everything. Thus, much of scientific inquiry today is not an objective search for the underlying rules that govern nature, but an attempt to prove or disprove a proposed hypothesis.

The form of induction adopted by science relies heavily on two premises: a) that which has happened consistently in the past will continue to happen in the future, and b) that which happens under observation happens even when unobserved.

However, neither of these premises can ever be proven absolutely correct. In fact, both of them are forms of inductive reasoning themselves, and as such, they can only be refuted. For this reason, induction has been called "the mother of all problems." Belief takes over where rational proof comes to an end.[3] So, although most scientists would blatantly deny it, the fundamental belief system of science is induction.

Ultimately, the purpose of science is to deduce everything from first principles. However, by necessity, it must pay homage to induction, which locks its aspiration to rely on deduction in a vicious cycle.

By contrast, Torah study is founded on humble submission to the Creator, and the belief that all of creation can be deduced from His Torah. Nonetheless, the sages reason inductively: first, they study the verses pertaining to the matter at hand together with the sayings of the sages who preceded them, and then they generalize. Each sage offers new information from Torah sources and the principle is adapted accordingly. From the general conclusion, new particulars can be deduced. Thus, Torah reasoning relies on both induction and deduction, but *necessarily* in that order: first, induction (the essence of Torah reasoning), then deduction (logical analysis of principles in order to yield particular statements).

In short, science has faith in induction, but aspires to a state of deduction, which breeds disparity. Judaism is founded upon faith in God, and so it aspires to induction and unity. In this sense, scientific wisdom is diametrically opposed to Torah wisdom.

Furthermore, because science is based on human logic, it is fallible. Even Albert Einstein made mistakes. By his own admission, his greatest blunder was that he denied the implications of his own general theory of relativity and the inescapable conclusion that the universe is expanding. This proves the risks that even the greatest scientist takes when working with a particular theory. By contrast, a great Torah scholar may be mistaken in his reasoning, but the accepted tradition of the Torah is always triumphant.▸

▸ Any innovative interpretations of the Torah that do not stand in line with accepted tradition must be avoided and rejected at all costs.

From a different angle, science reflects the attribute of judgment. Like a suspect in a crime who is "innocent until proven guilty," scientific theories are constantly subjected to harsh examinations until the "prosecution" collects enough evidence against them to enable their refutation. Similarly, scientific knowledge aspires to be exact and precise, like the attribute of judgment.

By contrast, the Torah is "the Torah of loving-kindness." It begins with clothing the naked[4] and concludes with burying the dead,[5] both exemplary acts of loving-kindness.[6] The Torah teaches us to always judge people favorably, against all odds. Even when the evidence seems obviously incriminating, we must allow for the most unlikely extenuating circumstances.[7]

Such dramatic differences seem to negate any possible connection between the two domains. Torah and science appear so diametrically opposed to one another that reconcilitation between the two seems highly improbable.

Yet, science sincerely desires to discover the unity inherent in nature.

And the ultimate purpose of the Torah is the realization of the verse, "…on that day, God will be one and His Name will be one,"[8] as Jewish people have proclaimed since time immemorial, "*Havayah* is our God, *Havayah* is one." The pursuit of unity is thus common territory where Torah and science can meet. The union of Torah wisdom and scientific wisdom would be a giant step in the right direction for both.[9]

The Marriage of Torah and Science[10]

> God said, "Let there be a firmament in the water and let it separate waters from waters." God made the firmament and separated the waters beneath the firmament from the waters above the firmament.[11]

The Torah verses describing creation contain profound mysteries that conceal the essence of Divinity. Kabbalah uncovers those mysteries using metaphors, allegories and parables. In Kabbalah, water is a metaphor for wisdom. Read from this perspective, these Torah verses teach us that, at the beginning of creation, all the water/wisdom in the universe was in an elemental state. This initial stage represents the primeval continuum of wisdom—the primordial Torah.

From our limited intellectual perspective, the primordial Torah has divided in two: the Divine wisdom of the Torah (as revealed to the millions of Jews who received it at Mt. Sinai), and secular wisdom, which defines the laws of nature. The higher waters represent Divine wisdom (Torah) and the lower waters represent mundane wisdom (science).[12]

These two types of wisdom were not created equal. The prescriptive commandments of the Torah sanctify and elevate mankind, bringing us to a closer connection with God. By contrast, the lower waters—the secular wisdom of science—define reality. However, they lack the ability to perfect life morally, psychologically and spiritually. In addition, their very presence conceals the Almighty and even promotes denial of His existence. The Zohar describes metaphorically how the lower waters, scientific wisdom, weep before God, crying, "We too want to stand before [i.e., be able to serve] God."[13]▸

This division between Torah and science is temporary. Science too will eventually contribute to perfecting mankind's moral and spiritual state and become a tool for serving God.

The Zohar[14] states further that the higher waters are masculine and the lower waters are feminine. In order to become productive, they must unite in "marriage."

▸ The Modern Hebrew word for "science" (מַדָּע) is a permutation of the word for "weeping" (דְּמַע). This illustrates an image of science weeping because it has been severed from the Divine.

Making the Match

The Torah is refined wisdom that emanates directly from the ultimate level of spiritual unity. By contrast, scientific wisdom stems from mundane reality. It relies on constant proofs to clarify its findings. Science is constantly being revised and updated, as old theories are replaced by newer ideas. Nonetheless, science does not discard the old theories, but evolves by developing new theories that incorporate and unify previous theories. This is the refining process that science must undergo until it will eventually perfectly reflect its source in the primordial Torah.[15]

Because science is in many ways the very antithesis of Torah, many great Torah scholars shy away from it, unless it is utilized for purposes of livelihood, or is beneficial to one's service of God.[16] Yet, the Torah bestows upon science—in certain areas at least—validity much greater than contemporary science claims for itself. In many instances *halachah* (Jewish law) accepts scientific findings not as possible or probable, but as certain and true.[17] Similarly, science and nature have always been an important element in Jewish philosophy. Talmudic literature is replete with parables from nature. For example, the Jerusalem Talmud describes how two sages observed daybreak in the Arbel Valley. One of the sages turned to his colleague and said, "This is how the redemption of the Jewish People will be, at first it will be gradual, but the more it progresses, the stronger it will become."[18] Another apt parable in the Midrash describes how living to a ripe old age is like plucking a ripe fig: "When a fig is picked in its time, it is good for the fig and good for the tree. But, when it is picked unripe, it is bad for it and bad for the tree."[19]

Nature and science have served as a source of parables to explain profound spiritual insights throughout Jewish history. This is even more so since the founding of the Chassidic Movement by the Ba'al Shem Tov, and particularly in the literature of Chabad Chassidut. The general Chabad philosophy is that when meditating on esoteric concepts, there is no direct way to experience them. In this case, parables are the foundation upon which truth can rest. These parables allow us to grasp the profundity of a Divine world that is hidden from view. The general teaching of Chassidut is that, for every Divine concept, there is a tangible parable from the realm of our physical senses, and also a spiritual parable from the psychological realm. Moreover, Kabbalah and Chassidut teach us that within the outer casing of secular wisdom lie sparks of holy wisdom that are waiting to be redeemed. Our task is to refine those sparks by entering the realm of science and extricating them to make use of them in God's service.

Modern science is becoming more and more enigmatic, while the most profound Torah teachings of Kabbalah—particularly when dressed in their Chassidic garb—are becoming more and more accessible to humanity. The path to uniting the two is thus shorter than ever before. ◄

The union between Torah and science is a two-stage process. At the first stage, Torah scholars must descend to study science until they understand its basic principles. They must also familiarize themselves to a satisfactory degree with new discoveries that lie on the frontiers of science. By examining the scientific concepts from the perspective of their Torah knowledge, they will see how these concepts serve as precise parables for topics discussed in the Torah, Kabbalah and Chassidut. At this point, they will perceive direct correspondences between the two topics, and discover the underlying model that unites them.

The second stage further divides into two levels. The first

▶ There are four spiritual Worlds. Science deals with the three lower worlds, *Asiyah* (the World of Action), *Yetzirah* (the World of Formation) and *Beriah* (the World of Creation). The highest world is *Atzilut* (the World of Emanation). This is the world of absolute unity to which both science and the Torah ultimately aspire.

level—which will become possible only once the first stage is complete—is to find ways in which the Torah can contribute to science. Indeed, it may become possible to find Torah insights that will direct scientists to discover new scientific principles (some tentative suggestions are made in this book). The second level of the second stage is when scientific knowledge automatically becomes a parable for the issue at hand in the Divine realm. Thus, the new insights contributed by the Torah cross-pollinate between Torah and science to produce a fruitful union.

The Torah reflects God's opinions and desires. When God gave the Jewish People the Torah, He gave them His essence[20]—His very consciousness, so to speak. So, instead of relying on the as-yet unrefined imagination of scientists (see Chapter 3), as creative as it may be, the optimal method to arrive at a new law of nature is to look into God's blueprint of the world—the Torah.[21] Ultimately, this will rectify the problems that plague the scientific method.

The first step in the matchmaking process is falling in love with both Torah and science (see Chapter 11). We need to contemplate the beauty inherent in each of them and head for the horizon to reveal the most profound mysteries of them both. As we approach the horizon, we see the point where their two different types of beauty are essentially one. There, the higher waters of Torah meet the lower waters of science at their source in the primordial Torah. The innate harmony that can be perceived at this point is the intellectually esthetic realm into which we delve in this book. The beauty of both of them draws us closer to perceive their origin. At this level, the more correspondences that are discovered between the two realms, the closer we will come to confirming their single source.

The Physical Constants

In science, physical phenomena are analyzed by calculating the interactions between objects and by quantitative measurement of the effects of the forces that impact them. Such analysis relies on certain constants that have been accurately measured by scientists over the decades, such as the atomic mass of an element (m_a), the speed of light (c), or the earth's gravity (g).

Most constants in the physical world are measured in terms of arbitrary, man-made units such as centimeters, grams or seconds, etc. The numbers that appear in these constants depend on the units of measurement. For instance, measuring the distance of the earth from the sun in kilometers results in a significantly different number than when measuring the same distance in miles. However, there are a number of physical constants that are pure numbers. They remain the same under all circumstances. One common example of such a dimensionless number is *pi* (π), the ratio between the circumference and the radius of a circle. The dimensionless constants present in the equations are the key pieces in the quantum puzzle.

The standard model of modern physics requires 25 such dimensionless constants. One more constant is required for cosmology (relating to gravitation and general relativity), bringing this total to 26.[22]◄

The constant that heads the list of physical constants is "the fine-structure constant." It is referred to by the Greek letter, *alpha* (α).

Alpha is also the electromagnetic coupling constant. It is defined by the ratio between the speed of the electron (in a hydrogen atom) and the speed of a photon (i.e., the speed of light). As such, this constant defines all interactions between these two particles. In particular, the probability amplitude of a charged particle is proportionate to the square root of the fine-structure constant.

► 26 is the numerical value of the essential Name of God, *Havayah*.

$$\alpha = \frac{e^2 l \hbar c}{4\pi\varepsilon_0} = \text{approximately } 1/137$$

One intriguing idea in modern science is that there is a limit to infinity and a finite limit to size. The speed of light (c) is relative infinity, because movement at a speed that is greater than the speed of light is inconceivable. Similarly, Planck's constant (h) defines the smallest possible units of measurement, i.e., relative zero. The one pure dimensionless number that connects these two in the equation above is the fine-structure constant, *alpha*.

During the past century, many of the world's most prominent physicists, mathematicians and philosophers have made repeated attempts to derive the value of the fine-structure constant mathematically from first principles. Although some close matches have been achieved, none explains the reason why *alpha* should be the value that it is. As physicist Max Born wrote:[23]

> *There seems to be little doubt that the existence of this dimensionless number, the only one which can be formed from e, c and h, indicates a deeper relation between electrodynamics and quantum theory than the current theories provide, and the theoretical determination of its numerical value is a challenge to physics. The solution to this problem seems to be closely connected with a future theory of elementary particles in general. [However] all attempts have so far been in vain.*

Born's surmise that *alpha* would be closely connected with a future theory of elementary particles has been fully realized in quantum field theories. *Alpha* appears as the coupling constant that determines the observed strength of the interaction between electrons and photons at low energies. Similarly, in the electroweak theory—which unites electromagnetism and the weak force in the atom—*alpha* is absorbed into two other coupling constants associated with electroweak gauge fields.

Of all dimensionless constants, *alpha* is the most enigmatic. It crops up unexpectedly in many places for no apparent reason. The fine-structure constant has at least ten different physical interpretations:

- The square of the ratio of the elementary charge to the Planck charge
- The ratio of two energies: (i) the energy needed to overcome the electrostatic repulsion between two electrons a distance of

d apart, and (ii) the energy of a single photon of wavelength $\lambda = 2\pi d$ (or of angular wavelength d)

- The ratio of the velocity of the electron in the first circular orbit of the Bohr model of the atom to the speed of light in vacuum.[24] This is Sommerfeld's original physical interpretation. Then the square of α is the ratio between the Hartree energy (27.2 eV = twice the Rydberg energy = approximately twice its ionization energy) and the electron rest mass (511 keV).

- The two ratios of three characteristic lengths: the classical electron radius (r_e), the Compton wavelength of the electron (λ_e), and the Bohr radius (a_0):

- In quantum electrodynamics, α is directly related to the coupling constant determining the strength of the interaction between electrons and photons.[25] The theory does not predict its value. Therefore, α must be determined experimentally.

- In the electroweak theory unifying the weak interaction with electromagnetism, α is absorbed into two other coupling constants associated with the electroweak gauge fields. In this theory, the electromagnetic interaction is treated as a mixture of interactions associated with the electroweak fields. The strength of the electromagnetic interaction varies with the strength of the energy field.

- Given two hypothetical point particles each of Planck mass and elementary charge, separated by any distance, α is the ratio between their electrostatic repulsive force and their gravitational attractive force.

- In the fields of electrical engineering and solid-state physics, the fine-structure constant is one fourth the product of the characteristic impedance of free space, $Z_0 = \mu_0 c$, and the conductance quantum, $G_0 = 2e^2/h$:

- The fine-structure constant gives the maximum positive charge of the central nucleus that will allow a stable electron-orbit around it. For an electron around the nucleus with atomic number Z, $mv^2/r = \frac{1}{4}\pi\varepsilon_0 (Ze^2/r^2)$. The Heisenberg Uncertainty Principle momentum/position uncertainty relationship of such an electron is just $mvr = \hbar$. The relativistic limiting value for v is c, and so the limiting value for Z is the reciprocal of fine-structure constant 137.

♦ Graphene, a two-dimensional layer of carbon atoms, absorbs almost exactly $\pi\alpha \approx 2.3\%$ of low-frequency light that hits it.

♦ *Alpha* is related to the probability that an electron will emit or absorb a photon.

Because *alpha* is a dimensionless number, scientists on any planet on the universe with advanced intelligent life would recognize it, no matter in what units they had formed their equations.[26]

James G. Gilson, a contemporary scientist, explains the significance of *alpha* in detail:

> *The α mysteries… are fundamental and pervade the whole of theoretical physics from the smallest systems, elementary particles, to the very largest one, the universe. This is no mere speculative fantasy or science fiction extrapolation of imagination about the importance of α. The fine-structure constant impacts on our every day life here and now … If α were to suddenly be switched off, here on earth—out there in those massive astronomical objects and indeed everywhere, atomic systems would shed all their orbiting electrons as the Coulomb attraction reduced to zero … All atomic, molecular, biological systems would be destroyed in the process. All life including our own, all human aspirations, society and institutions would be consumed in the instant catastrophic fireball and there would be no record left that they had ever existed.* [27]

Because of its phenomenal significance, physicists regard *alpha* as a most important link in the discovery of a Grand Unified Theory (which is one step on the way to the Theory of Everything). If the essence of this number were known to scientists, it would greatly aid in putting all the parts of the puzzle together into one complete picture. This would be a momentous breakthrough and a significant milestone in the history of quantum physics.

Of all the dimensionless constants that have been discovered over the centuries, *alpha* has another unique quality. It is uncannily close to the inverse of a whole prime number: 137. This fact makes *alpha* even more intriguing to scientists and philosophers alike.

Precision is of utmost importance for the practical application of numbers in technology and research. The mathematician would like to know the value of *alpha* to an exact number of decimal places. Yet, modern physics holds that the entire physical world is based on single units of energy. This suggests that at some level, the natural world is founded on *whole* numbers.

The Torah in general takes into consideration only whole, positive integers and their quotients. Kabbalah in particular relates to numbers as they are generated by the Hebrew letters and various Names of God. As such, they serve as a tool for analyzing numbers qualitatively. Kabbalists incorporate number theory into their meditations by rounding off numbers and removing the decimal point. For instance, using this system, *pi* (π) equals 314.[28] Similarly, according to this method, the golden ratio, *phi* (ϕ), is 1618. This is why we consider the inverse of *alpha* to be exactly 137.

137 is the numerical value of Kabbalah (קַבָּלָה), which means "parallel" or "corresponding." The wisdom of Kabbalah is to find correspondences between the mundane and spiritual levels of reality, thus bridging the gap between them. Similarly, *alpha*, because of its natural proximity to a whole number, serves to bridge the gap between the preciseness of science and the quantization of Torah. Science is related to the natural cyclic world, which aspires to exactitude but can never truly reach unity. Torah is the direct light that shines into the world and inspires us toward true oneness.

As quantum physicist, Wolfgang Pauli wrote to his colleague,

Werner Heisenberg, in 1934, "everything will become *beautiful* when [*alpha*] is fixed."[29]

Language of Science; Soul of Kabbalah

"To those who do not know mathematics it is difficult to get across a real feeling as to the beauty, the deepest beauty, of nature ... If you want to learn about nature, to appreciate nature, it is necessary to understand the language that she speaks in" (Richard Feynman).

Let's be honest, we can't get by without mathematics. Whether you want to double the recipe to bake a cake, or calculate how long it will take you to save up for a new car, mathematics is a necessity. As history has developed, our application of mathematics has become more and more sophisticated. So true is this that many mathematicians and physicists conclude that science today is taxing the upper limit of human intelligence.

Every language has its alphabet and the alphabet of mathematics is numbers, a convention accepted by all of civilization. This makes mathematics a more general and a more secular pursuit than any other subject.

Sefer Yetzirah (the "Book of Formation"), the earliest Kabbalistic text, states that there are thirty-two pathways of wisdom through which God created the universe.[30] These pathways are the ten *sefirot* and the twenty-two Hebrew letters.[31] In this Kabbalistic statement, the number ten, the base of our decimal system, precedes the twenty-two letters of the Hebrew alphabet.▸ Giving numbers top priority puts them on a higher spiritual plane than the letters of the Holy Tongue. In some mysterious way, numbers are the most holy language of all.

The Jewish People are referred to as the "People of the Book," which indicates the Jewish penchant for studying the written word.▸▸ There is also something essentially Jewish about counting. Like a gem collector counting precious jewels, scribes continuously count the letters and words of the Torah. In fact, the Hebrew word for "scribe"

▸ In secular education, the decimal system is considered an arbitrary choice. However, Judaism holds that decimality is innate in creation. One of the most basic precepts of Kabbalah is the ten *sefirot*. These are the ten channels of Divine energy through which God continuously re-creates the universe. Each of these ten channels is holistic in character, including within it each of the characteristics of the other nine. As such, every facet of creation manifests and reflects decimality, making base ten the natural and simplest choice to use.

▸▸ "Book" (סֵפֶר), "scribe" (סוֹפֵר) and "numbers" (מִסְפָּרִים) all share the same Hebrew root (ספיר). Another word with the same root is "counting" (סְפִירָה), which is also the spelling of *sefirah*. The ten *sefirot* are the ten mystical channels of Divine light with which God created the world.[32]

(סוֹפֵר) also means "counter."[33] The three-lettered root (ס־פ־ר) is also the root of *sefirah* (סְפִירָה). One of the ways to analyze and interpret the Torah text is by counting its holy letters. Contemplating its words quantitatively is one of many ways to express one's love of the Torah.

A general Chassidic principle states that anything that emanates from a higher spiritual source descends lower into the physical realm.[34] Because the origin of numbers is so high, they have descended lower and have become more secular. This is why they are usually more pertinent to science than to Torah. However, in Kabbalah and Chassidut, the Torah's inner dimension, numbers have primary significance.

One opinion is that the *sefirot* are the ten numbers from 1 to 10.[35] The firmament that divides the two types of water (as noted above) represents the 22 letters.[36]

This is illustrated in the form of the letter *alef*, the first letter of the Hebrew alphabet, which represents absolute unity.

א

As can be readily seen here, the form of the *alef* is an upright *yud* (י) on the right, an upside down *yud* on the left, and a letter *vav* (ו) between them. The numerical value of the letter *yud* is 10. The upper *yud* represents the ten *sefirot* as they descend from their source in the Divine. The lower *yud* represents the ten numbers of mathematics. The numerical value of one spelling of the *vav* (ויו) that connects them is 22.

The ten *sefirot* (or numbers) are lights. As we shall see, light preceded water, and it includes both the upper waters and the lower waters (the wave and particle aspects of water). Thus, numbers reflect the wisdom of the primordial Torah. They are the perfect tool to facilitate the union between the two types of wisdom.

Lights in Vessels

A most basic relationship between the ten *sefirot* and the twenty-two Hebrew letters is observed in a 1 by 10 rectangle. The perimeter of a rectangle is defined by the algorithm $2(l \perp h)$, where l is its length and h is its height. In this case, the perimeter is 22 units. The area of a rectangle is $l \cdot h$. In this case, the area is 10 units^2.

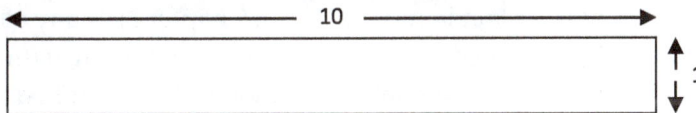

As a body is to a soul, so the twenty-two letters of the Hebrew alphabet are the relatively material vessels (the perimeter, the confining limit) that contain the spiritual lights of the ten *sefirot* (the area, the content of the rectangle).▸

In Kabbalah, vessels represent quantity, and light (or energy) represents quality. Our purpose is to contain the great lights of the inner, qualitative aspect of the Torah within the rectified, quantitative vessels of modern science. We begin doing so by discovering the links that connect them through their common language: mathematics.

Many philosophically-minded scientists have searched for a connection between the cosmic number 137 and various mystical phenomena. None have yet delved into the Torah and Kabbalistic texts in an attempt to discover its true essence. By throwing light on 137 as it manifests in the Torah, we will use it as a coupling constant between Torah and science.▸▸

Every verse in the Hebrew Bible consists of Hebrew letters. The inner soul of every letter is its numerical value, its *gematria*. We might say that the *gematria* of a letter or word translates it into a dimensionless constant. The most commonly known *gematria* system is the absolute/normative value, whereby the value of each Hebrew word is calculated by the specific number value associated with each of its letters (see Appendix D). The innate holiness of numbers becomes tangible

▸ The ten *sefirot* (numbers) are silent.[37] By contrast, the twenty-two letters of the Hebrew *alef-bet* are infused with meaning at every level of expression. Meaning hovers above individual letters, in their names and spellings, in their forms and in their numerical values. Explicit ideas dwell within whole words. The significance of the Hebrew letters becomes clearly comprehensible in complete sentences.

▸▸ The Torah is a prophecy from God; science is the wisdom of mankind. The sum of the numerical values of "prophecy" (נְבוּאָה) and "wisdom" (חָכְמָה) is 137. The initial letters of "wisdom" (חָכְמָה) and "prophecy" (נְבוּאָה) spell "grace" (חֵן). This word also implies "symmetry." Kabbalists are often referred to as "knowers of symmetry" (יוֹדְעֵי חֵן).

17

▶ Numbers descend from the lights of the ten *sefirot* in *Atzilut* (the World of Emanation).

when it is garbed in the Hebrew letters and is conveyed to us in the language of the Torah.◀ Science attempts to discover the mysteries of creation by defining physical reality through numbers. The Torah text reveals those mysteries through its hidden numerical code.

There are many ways in which numbers are significant in Kabbalistic teachings. Examples of such are figurate numbers and numerical series derived from the numbers of letters or words in a text, or their values (as we shall explain in more detail in Chapter 3). Many of these involve more sophisticated calculations, some of which are similar to those used in modern mathematical methods. When applied to the Torah text, these forms of wisdom reveal innumerous instances of self-reference, infusing the text with multi-dimensional richness. This new and inspiring quality resonates with the numerical definitions of science. Analyzing the Torah's verses numerically produces a melody that blends with the mathematical analysis that is the basis of modern science. Listening to the symphony that results from playing them together is an experience that can uplift the human soul to spiritual heights never reached before. By studying the ideas in this book, you are invited to be inspired by the beauty of this symphony. Herein lies the link between these two fields of wisdom that have been isolated from one another for so many generations. It opens up a two-way channel that infuses quality into the quantitative discoveries of science and inspires new associations for Kabbalistic meditations that deepen our understanding of the secrets of creation.

Our hope is that the ideas presented in this book will take root in the minds and hearts of all people around the world. Perhaps this will lead to putting all the pieces of our enigmatic puzzle into place.

This book is based on a series of sixteen lectures presented by Rabbi Yitzchak Ginsburgh in New York and Toronto in the summer of 5770 (2010). More material has been gleaned from other books and lectures by Rabbi Ginsburgh.

As preliminary reading, we highly recommend that the novice in Kabbalistic concepts first peruse our book *What You Need to Know about Kabbalah*, which gives the beginner a basic understanding of Kabbalistic terminology that will be of assistance throughout this book. Included at the end of this book is a comprehensive glossary of Hebrew terms, which will also facilitate those unfamiliar with these concepts.

All the words and phrases in the Torah that allude to 137 that are mentioned in this book, together with many more, can be found on a special web-page at http://www.inner.org/torah_and_science/allusions-of-137. The page will be updated periodically as new allusions are discovered.

Despite the care taken to prevent mistakes, the complexity of the book's content and its design and the simple human nature to err, will probably have left their mark. Please send any corrections, comments or questions that you might have regarding this book to 137riddle@gmail.com. This will enable us to improve following editions.

R. Gordon

11th Sivan 5778▸

web page for allusions
in the Torah to 137

▸ The numerical value of 11th Sivan (יא סיון) is 137.

19

REFERENCE NOTES FOR INTRODUCTION

1. Brian Greene, *The Elegant Universe* p. 123

2. Many discussions on the nature of reality are apparent in the Talmud. In particular there are the discussions of the sages with the Elders of Athens (*Bechorot* 8b etc.).

3. *Likutei Moharan* II, 8:7.

4. Genesis 3:21.

5. Deuteronomy 34:6.

6. *Sotah* 14a.

7. *Avot* 1:6; *Shabbat* 127b.

8. Zachariah 14:9.

9. See our book, *Rectifying the State of Israel*, pp. 111 ff.

10. For a more detailed discussion on this topic, see our book *Wisdom: Integrating Torah and science.*

11. Genesis 1:6.

12. See *Zohar Bereishit* 117a. See also the explanation of *Ashmoret Haboker* on the *Zohar*.

13. *Tikunei Zohar, Tikun* 5 19b. See also ibid *Tikun* 40; *Etz Chayim, Sha'ar Haklalim, Klal* 2 & 4.

14. *Zohar Bereishit* 17b-18a.

15. See at length, Rebbe Isaac of Homil, *Ma'amar Yetziyat Mitzrayim* chs. 12-13.

16. See for example, *Tanya* Chapter 8.

17. Such is the case with reference to certain laws pertaining to *niddah*, in particular. See Rabbi Neriyah Gutel, *Hishtanut Hatevaim*.

18. *Yerushalmi Berachot* 1:1.

19. *Bereishit Rabah* 62:1.

20. *Midrash Tanchuma, Parashat Yitro.*

21. *Zohar Shemot* 161a.

22. John Baez, *How Many Fundamental Constants Are There?*

23. Max Born, *Atomic Physics*, 8th edition.

24. http://physics.nist.gov/cuu/Constants/*alpha*.html.

25. Riazuddin, Fayyazuddin. *A Modern Introduction to Particle Physics* (Third ed.). World Scientific. p. 4.

26. Professor Laurence Eaves, http://www.sixtysymbols.com/videos/finestructure.htm.

27. James G. Gilson, Strong Quantum Coupling and Relativity, July 2002.

28. For more on π as a whole number, see Part 1 of the article "The Story of π", on our website:

    http://www.inner.org/torah_and_science/mathematics/story-of-pi-1.php.

29. Arthur I. Miller, Early Quantum Electrodynamics: A Sourcebook, p. 62.

30. *Sefer Yetzirah* 1:1.

31. Ibid 1:2.

32. *Sefer Yetzirah* 1:1-3.

33. *Kidushin* 30a.

34. Ma'amarei Admor Hazakein, 566 part 1, p139.

35. Rabeinu Saadiah Gaon's interpretation on *Sefer Yetzirah* 1:1; Ibn Ezra's interpretation ibid.

36. *Tanya, Sha'ar Hayichud Vehaemunah* 5.

37. *Sefer Yetzirah* 1:1.

KABBALAH AND CREATION

Let There Be 137

Light, life and love are those three elusive qualities that we all take for granted, yet their source remains cryptically intangible. Physicists delve into the undiscovered realms of quantum mechanics to fathom the source of light. Biologists have yet to decipher the secret of life and the riddle of human consciousness. Psychologists search for the hidden source of sound mental health that is nurtured by unconditional love.▸ It seems that ever since the human race first appeared on the face of the universe, man's intellect has been determined to investigate the origin of these mysterious secrets of creation.

In the Torah, light was created on the first day of creation, biological life on the third day, and human life appeared on the sixth day. Love is the power that motivates all of creation, light and life included.▸▸

Perhaps light is so intriguing to humans because it has such primordial roots. Indeed, whether it is rays of sunlight sparkling on a freshwater stream, or the glimmer of city lights as you approach it at night, we humans are fascinated by light in all its forms.

Two ancient Greek philosophers, Aristotle and Democritus, were the first to argue about the properties of light. Aristotle proposed the theory that light was a disturbance of air and behaved like waves. Democritus believed that like everything in the universe, light was formed of indivisible subcomponents that he called "atoms."

When Isaac Newton came on the scene, he favored the corpuscular hypothesis of light, arguing that only particles could travel in such straight lines as a beam of light. Yet, two of Newton's contemporaries, and others later, demonstrated that the refractive properties of light could be easily explained by wave propagation.

▸ Conditional love is unsound; unconditional love is stable.[1]

▸▸ The Zohar[2] associates the creation of light on "day one" to the light of Abraham, as in the phrase "One was Abraham." Abraham is the archetypal soul associated with love. Indeed, God calls him, "Abraham, my lover."[3] The numerical value of Abraham (אַבְרָהָם) is 248. The numerical value of the essential Name of God, *Havayah* is 26. Their sum is 274, thus the average numerical value of these two is 137.

One way to determine which hypothesis is correct is by means of the "double-slit" experiment. Light is shined through a double slit. If light is a particle, then only the rays of light that hit exactly where the slits are located will pass through them. The effects in this case would be like spraying paint from a spray can through the openings. If light is composed of tiny particles, two lines will appear on the viewing screen; one opposite each of the slits.

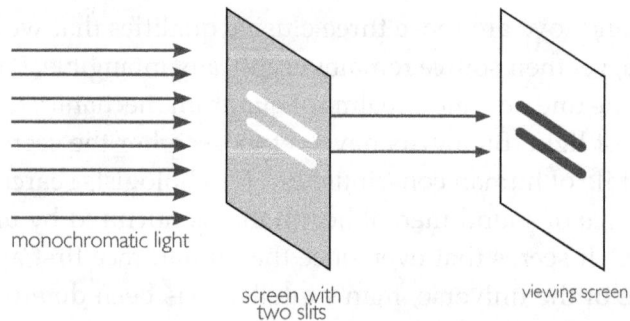

monochromatic light

screen with two slits

viewing screen

However, if light is a wave, the results of this experiment are very different. There are still only two light rays that actually pass through the slits, but since they are waves, as soon as they traverse the screen, they begin to bend and interfere with one another. Thus, an interference pattern is observed on the viewing screen.

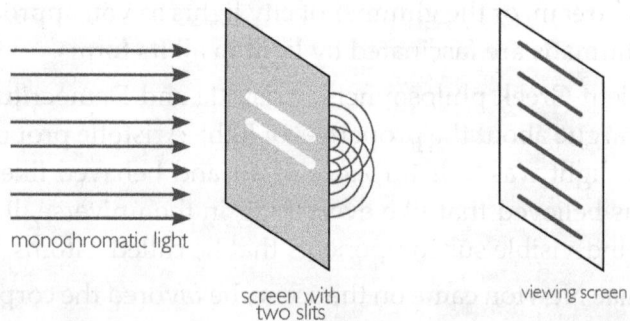

monochromatic light

screen with two slits

viewing screen

When 18th century scientist, Thomas Young, carried out this experiment, he discovered that double-slit interference does exist. Like waves in a pond, the ripple effects of light passing through a slit do indeed interfere with one another. This proved that light is a wave. The wave theory of light had won the day.

Thus, light/energy was understood to behave like waves. Like water flowing in a river it could be emitted or absorbed continuously in arbitrary amounts. James Maxwell subsequently discovered the wave formula that resulted in an accurate calculation of the speed of light. It seemed that the particle theory of light was doomed forever. Or was it?

The classical wave perception of energy reaped good results in most cases. However, some questions remained unsolved. For example, when certain elements are irradiated by light, electrons are emitted from their surface. This phenomenon is called the photoelectric effect. According to the wave theory, the average energy carried by an ejected (photoelectric) electron should increase with the intensity of the incident light. But this is not so. The energies of the emitted electrons are independent of the intensity of the incident radiation. This was one phenomenon that could not be explained by wave theory.

Another enigma was black body radiation. All normal matter absorbs electromagnetic radiation to some degree. An object that absorbs all radiation falling on it at all wavelengths is called a black body. When a black body is at a uniform temperature, its emission has a characteristic frequency distribution that depends on the temperature. Its emission is called black body radiation. Nobody had an acceptable hypothesis to explain why this type of radiation exists.

Physicists continued their quest to discover the solutions to these problems. At the turn of the 20th century, Max Planck finally reaped success. He derived a function that correctly described the effect of black body radiation. He did so by abandoning the standard idea that light energy behaved like a liquid. Instead, he proposed that energy is not "a continuous, infinitely divisible quantity, but a discrete quantity composed of an integral number of finite equal parts."[4] He called these parts, "quanta."

Although Planck's quantum hypothesis "worked" to solve the black body enigma, physicists of his day (including Planck himself) agreed that viewing light energy as quanta was a flaw in his model. It took Einstein to notice that Planck's black body model also produced an effective solution to the inquiry regarding the photoelectric effect. This revelation is alluded to in the Book of Daniel: "The wise will glow like the glow of the firmament."[5]

Einstein's predictions were confirmed by experiment in 1915. He received the Nobel Prize in 1921 for the simple, yet revolutionary suggestion of quantized light. These light quanta were later called "photons." They represent the quintessential example of wave–particle duality. Electromagnetic radiation propagates following linear wave equations, but can only be emitted or absorbed as discrete elements. In terms of the double-slit experiment, projecting single particles through the apparatus results in single particles appearing on the screen, as expected. However, as more of these single particles are projected through the slits, an interference pattern emerges, suggesting that the particles, acting like waves, go through both slits simultaneously! The paradoxical conclusion that physicists have reached is that light is neither wave nor particle, but both, at one and the same time.◄ This enigma reinforces the human fascination with light at the subatomic level.

Light, Water, Firmament

Before the photon collides with another particle, it is a wave of energy. When it hits an atomic particle, it increases the particle's energy by one quantum.

This short description is the basis of quantum theory. It describes three levels: 1) light as a wave of energy; 2) light as a particle (photon); 3) the photons, drops of light so to speak, hitting and being absorbed by one of the particles of the atom, raising its energy level by one quantum.

In Kabbalah, the secret of these three states is one of the most significant mysteries of what is known as the "Workings of Creation." It is

► Wave-particle duality has been demonstrated in photons, electrons, atoms and molecules. It has been experimentally confirmed even in large carbon molecules ("buckyballs").

Professor Aharonov from Tel Aviv University disputes the wave interpretation, proposing a mechanism with non-local correlation where each particle goes through only one of the two slits.

reflected in the root *amar* (א־מ־ר), "say." The sages state that "the world was created with ten sayings."[6] The general process of all creation is the mystery of how God's "saying" created reality.

The famous 16th century Kabbalist, Rabbi Isaac Luria (the Arizal)[7] expounds that the root *amar* (א־מ־ר) is composed of the initial letters of the words for "light" (אוֹר), "water" (מַיִם) and "firmament" (רְקִיעַ). Each of these words appears five times in the account of the first two days of creation. On the first day of creation "light" appears five times: "And God said, 'Let there be *light*!' and there was *light*"; and God saw the *light* that it was good and God separated the *light* from the darkness"; "and God called the *light* day and the darkness He called night and it was evening and it was morning, one day."[8]▶

On the second day of creation "water" (מַיִם) and "firmament" (רְקִיעַ) appear five times each: "God said, let there be a *firmament* within the *water* and it shall separate between *water* and the *water*: and God made the *firmament* and He separated between the *water* that was below the *firmament* and the *water* that was above the *firmament* and it was so. And God called the *firmament* heaven and it was evening and it was morning, second day."[9]

Although there were many more entities created during the six days of creation, in Kabbalah these three—light, water, firmament—dominate and form an all-inclusive rule, as we shall see.

Just as this teaching of the Arizal relates to creation, it also relates to the formation of new life through the procreation process. The mystery of *amar* relates to the transition from electrical impulses to physical seed in the process of creation and birth.[11] Biological procreation begins with a manifestation of energy in the mind. At this stage, it is a series of electrical impulses that pass through the nervous system, but do not yet manifest as physical seed. The individual becomes "electrified" at different levels as waves of energy flood his psyche. The energy gradually becomes more physical. It is still inside the male initiator but has no essential form. This is the second stage in the transition from light-water-firmament that parallels droplets

▶ The Arizal's exposition of this process is based on the phrase, "Day to day expresses speech, and night to night experiences knowledge" יוֹם לְיוֹם יַבִּיעַ אֹמֶר וְלַיְלָה לְּלַיְלָה יְחַוֶּה דָעַת.[10] The numerical value of "day... and night" (יוֹם...וְלַיְלָה) is 137.

▶ All the stages of light, water and firmament, also occur in the female procreative process. "Let there be light" refers to energy in the mind of the male, "and there was light" refers to the energy in the mind of the female. The energy then manifests as "male waters" and "female waters." The firmament within the waters relates to the mother's womb.

▶▶ Elsewhere,[12] we have explained how these three stages of procreation correspond to three different states of consciousness within the pre-intellectual and intellectual faculties of the mind:

Sefirah	Transition State	State of Consciousness	Physical Response
Keter	Light	Totally unconscious	Will to inspire or impregnate
Chochmah	Water	Transition between unconscious and conscious	Volition towards consummation
Binah	Firmament	Initiation of consciousness	Initiation of physical sensation

(particles) of water. This stage is implemented through the realization that it is impossible to impregnate reality while stagnant. This creates a vector force of liquid motion that flows towards consummation. This liquid is composed of droplets, the physical seed. Its wavelike property is the presence of light, its spiritual source, within it. ◀

The third stage begins once the droplets are transferred to the female receiver. This stage is called firmament; it is the stage of physical manifestation, the initial stage of conception. ◀◀

The first stage of psychological energy, before it becomes water, is the wave state. The second, water stage is composed of droplet-like photons, light particles. The third stage, the firmament, manifests once it transforms into a hyper-energetic electric state within the electrons, which are the components that receive the extra energy, as we shall explain.

The particle aspect of light becomes apparent at the stage when light energy is intercepted by a receiving object. If light does not come into contact with an object, there is no need for us to suppose that there are photons. But the moment that light is radiated outward to something else and the object receives the light's influence, we must suppose that light is a particle and its wave function no longer suffices. The particle of light adds one quantum of energy to the object that it hits, and it begins to glow. This is the process of giving life. This is also the process of creation.

Naming the Particles in Hebrew

The first chapter in Ezekiel is the most esoteric chapter of the entire Hebrew Bible. It is even more profound than the "Workings of Creation" described in Genesis. Referred to by the sages as "the Workings of the Chariot," it contains secrets that are taught only to select Torah scholars who are deemed morally worthy of contemplating them.

In his mystical vision of fire and light, Ezekiel describes the cyclical movement of the wheels of the Divine Chariot: "The appearance of the wheels ... and their workings were as a wheel would be within a wheel."[13]▶ Rashi, a medieval Biblical commentator, explains that the wheels of the Divine Chariot could travel in any direction.[14]

The electron, which orbits the nucleus of the atom in concentric shells, can also be described as "a wheel within a wheel." We suggest that this word—*ofan* (אוֹפָן)—be adopted as the official Hebrew translation of "electron." The numerical value of the Hebrew word for "wheel" (אוֹפָן) is 137.

Like *ofan* for "electron," the Hebrew word *egel* (אֶגֶל) is most fitting as a translation of "photon." It expresses the dual character of light as both particle and wave.▶▶ It is found in a unique expression in the Book of Job, *eglei tal* (אֶגְלֵי טָל), which many commentaries translate as "drops of dew."[15] However, Rashi states that the root of *egel* is *gal* (גַּל) meaning "wave." Dew is a wave that passes across the face of the earth each morning, before settling on the ground. Elsewhere in the Bible, we find dew directly identified with light, as in the verse from Isaiah: "Your dew is a dew of lights."[16] Thus, the idealized concept of dew is water and light, wave and particle, combined into a single droplet. *Egel* captures both the wave and particle phenomenon of the photon.

A photon is formless energy. An electron is the most elementary particle of matter. When a photon is absorbed by an electron, it transforms from energy into matter.▼▼▼ The maximum speed of an electron is $\frac{1}{137}$th times the speed of light (the speed of a photon).

▶ The reduced value of the concluding phrase of this verse, "and their appearance and their workings were as a wheel would be within a wheel" (וּמַרְאֵיהֶם וּמַעֲשֵׂיהֶם כַּאֲשֶׁר יִהְיֶה הָאוֹפַן בְּתוֹךְ הָאוֹפָן) is 137. The reduced value of the first four words (וּמַרְאֵיהֶם וּמַעֲשֵׂיהֶם כַּאֲשֶׁר יִהְיֶה) is 73 [the numerical value of "wisdom" (חָכְמָה)] and the reduced value of the final three words (הָאוֹפַן בְּתוֹךְ הָאוֹפָן) is 64 [the numerical value of "prophecy" (נְבוּאָה).

▶▶ "Wave" (גַּל)—refers to either a light wave, or an ocean wave. Other words with the same root are "revelation" (גִּלוּי) and "wheel" (גַּלְגַּל). These two words reflect the two types of wave. A light wave is a form of revelation. An ocean wave is cyclic, like a wheel.

Another Hebrew word for "wheel" (אוֹפָן) has a numerical value of 137.

▼▼▼ The matter property of an electron is reflected in the filled numerical value—310—of its Hebrew name, *ofan* (אלף וו פא נון) which is also the numerical value of "something" (יֵשׁ). The photon is formless energy, relating to "nothing" (אַיִן). The numerical value of the common phrase, "something from nothing" (יֵשׁ מֵאַיִן) is 411, which equals 3 times 137.

The numerical value of the phrase, "or who gave birth to drops of dew" (אוֹ מִי הוֹלִיד אֶגְלֵי טָל) is 195, which equals 5 times 39, the numerical value of "dew" (טָל). This is the average numerical value of each word.

When the value of each letter is considered as the triangle of its numerical value, the sum is 2,500, which is equal to 50^2. This relates to the revelation of the complete inter-inclusion of the fiftieth gate

of understanding. In this way, the average numerical value of each word is 500, which is equal to the numerical value of "be fruitful and multiply" (פְּרוּ וּרְבוּ),[17] which alludes to marriage.

When the value of each letter is considered as the square of its numerical value, the sum is 4,805, which is equal to 5 times 961 (31^2), i.e., the average numerical value of each word is 961. This relates to the sum of the numerical values of the names of the first family of mankind, Adam, Eve, Cain, Abel and Seth (אָדָם, חַוָּה, קַיִן, הֶבֶל, שֵׁת) and also the first Jewish family, Abraham, Sarah and Isaac (אַבְרָהָם, שָׂרָה, יִצְחָק). 31 is the numerical value of "not" (לֹא) and the numerical value of "towards" (אֶל). Adam and his family relate to "not" and Abraham and his family relate to the vector force, "towards," as explained elsewhere.

The numerical value of *egel* (אֵגֶל) is 34. Beginning from 1, 137 is the 34th prime number. This phenomenon reflects the relationship between the photon, *egel* (אֵגֶל) and electron, *ofan* (אוֹפָן), which has a numerical value of 137.▼

Light Means "Mystery"

Over the years, physicists have observed the behavior of the photon, the basic unit of electromagnetic radiation (which includes visible light). They have determined that the coupling constant (depicted by the Greek letter *alpha, α*) that defines the strength of the interaction between electrons and photons, is closely equal to the inverse of a whole number: 137.

Since the beginning of the last century, *alpha* has baffled the greatest physicists.

In 1935, Max Born, a Jewish, German-born physicist, wrote a paper entitled, "The Mysterious Number, 137," stating: "It is clear that the explanation of this number must be the central problem of natural philosophy [i.e., science]."

▸ The average numerical value of *egel* (אֵגֶל) and *ofan* (אוֹפָן) when both are spelled in full (אלף גימל למד אלף וו פא נון) is 289, which equals 17^2. This is the numerical value of the phrase, "God created" (בָּרָא אֱלֹהִים). The numerical value of "good" (טוב) is 17, thus, 17^2 refers to the interinclusion of all goodness.

Egel (אֵגֶל) appears in the verse in the plural—*eglei tal* (אֵגְלֵי טָל)—with an additional *yud* that connects it with the following word, *tal* meaning "dew" (אֵגְלֵי טָל). The complete phrase reads "or, who gave birth to droplets of dew" (אוֹ מִי הוֹלִיד אֵגְלֵי טָל). When all fifteen letters of this phrase are written in the form of a triangle, *egel* appears in the singular as the corners of the triangle.

```
              א
          ו       מ
      י       ו       ה
  ל       י       ד       א
ל     י     ט     י     ל     ג
```

Here, we see that *egel* (אֵגֶל) is divided into *alef* (א) at the top, with a numerical value of 1, and "wave" (גֵל), which has a numerical value of 33. This is in keeping with the opinion of modern mathematicians, who conventionally prefer 2 as the first prime number,[18] in which case 137 is the 33rd prime number. From our Kabbalistic perspective, the number 1 is the all-inclusive root of the prime numbers.

One way to "read" such a triangle is by relating to each layer of letters from the outermost layer to the innermost layer (the center of the bottom line).

```
              א
          ו       מ
      י       ה       ו
  ל       י       א       ד
ג     ל     י     ט     ל
```

In this case, the numerical value of the entire middle layer is 58, which is the numerical value of "symmetry" (חֵן). The sum of the numerical values of the first (127) and third layers (10) is 137.

Another 20th century Jewish physicist, Richard Feynman, said, "All good theoretical physicists put this number up on their wall and worry about it." No matter how much we know about physics, nobody knows why God chose 137 to create light. In Feynman's words, "Where does this number come from? … We would like to have a little clue to how He [God] thinks to make a number like this."

One contemporary scientist, James G. Gilson, has found a mathematical formula for *alpha* that agrees with experimental data to ten decimal places. This formula treats 137 as a whole number:

$$\alpha = \alpha(137, 29) = 29\cos(\pi/137)\tan(\pi/(137 \cdot 29))/\pi$$

Gilson writes:

> *The fine-structure constant has been for many years a source of scientific and philosophical questions regarding its value and significance and, as a fundamental object of theory, it has an unavoidable place in our attempted explanations of those extraordinary images of distant galaxies returned by the Hubble telescope. Thinking persons cannot ignore those new spectacular views of our world such as that of the Eagle Nebula which have recently been returned by Hubble [1995]. They jolt us back to the age old question of what is it all about?*

Kabbalah, the inner dimension of the Torah, proposes to answer "what it is all about." In its literature, where the mysteries of creation and the human psyche are discussed through allusions and mystical allegories, many revelations of Divinity are referred to as light.

The *Sefirot* and 137

It is no coincidence that the numerical value of the Hebrew word *Kabbalah* (קַבָּלָה) is 137. But, besides this straightforward clue, one of the most convincing correspondences between Kabbalah and the number 137 appears in the ten *sefirot*, the spheres or channels of Divine energy that permeate all aspects of creation.

In their manifestation in the psyche, the *sefirot* are divided into three strata that relate to the super-conscious, the intellectual, and the seven emotive/instinctive powers of the soul, as follows:

super-conscious		*keter* (crown) — faith, pleasure, will
intellectual	*binah* (understanding) joy	*chochmah* (wisdom) selflessness
	da'at (knowledge) connection	
emotive	*gevurah* (might) fear	*chesed* (loving-kindness) love
	tiferet (beauty) compassion	
	hod (acknowledgment) sincerity	*netzach* (victory) confidence
	yesod (foundation) truth	
	malchut (kingdom) lowliness	

The manifestation of a subset of the *sefirot* within each individual *sefirah* is called inter-inclusion. For example, the *sefirah* of *chochmah* (wisdom) contains all ten *sefirot* within it (i.e., *chochmah* within *chochmah*, *binah* within *chochmah*, *da'at* within *chochmah*, etc.). The Arizal taught that each of the three levels (super-conscious, intellectual, emotive) exhibits a different level of inter-inclusion of the ten *sefirot*.[19] Each of the seven lower *sefirot* is considered a single unit (total: 7); the three intellectual powers of the soul are each inter-included with 10 *sefirot* (total: 30). Finally, the highest *sefirah* of *keter* (the super-conscious crown) is inter-included with 10^2 *sefirot* (total:

100). The total number of *sefirot* when counted in this way is thus 137.▼

Chassidut teaches us that each of the *sefirot* also has an inner motivating power (see table above). Coupling, or connection, is the inner energy of *da'at* (the *sefirah* of knowledge). *Da'at* alludes to intimate knowledge of another, as illustrated by the first appearance of this verb in the Torah, "And Adam knew Eve, his wife."[21]

In the *sefirot*, *da'at* connects between *chochmah* and *binah*.▶▶ This is *da'at elyon* (upper knowledge). It also connects between the intellectual powers of the soul and the emotive powers. This connection is *da'at tachton* (lower knowledge). *Da'at* also connects the super-conscious with the conscious mind. Like wave-particle duality, only one of these two manifests at any moment. Thus, *da'at* is the *sefirah* that most suggests a connection to the electromagnetic coupling constant.

Each of the *sefirot* is associated with a Divine Name. The Name associated with *da'at* is *Akvah* (אהוי-ה).[22] The numerical value of this Name is 17, which is the numerical value of "good" (טוב). It is therefore sometimes referred to as the Goodly Name. It signifies the Divine power to

▶▶ Proverbs (3:20) connects these three *sefirot* in one phrase, "*Havayah* founded the earth with wisdom, establishes it with understanding. With His knowledge the depths were split..." The numerical value of the entire statement is 3,014, which equals 22 times 137.

▶ The sum of the numerical values of *binah* (בִּינָה), *tiferet* (תִּפְאֶרֶת) and *malchut* (מַלְכוּת) is 1,644, which equals 12 times 137. 1,644 is also the numerical value of the entire verse, "And you shall love *Havayah* your God with all your heart and with all your soul and with all your might" (וְאָהַבְתָּ אֵת הוי' אֱלֹהֶיךָ בְּכָל לְבָבְךָ וּבְכָל נַפְשְׁךָ וּבְכָל מְאֹדֶךָ). [In this context, "with all your heart" corresponds to *malchut*, "with all your soul" corresponds to *tiferet*, "and with all your might" corresponds to *binah*.]

In Chapter 6, we will explain that *Keter* contains three different levels: *emunah*, *ta'anug* and *ratzon* (faith, pleasure and will). When these are added to the ten *sefirot*, the total is thirteen. The number 13 is a particularly significant number in Kabbalah. It is the numerical value of "one" (אֶחָד), and "love" (אַהֲבָה).

The sum of the numerical values of all thirteen names of the *sefirot* (אֱמוּנָה תַּעֲנוּג רָצוֹן חָכְמָה בִּינָה דַּעַת חֶסֶד גְּבוּרָה תִּפְאֶרֶת נֶצַח הוֹד יְסוֹד מַלְכוּת) is 3,699, or 27 times 137. This array of the ten *sefirot* together with the three heads of *keter* serves as the basis for the entire wisdom of Kabbalah.

The sum of the numerical values of *keter* (כֶּתֶר), *tiferet* (תִּפְאֶרֶת) and *yesod* (יְסוֹד) is 1,781, which equals 13 times 137. Replacing *keter* with

chesed (חֶסֶד) in this equation yields 1,233, which equals 9 times 137 [i.e., the difference between *keter* (620) and *chesed* (72) is 548, which equals 4 times 137]. *Keter* is the first of the three intellectual faculties (when *da'at* is not included) and *chesed* is the first of the emotional powers of the soul. The relationship between *keter* and *chesed* is that *keter* relates to God's intention in creating the world, which is His desire to reveal His loving-kindness (*chesed*).

The verse in which the seven emotional attributes of the soul appear is "To You is the greatness (i.e., *chesed*) and the might (i.e., *gevurah*) etc." These first two attributes also include the five that follow them. The numerical value of "the greatness and the might" (הַגְּדֻלָּה וְהַגְּבוּרָה) is 274, which equals 2 times 137, i.e., the average numerical value of each of these two inclusive attributes is 137.

Since each of the intellectual powers of the soul is inter-included within ten *sefirot*, the ten *sefirot* inter-included in *da'at* are represented by the ten numbers from 128-137. The value of α depends upon the energy at which it is measured. It increases with increased energy. Thus, although *alpha* is usually considered $\approx 1/137$ (at zero energy levels), when measured at higher energy levels it can rise to $\approx 1/128$.[20]

▸ When spelled with the filling corresponding to *Sag*, the numerical value of the filling of the Name *Ekvi* (אלף הי ואו יוד) (associated with the *gevurot* of *da'at*), is also 137 (לף יאו ד).

The numerical value of another spelling of *Akvah* (אלף הי ויו הי) is 163. The numerical value of one spelling of *Ekvi* (אלף הה ויו יוד) is also 163. 163 is the sum of the numerical values of *netzach* (נצח) and *hod* (הוד), the two *sefirot* that are the final extension of the two sides of *da'at*.

▸▸▸ A close mathematical connection between 37 and 137 is in the series formed by the general equation $f(n) = n^2 \pm 1$. The sixth number of the series is 37 and the 37th number ($37^2 \pm 1$) is 1,370, which equals 10 times 137.

n: 1 2 3 4 5 6 7 8 9 10 …37

f(n): 2 5 10 17 26 37 50 65 82 101 … 1,370

The series includes the numerical numbers of two significant Names of God, *Havayah* (26; $5^2 \pm 1$) and *Adni* (65; $8^2 \pm 1$). The Name *Adni* is used as an alternative pronunciation when reading the Name *Havayah* in holy texts outside of the Temple.

▸▸ The four most significant fillings of the essential Name of God, *Havayah*, are referred to by their numerical values: *Ab* (72, i.e., the numerical value of *Havayah*, 26, plus the filling letters with a numerical value of 46), *Sag* (63, i.e., 26 plus 37), *Mah* (45, i.e., 26 plus 19) and *Ban* (52, i.e., 26 plus 26). These four Divine Names further correspond to the four spiritual Worlds, in descending order. *Ab* corresponds to *Atzilut* (the World of Emanation); *Sag* corresponds to *Beriah* (the World of Creation); *Mah* corresponds to *Yetzirah* (the World of Formation), and *Ban* corresponds to *Asiyah* (the World of Action).

Ab	Atzilut
Sag	Beriah
Mah	Yetzirah
Ban	Asiyah

Ab is associated with total unity, *Sag* is associated with the chaotic lights that caused the original seven vessels of *Mah* to shatter, thus separating *Ban* (*malchut*) from the other three Worlds. Rectification begins by emanating a new *Mah* from *Ab*, that reunites with *Ban*. The reunion of *Mah* and *Ban* in a rectified manner is alluded to by the numerical value of *Meheitavel* (מְהֵיטַבְאֵל), mentioned later in this chapter, which is 97, the sum of 45 (*Mah*) and 52 (*Ban*).

unite spirituality (heavens) and physicality (earth).[23] From a scientific perspective, this Name could be the missing link necessary to connect the quantum world with gravitation. When spelled in full (אלף הי ואו הי) the numerical value of the filling letters of this Name (לף י או י) equal 137. ◂

This spelling corresponds to the Divine Name *Sag* (סג), which has a numerical value of 63. This Name refers to the numerical value of the filling of God's Name *Havayah* (יוד־הי־ואו־הי). ▾▾ The numerical value of the filling letters of *Sag* (וד י או י) is 37. The connection between the numerical value of the filling of the Name *Akvah* (137) and the filling of the Name *Sag* (37) is our first clue to the significance of 137 and 37 as companion numbers, as we shall see on numerous occasions throughout this book. ◂◂◂

More simply, we see the relationship between 137 and 37 when the inter-included *sefirot* of *keter*, which total 100, are omitted, i.e., when *da'at tachton* is the governing factor. Thus, 137 emerges as the connecting force between Kabbalah (the super-conscious blueprint of creation,[24] corresponding to *keter*), and science (man's conscious attempt to grasp creation with the human mind, corresponding to *da'at*).

Behind *Bereishit* (Genesis)

What could be more intriguing than to find the fine-structure constant in the words God spoke to create the universe?

The first Hebrew word in the Torah is *Bereishit* (בְּרֵאשִׁית), "In the beginning." In addition to the Hebrew text of the Torah there is an Aramaic translation, which is also considered a holy text. In his teachings, the Arizal made a point of calculating not only the numerical values of the words in the original Hebrew text, but also the numerical values of the Aramaic translation of the Torah. He taught that the translation is like a "face on the back" that can also be contemplated. Like the back of a beautiful tapestry that reveals how the craftswoman created her work of art, so the Aramaic translation of the Torah text reveals God's exquisite craftsmanship.

There are three principal Aramaic translations of the Torah: Targum Unkelus, Targum Yonatan and Targum Yerushalmi. Although generally similar to one another, their interpretations sometimes differ. This is true of the first word of the Torah, *Bereishit* (בְּרֵאשִׁית). Targum Unkelus, the prevailing translation, uses the word, "Before" (בְּקַדְמִין), in the sense of "primordial existence." Targum Yonatan offers the most literal meaning, translating *Bereishit*, "In the beginning" (מִן אוּלָא), which is the version generally adopted by English translations. The third Aramaic translation, Targum Yerushalmi, translates *Bereishit* as "With wisdom" (בְּחוּכְמָא). ▶This relates to the verse in the Book of Psalms, "How great are Your acts, God! You have made them all with wisdom."[25]▶▶ It also corresponds to the expression, "The beginning of wisdom" (רֵאשִׁית חָכְמָה),[26] which synthesizes the latter two translations into one phrase.

The sum of the numerical values of these three different translations of *Bereishit* (בְּרֵאשִׁית) is 411. Appropriately, the numerical value of "something from nothing" (יֵשׁ מֵאַיִן) is also 411, which equals 3 times 137. Thus, the average value of each translation of *Bereishit* is exactly 137.

Name of translation	Translation of *Bereishit*	Meaning in English	Numerical value ▶▶▶
Unkelus	בְּקַדְמִין	"Before"	206 (the mid-point of 411)
Yonatan	מִן אוּלָא	"In the beginning"	128
Yerushalmi	בְּחוּכְמָא	"With wisdom"	77
Total numerical value			411
Average numerical value			137

▶ From the perspective of Kabbalah, each alternative translation corresponds to one of three different spiritual dimensions.

"Before" (בְּקַדְמִין), refers to the inner dimension of the unconscious beginning, or the "conception" of the world. "In the beginning" (מִן אוּלָא) refers to the external dimension of the unconscious beginning, relating to "pregnancy." "With wisdom" (בְּחוּכְמָא) refers to the conscious beginning of creation that corresponds to "birth."

▶▶ A further indication that *Bereishit* (בְּרֵאשִׁית) is related to wisdom is that the numerical value of the entire first verse of the Torah equals 2,701, which is the triangle of 73 (i.e. the sum of all numbers from 1 to 73), the numerical value of wisdom (חָכְמָה).

▶▶▶ Taking the finite differences between the numerical values of each translation, we construct the following series:

53 **77 128 206** 311 443 602

24 **51 78** 105 132 159

27 **27** 27 27 27

The sum of the first seven integers in the series is 1,820 (70 times 26, the numerical value of God's essential Name, *Havayah*). The sum of the first two integers (53 and 77) is 130 (5 times *Havayah*). The sum of the remaining five integers is 1,690, which equals 26 (the numerical value of the Name *Havayah*) times 65 (the numerical value of the Name *Adni*).

The Torah opens, "In the beginning, God created the heavens and the earth" (בְּרֵאשִׁית בָּרָא אֱ-לֹהִים אֵת הַשָּׁמַיִם וְאֵת הָאָרֶץ). This entire first verse of creation contains eleven different letters, exactly one half of the twenty-two letters of the Hebrew *alef-bet*:

א ב ג ד ה ו ז ח ט י כ ל מ נ ס ע פ צ ק ר ש ת

Just as the Aramaic translation reveals the "face on the back," there are eleven letters that are hidden from our direct view. The sum of the numerical values of these eleven letters is also 411—again, 3 times 137.

The eleven revealed letters correspond to direct light, while the eleven concealed letters correspond to returning light. When the letters are lined up "directly" and "returning," the numerical value of their dot product is 6,950, which equals 25 [the numerical value of "Let there be" (יְהִי)] times 278 [the numerical value of, "concealed light" (אוֹר הַגָּנוּז)].

א	ב	ה	ו	י	ל	מ	צ	ר	ש	ת	Total
ק	פ	ע	ס	נ	כ	ט	ח	ז	ד	ג	
1·100	2·80	5·70	6·60	10·50	30·20	40·9	90·8	200·7	300·4	400·3	
100	160	350	360	500	600	360	720	1,400	1,200	1,200	6,950

The numerical value of the first verse of creation is 2,701, which equals Abel (37, הֶבֶל), times "wisdom" (73, חָכְמָה), relating this verse to the father principle.◄ However, when the entire verse is rewritten by exchanging the letters for these "hidden" letters, the numerical value is 1,273, which is the numerical value of the phrase, "the hidden things are for *Havayah*, our God" (הַנִּסְתָּרֹת לַהֲוָי' אֱ-לֹהֵינוּ). This number also equals "Eve" (19, חַוָּה) times "understanding" (67, בִּינָה), both of which relate to the feminine, mother figure. The sages state that women are endowed with greater understanding than men.[28] Like the revealed and hidden letters when examined in this way, Adam and Eve were created back-to-back. In Chassidut, Eve is the unconscious aspect of Adam's psyche.[29]

Primordial Chaos and Rectification

Continuing our reading of the Torah, we soon come across a simple

► The normative value of "wisdom" (חָכְמָה), mentioned here, is 73. However, the ordinal value is 37 and the reduced value is 19. In each case, as we reduce the word further, the next number is the mid-point of its predecessor.[27]

word that has a numerical value of 411. This is the tenth word of the Torah, "chaos" (תהו). This is a very significant word that we will meet many more times before the end of this book. "Chaos" describes the state of creation before it was rectified by the creation of light. Since this word has three letters, the average numerical value of the letters in the word is 137.

There is a general rule in the Torah that states that "The tenth is holy to God."[►30] "Chaos" is the tenth word in the Torah. Does this mean that chaos is holy? By standards of Jewish law, a fact is established by three occurrences.[31] If chaos aspires to be holy, its connection to the number 10 needs to be present at least twice more, as indeed it is:

- "Chaos" is the tenth word of the Torah
- There are exactly ten appearances of the word "chaos" (תהו) in the Hebrew Bible without a prefix letter[32]
- The word "chaos" (תהו) appears ten more times with a prefix letter (or letters)[33▼▼]

Thus, we have three different phenomena that relate "chaos" (תהו) to the number 10, which indicates that it has particularly holy significance.

The statement in *Sefer Yetzirah*, "There are ten *sefirot* … ten and not nine, ten and not eleven"[34] is true specifically in the World of Chaos.[35] In the World of Rectification, there may be *partzufim* (inter-included personae) that contain nine, or eleven *sefirot*. This means that ten is typically a property of chaos. In the World of Rectification, either the energy is limited (nine *sefirot*), or the vessels are proportionate to the energy input (eleven *sefirot* contained in twenty-two vessels, or letters).

In the World of Chaos there is an abundance of energy and an unproportional supply of unyielding vessels. The primordial World

▶ The foremost verse where this phrase appears relates to the tithing of cattle. This represents the rectification of the chaotic-animalistic aspect of the world.

▶▶ The sum of the numerical values of all ten prefixes (11 letters) is 83. The sum of the numerical values of all ten words with the prefix is 8,303, which equals 23 [the numerical value of Chayah (חַיָּה), as the first woman was called before the primordial sin] times 19^2 [the squared value of Eve (חַוָּה), the name given to her after the sin]. This implies that before the sin, Eve possessed the great lights and vessels necessary to rectify all chaos in reality.

The companion word of "chaos" (תהו) is "void" (בהו), which appears three times in the Bible. It appears twice as "and void" (ובהו) (numerical value of 19) and once without a prefix (13).

Together there are 23 (חַיָּה, Chayah, rectified Eve) words with 13 [the numerical value of (בהו)] prefix letters. The sum of the numerical values of all 13 prefix letters is 95, which equals 5 times 19 [Eve (חַוָּה)].

▶ The numerical value of Hadar (הֲדַר) is 209, which is the numerical value of "the righteous one" (הַצַּדִּיק), implying Hadar's association with yesod (the sefirah of foundation) relative to the other kings. This reflects the fact that Hadar is the initial point of rectification.

The numerical value of Meheitavel (מְהֵיטַבְאֵל) is 97, which is the sum of the two Divine Names Mah and Ban (with numerical values of 45 and 52, respectively). The Name Mah (מה) is derived by filling the letters of the essential Name of God, Havayah with the letter alef (יוד-האיואו-הא). When the numerical values of the three alefs that fill this Name are considered to be 1,000 each, the total numerical value of Havayah is 3,042, which is equal to 2 times 39^2 [where 39 is the numerical value of "God is one" (הוי אֶחָד)], and is also a multiple of 26 (the numerical value of Havayah).

▶▶ The numerical value of Hadar (הֲדַר) when spelled in full (הי דלת ריש) is 959, which is 7 times 137. This spelling has eight letters (remember, Hadar was the eighth king). The numerical value of Meheitavel (מְהֵיטַבְאֵל) is 97, which is the numerical value of "time" (זְמַן). When the numerical value of the alef (א) in Meheitavel (מְהֵיטַבְאֵל) is considered to be 1,000 [one meaning of the letter alef is "one thousand" (אֶלֶף)], the numerical value of her name is 1,096, which equals 8 times 137. Similarly, there is a method of filling the letters of Meheitavel (מְהֵיטַבְאֵל) that yields a total of 1,096. Meheitavel empowered Hadar with an additional force that enabled him to rise above the failure of the preceding kings, and become the eighth king. Indeed, 1,096 is also the numerical value of the phrase "a wise woman"[38] אִשָּׁה מַשְׂכֶּלֶת), a subject that we will expand upon later (Chapter 8).

of Chaos contained seven vessels that shattered one after the other. These vessels were unable to accommodate the chaotic high-energy "lights" that emanated from the three higher *sefirot*. Imagine what happens when too much power floods through the filament of an electric light bulb. The filament bursts, and the light bulb explodes from the inside out. When there's too much energy, vessels shatter. This is exactly what happened.

The seven vessels of chaos are symbolized by the seven Edomite kings mentioned in Genesis.[36] Each of their names is accompanied by the phrase, "and he reigned … and he died." A short lifespan is a manifestation of chaos. Like volatile elementary particles, each of the seven kings died shortly after his coronation. Having listed the names of these seven kings, whose untimely deaths disclose their chaotic origins, the Torah then mentions Hadar [הֲדַר; lit: "Splendor"], the eighth king. ◀

Hadar was obviously different from the other seven kings. The Torah states that he reigned, but does not state that he died. He corresponds to a more stable type of particle with a longer lifespan. In Kabbalistic terms, his vessel did not break. Hadar's appearance marks the transition from the World of Chaos to the World of Rectification. This new world aspires to build refined vessels that are able to contain the great chaotic lights. Hadar is thus the kernel of rectification within chaos.

A second difference between Hadar and all the other kings of the World of Chaos is that he was married. His happy marriage to Meheitavel was the secret of his rectification. Kabbalah teaches us that the success of their marriage manifested in their ability to achieve a rectified state of inter-inclusion, by which each component of the dual state successfully integrates the reflection of its counterpart.[37] Hadar integrated Meheitavel into his psyche and vice versa. ◀◀ Hadar's marriage to the wise Meheitavel thus alludes to the union of Torah and science (as mentioned in the Introduction).

Bachelorhood is a high-energy state of chaos that is doomed to break. Unmarried individuals are liable to experience in their consciousness—and especially in their dreams—the dire effects of vessels breaking. Even a married person who has not perfected a faithful attitude to his or her spouse, may still belong to this chaotic realm. A loving marriage is the secret of a good and long life, as the verse in the Book of Ecclesiastes states, "See life with the wife whom you love."[39] The lesson to learn from this is that living in chaos means taking the risk of being jolted by high voltage energy, leading to spiritual breakdown. (Of course, the other option is to live a happily married life.)▶

We can conclude that chaos is holy when it is coupled to a stable force. In a rectified state, the holiness of chaos is contained within limits and infuses the world with vital energy.

The Creation of Light

Advancing to the third verse of Genesis, we come to the description of God's creation of light: "'Let there be light!' and there was light" (יְהִי אוֹר וַיְהִי אוֹר).[41] When each of the 13 letters of this phrase are spelled in full, we discover a remarkable phenomenon.▶▶

Letter of phrase	Spelling	Number of letters	Numerical value of spelling
י	יוד	3	20
ה	הא	2	6
י	יוד	3	20
א	אלף	3	111
ו	וו	2	12
ר	ריש	3	510
ו	וו	2	12
י	יוד	3	20
ה	הא	2	6
י	יוד	3	20
א	אלף	3	111
ו	וו	2	12
ר	ריש	3	510
Total		**34**	**1,370**

▶ The sum of Hadar's name, written in full (959), and Meheitavel when written in this way (1,096), is 2,055, which is 15 times 137. 15 is the numerical value of the Name *Kah* (יה), which implies the presence of the *Shechinah* in the union between man and wife.[40]

The numerical value of Hadar is 959 when the letter *hei* is filled with a *yud* (הי). The regular spelling of *hei* is with an *alef* (הא). Filling it with a *yud* (הי) relates to the Name *Sag*, which is the root of the World of Chaos. The numerical value of Meheitavel with the same filling (ממ הי יוד טית בית אלף למד) is 1,131. The sum of this number and 959 is 2,090, which equals 10 times 209, the numerical value of Hadar (הֲדַר). The average numerical value of each letter when spelled in full is 209 (הֲדַר).

▶▶ There are 13 letters in the phrase "'Let there be light!' And there was light" (יְהִי אוֹר וַיְהִי אוֹר)—13 is an inspirational number, i.e., the sum of two consecutive square numbers (in this case, 2^2 and 3^2).[42] Inspirational numbers can be represented by drawing the smaller square inside the larger square. Thus, the 13 filled letters of this phrase can be written:

יוד   הא   יוד

וו   אלף

יוד   וו   ריש

הא   יוד

אלף   וו   ריש

This reveals that the sum of the midpoints of the four sides is 548 (4 times 137) and the remaining diagonals add up to 822 (6 times 137).

The sum of the numerical values of the four corners is 661 (the numerical value of Esther, אֶסְתֵּר). The sum of the midpoints of the four sides is 548 (twice the numerical value of Mordechai, מָרְדְּכַי).

The entire phrase now contains 34 letters, alluding to the 34th prime number, 137.▼ The total numerical value of these letters is 1,370, which equals 10 times 137. This principal phrase relating to the creation of light thus has a numerical value equal to 137 multiplied by the entire array of God's ten channels of Divine light. This is one reference to the relationship between the Hebrew names we suggested for the photon (*egel*) and the electron (*ofan*), which have numerical values of 34 and 137, respectively, as mentioned.

Also as mentioned in the Introduction, one of the earliest and most mystical forms of Torah analysis is counting the verses, the words and the letters of the Torah.

Like the 34 letters in the phrase relating to the creation of light, there are 34 verses in the account of the seven days of creation.◄◄

Now, let's count the number of words in the account of the seven days of creation. There are 469.◄◄◄

If we categorize all 469 words into groups according to their number of letters, we find that there are seven groups, from short words containing two letters to the longest word containing eight letters.▼▼▼▼

▶▶ Of these 34 verses, the first verse begins with the letter *bet* (2), the first letter of *Bereishit* (בְּרֵאשִׁית). All the other 33 (!) verses begin with the letter *vav* (6). The sum of the numerical values of the initial letters of these 34 verses is exactly 200, the numerical value of the letter *reish* (ר), the letter that follows the letter *bet* (ב) in the word *Bereishit* (בְּרֵאשִׁית). The sum of the numerical values of the remaining initial letters in these 34 verses is 1,781, which equals 13 times 137.

▶▶▶ 469 is the sum of the numerical values of "light" in Hebrew (אוֹר) and in Aramaic (נְהוֹרָא). It is also the 13th Shabbat (hexagonal) number.

▶ Prime numbers are positive, non-zero numbers that have exactly two factors—no more, no less. A prime number is divisible only by 1 and itself. The number 1 fulfills these conditions and is therefore a prime number.

There is no way to predict what number in the series of primes a particular prime number will be. Prime numbers correspond to the unknowable head of mathematics.

The 34th prime number (137) alludes to the 34 letters of the filling of the filling of the Name *Adni* (אלף למד פא דלת למד תו נון ואו נון יוד ואו דלת).[43] An alternative way of presenting this Name is with 35 letters.[44] The numerical value of the filling letters alone (וד או וד לף ו יש וד או וד לף ו יש) is 900, which equals 30^2. The letter that has a numerical value of 30 is the letter *lamed* (ל). 30^2 represents the secret of the Jewish heart which is two *lamed*s facing one another to form the shape of a heart, (as taught by the medieval Kabbalist, Rabbi Avraham Abulafia).

This alludes to the love of a married couple who have rectified their initial back-to-back relationship by turning to face one another.

▶▶▶▶ The total number of three-lettered words in the first day of creation is 16 [the numerical value of "where" (אַיֵּה). The numerical value of this word when spelled in full is 137 (see Chapter 8)]. After "chaos" (תֹהוּ, the third three-lettered word of creation), there are 13 more words in the first day of creation. The numerical value of the last word, "one" (אֶחָד), is 13. This represents the conclusion of the "void" (בֹהוּ), which also has a numerical value of 13 (in the verse, "void" is preceded by a prefix letter and is thus not included in the list of three-lettered words). In addition, the sum of the numerical values of the 14 words from "chaos" (תֹהוּ) to "one" (אֶחָד) is 2,209, which equals 47^2. 47 is the numerical value of the phrase "that it was good" (כִּי טוֹב) with which God acknowledges the creation of light. It is also the numerical value of "selflessness" (בִּטוּל), the inner attribute of *chochmah* (the *sefirah* of wisdom).

Group	Number of words	Number of letters in words
1. 2 letter words	54	108
2. 3 letter words▶	137	411
3. 4 letter words▶▶	127	508
4. 5 letter words	123	615
5. 6 letter words	24	144
6. 7 letter words	3	21
7. 8 letter words	1	8
Total:	**469**	**1,815**

Analyzing the table above, we see that the number of three-lettered words in the entire account of creation is 137, the maximal number of words of any size. This is another outstanding example of an appearance of 137 in creation. The total number of letters in these 137 words is 411, which is the numerical value of "chaos" (תהו), as mentioned. The total number of four, five and six lettered words (127 ⊥ 123 ⊥ 24) is 274, which equals 2 times 137. The number of letters in the two, seven and eight lettered words (108 ⊥ 21 ⊥ 8) is also 137.

Seven Days of Understanding

God created the world with wisdom, i.e., with *chochmah* (the *sefirah* of wisdom). The inseparable companion to *chochmah* is *binah* (the *sefirah* of understanding). *Binah* takes the initial conceptual flash of wisdom and develops it into a complete picture.

The numerical value of *binah* (בִּינָה) is 67. As we saw above, the number of words in the account of creation is 469, which is the product of 7 times 67. This means that the average number of words for each of the seven days is 67 words per day.▶▶▶

During each of the six days of creation, the initial inspirational spark of *chochmah* expanded and developed through *binah* into a completed form, reaching its consummate state on the seventh day: Shabbat.

▶ The total number of three-lettered words in the first day of creation is 16 [the numerical value of "where" (אַיֵּה). The numerical value of this word when spelled in full is 137 (see Chapter 8)]. After "chaos" (תהו, the third three-lettered word of creation), there are 13 more words in the first day of creation. The numerical value of the last word, "one" (אֶחָד), is 13. This represents the conclusion of the "void" (בהו), which also has a numerical value of 13 (in the verse, "void" is preceded by a prefix letter and is thus not included in the list of three-lettered words).In addition, the sum of the numerical values of the 14 words from "chaos" (תהו) to "one" (אֶחָד) is 2,209, which equals 47^2. 47 is the numerical value of the phrase "that it was good" (כִּי טוֹב) with which God acknowledges the creation of light. It is also the numerical value of "selflessness" (בָּטוּל), the inner attribute of *chochmah* (the *sefirah* of wisdom).

▶▶ The number of four-lettered words is 127, the life-span of Sarah and the number of five-lettered words is 123, the life-span of Aaron (see Chapter 7).

Regarding the life-spans of the seven generations between Abraham and Moses, the relation between 137 and 127 is essentially male and female, respectively (see Chapter 7). Similarly, the relation between 3 and 4 is male-female. There are three Patriarchs and four Matriarchs (see Appendix B).

▶▶▶ The numerical value of the letter *zayin* (ז) is 7. When we spell the name *zayin* in full (זַיִן), its numerical value is 67. So, the number of words in the passage (469) equals 7 (ז) times 67 (זַיִן).

▶ The number of letters that remain after subtracting the three-lettered words (1,815-411) is 1,404, which equals 54 (the number of two letter words in creation) times 26 (the numerical value of God's essential Name, *Havayah*). 1,404 also equals 2 times 702, the numerical value of Shabbat (שַׁבָּת). This alludes to the teaching that if all the Jewish People would correctly observe two consecutive Shabbats, the world would immediately merit the complete redemption.[45] When the digits of the number 702 are reversed, they form the number 207, the numerical value of "light" (אוֹר), which equals 54 times 13, the numerical value of "love" (אַהֲבָה).

The letters that spell "love" (אהבה), are the initial letters of the phrase, "The light of the Holy One Blessed be He" (אוֹר הַקָּדוֹשׁ בָּרוּךְ הוּא).

The creation of human life is emphasized uniquely with the words "male and female [did He create them]" (זָכָר וּנְקֵבָה). The numerical value of this phrase is 390, which equals 30 times 13, the numerical value of "love" (אַהֲבָה). This teaches us that God not only granted mankind life, He created us with love.

The pinnacle of God's love is expressed to us in the creation of Shabbat. This relates to the three concepts of light, life and love, presented at the beginning of this chapter.

As we shall see, Shabbat, the seventh day of the week, is the culmination of this process. Shabbat also alludes to the to the time of the final redemption when chaos will be rectified, achieving the ultimate state of perfection. ◀

"Let There Be"

So far, we have discovered some relatively simple appearances of the number 137 in the Torah. These we found in the numerical values of words and phrases relating to the first three verses of Genesis describing creation, and in the number of letters and words in that narrative. Let's see what happens when we investigate other ways of counting the letters of the Torah text.

The Mishnah states that God created the world with ten sayings.[46] Nine of these ten sayings are the nine times that the phrase "God said" appears explicitly in the description of creation. The first saying is implicit in the phrase, "In the beginning God created."[47]

The most common statement of creation is "Let there be" (יְהִי), which is the phrase related to the creation of light. It appears three times in the ten sayings.

"Let there be light!"	יְהִי אוֹר
"Let there be a firmament in the water to separate between the [higher] waters and the [lower] waters."	יְהִי רָקִיעַ בְּתוֹךְ הַמָּיִם וִיהִי מַבְדִּיל בֵּין מַיִם לָמָיִם
"Let there be luminaries in the firmament of the heavens, to differentiate between day and between night; and they shall be signs for seasons and days and years. And they shall be for luminaries in the expanse of the heavens to illuminate the earth."	יְהִי מְאֹרֹת בִּרְקִיעַ הַשָּׁמַיִם לְהַבְדִּיל בֵּין הַיּוֹם וּבֵין הַלַּיְלָה וְהָיוּ לְאֹתֹת וּלְמוֹעֲדִים וּלְיָמִים וְשָׁנִים וְהָיוּ לִמְאוֹרֹת בִּרְקִיעַ הַשָּׁמַיִם לְהָאִיר עַל הָאָרֶץ

These three verses correspond to the three translations of the word *Bereishit* ("In the beginning") mentioned earlier.▸

- "Let there be light,"[48] corresponds to *chochmah* (the *sefirah* of wisdom), the father principle. *Chochmah* emanates from the inner dimension of *keter* (the *sefirah* of the super-conscious crown), often called "nothing" in Kabbalah because it is not consciously experienced in the soul. So, "Let there be" implies creating something (light) from nothing.▸▸

- The second saying corresponds to *binah* (the *sefirah* of understanding), the mother principle, which emanates from the external aspect of *keter*. It implies developing something from something that already exists, just as a woman develops the fetus in her womb.

- Finally, in the description of the fourth day of creation, we find the expression, "Let there be luminaries in the firmament of the heavens, to differentiate between day and between night; and they shall be signs for seasons and days and years. And they shall be for luminaries in the expanse of the heavens to illuminate the earth."[51] This corresponds to *da'at* (the *sefirah* of knowledge). *Da'at* includes both the power of union and the power of differentiation.

These three verses also relate to the stages of creation mentioned previously—light, water and firmament. The first verse clearly relates to light. The subject of the second verse is the creation of the firmament, which takes place entirely in the realm of water, dividing between the upper and lower waters. Finally, the subject of the third verse describes the creation of the light-giving luminaries situated in the firmament.▸▸▸

Like the procreation process alluded to in these three stages of creation, there is a definite progression in the number of letters in each phrase. The first phrase, "Let there be light," has 6 letters, relating to the letter *vav* (ו), the sixth letter of the Hebrew *alef-bet*, which has a

▸ These three verses correspond to charge (C), parity (P) and time (T), respectively. CPT symmetry is discussed in Chapter 14

▸▸ In the Shabbat morning prayers, we recite the liturgical poem beginning "*Kel Adon.*"[49] The 19th line of this poem is: "He called upon the sun and light shone forth" (קָרָא לַשֶּׁמֶשׁ וַיִּזְרַח אוֹר). The numerical value of the initial letters of this phrase (ק ל ו א) is 137. The numerical value of the first three words "He called upon the sun" (קָרָא לַשֶּׁמֶשׁ וַיִּזְרַח) is 1,202, which is the numerical value of "In the beginning God created" (בְּרֵאשִׁית בָּרָא אֱלֹהִים). Thus, the phrase alludes to "In the beginning, God created ... light."

▸▸▸ The total numerical value of all four verses of the fifth day of creation, on which the fish and the birds were created is 13,689, which equals 117^2. 117 is the discrete version of the square root of 137, which is 11.7047, which is why 13,689 is so close to 13,700 (100 times 137).

▶ "Let there be" refers to creation "something from nothing" and also to creation "something from something." The numerical value of "Let there be something from nothing, something from something" (יְהִי יֵשׁ מֵאַיִן, יֵשׁ מִיֵשׁ) is 1,096, which equals 8 times 137. As explained earlier, 1,096 is the numerical value of Meheitavel, according to two different renderings of her name. It is also the numerical value of "an intelligent woman" (אִשָּׁה מַשְׂכָּלֶת).[52]

▶▶ The dot-product of the three letters of the word, "Let there be" (יְהִי) is 500 (10·5·10), which is equal to the dot-product of the letters of "nothing" (אַיִן; 50·10·1), the 500th word of the Torah.

The numerical value of, "be fruitful and multiply" (פְּרוּ וּרְבוּ) is also 500. The power of procreation is the finite "something-ness" of man that derives from the infinite power of God, who appears to us to be "nothing," i.e., the power of procreation reflects God's creation of "something" from "nothing."

numerical value of 6. (The relationship between *vav* and the number 137 is explained in Chapter 9.)

"Let there be a firmament in the heavens to separate between the [higher] waters and the [lower] waters," relates to the splitting of the waters in the heavens, which created a dual reality in the world. This second phrase contains 34 letters, alluding to the splitting of spectral lines (to be discussed later), represented by the 34th prime number, 137.

The number of letters in the third phrase is 97, which is the numerical value of Meheitavel (מְהֵיטַבְאֵל), whom we met above with reference to rectifying chaos. It is also the numerical value of "time" (זְמַן), which is the general subject of this phrase.◀

The total number of letters in these three phrases (6 ⊥ 34 ⊥ 97), spoken by God, each of which begins, "Let there be," is 137.◀◀

The Colorful Constant

The first chapter in Ezekiel, mentioned previously with reference to the correspondence of the *ofan* to the electron, concludes with the verse that begins, "Like the appearance of a rainbow in a cloud on a rainy day, so was the appearance of the surrounding glow." The refraction of sunlight through "a cloud on a rainy day" is an explicit reference to the color spectrum.

Several centuries ago, Isaac Newton observed that when white sunlight is shone through a prism it refracts and disperses into a rainbow of colors. Upon closer inspection of the color spectrum produced by sunlight, scientists noticed that at certain points in the spectrum, dark lines appear that break its continuity. As a result of this study, scientists discovered that the chemical elements in the gaseous substance of the sun absorb the frequencies of light that would normally produce the missing colors. Further experimentation has confirmed that every chemical element absorbs different frequencies of light. This is due to the number and location of the electrons that orbit the nucleus in the atom. Since each chemical element has a specific number of electrons, each has its own unique set of spectral lines.

Physicists later observed that spectral lines have a fine-structure, meaning that they sometimes split into two. The fine-structure of an element can be measured experimentally. In 1916, Arnold Sommerfeld discovered the basic equation that defines the fine-structure of any particular element. He did so by introducing the fine-structure constant, which is approximately equal to $\frac{1}{137}$.

Sommerfeld's calculations were founded on the assumption that electrons orbit the nucleus of an atom elliptically, similar to the way the planets orbit the sun. The model of fixed atomic orbits was superseded a decade or so later by an innovative synthesis of quantum theory and special relativity. This new hypothesis—still accepted today—shows that the electron displays an elusive quality, described by the probability that the electron will be in a specific region anywhere around the nucleus, depending on its quantum energy level.

Like Sommerfeld's theory, this later theory provided an explanation for the fine-structure observed in the atomic spectra, but on the basis of a completely different phenomenon. Yet, despite the differences between the two methods of calculation, the new equations were almost identical to Sommerfeld's and also incorporated the fine-structure constant (α)! The mysterious impression left by this quirk of fate was enhanced by the fact that *alpha* is the only dimensionless quantity that can be formed from any combination of its basic components: electric charge (e), the permittivity of free space (ε_0), Planck's constant (h), and the speed of light (c).

As we shall discuss in Chapter 9, the splitting of spectral lines is the result of the spin property of the electron, which we have associated with the *ofan*, the spinning "wheel within a wheel" of the Divine Chariot. Returning to the abovementioned verse from Ezekiel, like the numerical value of *ofan* (אוֹפָן), the numerical value of "the surrounding glow" (הַנֹּגַהּ סָבִיב) is 137.

> ▶ Beginning with the first letter and skipping four letters each time, we discover the word "where" (אַיֵּה). The numerical value of this word when spelled in full is 137 (see Chapter 8).
>
> אוֹר חַיִּים אַהֲבָה

Back and Front

We began this chapter with the three words, "light" (אוֹר), "life" (חַיִּים) and "love" (אַהֲבָה). We conclude by contemplating these words as the basis for another meditation. ◀

The sum of the numerical values of these three words is 288. This number is most significant in Kabbalah, especially with relation to rectifying chaos.

As a result of the excessive energies in the World of Chaos, the seven vessels in that world shattered and fell from their elevated spiritual plane. They descended into the World of Rectification. Kabbalah teaches us that 288 sparks of light remained in the broken vessels. In order to rectify these sparks, it is incumbent upon us to separate them from the shards of broken vessels, refine them, and elevate them to their source.

Hebrew words can be analyzed by using another central Kabbalistic technique: the "back" and the "front" of a word. The "back" value of a word is computed by inflating the word by duplicating and adding the letters, and then calculating the total numerical value. If we take the English word "love" as an example, we would write it "l-lo-lov-love," to calculate the "back" value. The "face" value is computed by degenerating the word, depleting the letters one by one until the word disappears completely, e.g., "love-ove-ve-e." The back represents the original, chaotic attempt at approaching the destination while facing the origin. The front represents the rectified return to the origin with the face turned towards the destination, a state referred to as "face to face." In this case, we write the back and face of all three words "light" (אוֹר), "life" (חַיִּים) and "love" (אַהֲבָה)—like so:

א או אור אור ור ר

ח חי חיי חיים חיים ייم יم م

א אה אהב אהבה אהבה הבה בה ה

The numerical value of the entire array is 1,233, which equals 9 times 137. This means that the average value of each expression is 411, the numerical value of "chaos" (תֹּהוּ), the root of the 288 fallen sparks.▼

▶ The sum of 1,233 (the total numerical value of all three words spelled back and front) and 288 (the regular numerical value of the three words) is 1,521, which equals 39^2. 39 equals 3 times 13 and is the numerical value of "*Havayah* is one" (הוי' אֶחָד). The ultimate rectification of these three concepts is to return them to their source in the one God, who is the source of love, and the Creator of light and life. The numerical value of "love" (אַהֲבָה) is 13, which is also the numerical value of "one" (אֶחָד).

As seen in the table, the sum of the numerical values of the three words, "light, life, love" (אור חיים אהבה) together with the filling and the filling of the filling is 6,448, which is equal to 26 (the numerical value of *Havayah*) times 248 (the numerical value of Abraham, who is associated in particular with light and love). The number 248 is also the numerical value of "particle" (חֶלְקִיק). One possible unified theory of the universe is that there are exactly 248 different elementary particles (see Chapter 4).

Word with filling and filling of filling			N. value
word — אהבה	חיים	אור	288
filling — הא בית הא אלף	מם יוד יוד חית	ריש וו אלף	1,706
filling of filling — הא יוד אלף תו / בית / הא למד אלף פא	מם מם / יוד וו / יוד וו דלת / חית דלת	וו ריש יוד / וו שין / למד וו פא	4,454
Total numerical value			6,448

REFERENCE NOTES FOR CHAPTER 1

1. *Avot* 5:16.
2. See *Zohar Bereishit* 13a.
3. Isaiah 41:8.
4. *Annalem der Physik*, vol 4 (1901), p. 552 – translation in *Great Experiments in Physics* ed: M. H. Shamos.
5. Daniel 12:3.
6. *Avot* 5:1.
7. *Etz Chayim* 11:6.
8. Genesis 1:3-5.
9. Genesis 1:6-8.
10. Psalms 19:3.
11. *Etz Chayim* 11:6. This further corresponds to the process of *chash-mal-mal*. See Chapter 9.
12. See our article "The Nine Steps to Actualizing Potential." Available on request at www.inner.org.
13. Ezekiel 1:15, 16.
14. Ezekiel 1:16; Rashi ad loc.
15. Job 38:28.
16. Isaiah 26:19.
17. Genesis 1:28.
18. See our book, *913: The Secret Wisdom of Genesis*, p. xii.
19. *Etz Chayim* 13:5. See also *Likutei Torah, Vezot Haberachah* 93.
20. http://www.physics.nist.gov/cuu/Constants/alpha.html
21. Genesis 4:1.
22. See *What You Need to Know About Kabbalah* pp. 150-151.
23. Rabbi Yehudah Leib of Anipolya, a contemporary and close friend of Rabbi Shneur Zalman of Liadi, authored the book *Or Haganuz*. There, he interprets the significance of each of the implicit appearances of this Name in the Five Books of Moses.
24. *Zohar Shemot* 161a.
25. Psalms 104:24.
26. Psalms 111:10; Proverbs 4:7.
27. This phenomenon is explained further in our book *913: The Secret Wisdom of Genesis*, p. 32.
28. *Niddah* 45b.
29. See *The Mystery of Marriage*, Chapter 1, p.8-9.
30. Leviticus 27:32.
31. *Baba Metzia* 106b.
32. Genesis 1:2; I Samuel 12:21; Isaiah 24:10; 34:11; 44:9; 45:18, 19; 59:4; Jeremiah 4:23; Job 26:7.
33. Four times "in chaos" (בתהו, Isaiah 29:21; Psalms 107:40; Job 6:18; 12:23); twice "and chaos" (ותהו, Isaiah 40:17; 41:29); once "the chaos" (התהו, I Samuel 12:21); once "like chaos" (כתהו, Isaiah 40:23), once "for chaos" (לתהו, Isaiah 49:4), and once "and in chaos" (וּבְתהו, Deuteronomy 32:10).
34. *Sefer Yetzirah* 1:4.
35. *Etz Chaim, Sha'ar Hakelalim* Ch. 1.
36. Genesis 36:31-39.
37. *Kuntres Hechaltzu*.

38. Proverbs 19:14.

39. Ecclesiastes 9:9.

40. See our book *The Mystery of Marriage*, ch. 5.

41. Genesis 1:3.

42. We refer to this series of numbers $f(n) = n^2 \perp (n \perp 1)^2$ as "inspirational numbers." See our book in Hebrew, *Einayich Breichot Becheshbon* for a full explanation of an inspirational number.

43. *Etz Chayim* Gate 34, Ch. 2:42.

44. Ibid.

45. *Shabbat* 118b.

46. *Avot* 5:1.

47. *Rosh Hashanah* 32a; *Megillah* 21b.

48. Genesis 1:3.

49. The grand significance of this poem is mentioned in *Zohar Shemot, Parashat Terumah*, 132a-b.

50. Genesis 1:6.

51. Ibid 1:14-15.

52. Proverbs 19:14.

COMPLEMENTARITY AND UNCERTAINTY

"We have two contradictory pictures of reality; separately neither of them fully explains the phenomena of light, but together they do." (Albert Einstein)

The Paradox of Complementarity

Sophisticated machines, skyscrapers and all forms of motorized transportation are so much a part of our everyday lives that we tend to take them for granted. We rest assured that the roof above our heads will remain intact, that we can safely drive a car on a well-built road, and that the underground tunnel we are driving through will not cave in. We have faith in their precise construction according to the well-defined laws of classical mechanics.

In classical mechanics, everything is simple. Be it a single molecule or a planet, once we know the physical properties of an object and the forces that affect it at any moment, we can predict the outcome of its interactions with reliable precision. Newtonian mechanics has proven itself over decades to be accurate and consistent.

Yet, when science descended to the subatomic level, scientists were perplexed to discover that classical mechanics just didn't work in this realm.

In 1927 (a decade after the fine-structure constant was first discovered by Sommerfeld), in a combined attempt to define the behavior of subatomic particles, Werner Heisenberg, together with Niels Bohr and Wolfgang Pauli, proposed the uncertainty principle. This was the initiating insight that led to the development of modern quantum mechanics.

The uncertainty principle states that, contrary to Newtonian mechanics, it is impossible to determine simultaneously both the position and the momentum of an electron—or any other elementary particle—with any degree of accuracy. Sometimes it may be possible to switch back and forth between different views of a particle to observe these properties. But, it is impossible to view both at the same time. Despite their simultaneous coexistence in reality, one can either measure the momentum of an electron in motion, or its position; together, the two can never be clearly defined.

Like the paradoxical nature of light, which behaves either as a wave or as a particle, the uncertainty principle defies the preconceived notions of Newtonian mechanics. This was another mind-boggling innovation of quantum mechanics that scientists had to grapple with.

Niels Bohr and his colleagues explained the uncertainty principle by introducing the idea of complementarity.

Complementarity supposes that multiple properties of apparently contradictory phenomena all derive from a common origin. In other words, the two extremes of every complementary pair are united at their source. Because of the paradox implied by the concept of complementarity, many of Bohr's contemporaries, including Einstein, considered it too mystical to be scientific. Be that as it may, Einstein's discovery that the fundamental complementary pair of energy and mass are united by his classic equation, $E = mc^2$, was a giant leap ahead that may eventually lead to discovering the source that unifies the two.

At the subatomic level, the two most common complementary pairs are the position-momentum pair and the wave-particle pair. However, Bohr and his colleagues recognized that complementarity is a universal principle that shapes our everyday lives with other fundamental pairs, such as light and darkness, life and death, or love and hate. Each member of every such pair is mutually exclusive of the other. But, Bohr believed, they emanate from the same source.

Complementarity in the Torah

In English, every sentence starts with a capital letter. In contrast, large (and small) letters are a very rare occurrence in the Torah. The location of large and small letters in the Torah is a matter of tradition, passed down from generation to generation. Their appearance is therefore open to interpretation. The first and most significant appearance of a large-sized letter in a Torah scroll is the large letter *bet* (ב) that begins the first word of creation, *Bereishit*. The letter *bet* is the second letter of the *alef-bet* and has a numerical value of 2, signifying duality. Its appearance as the initial letter of the Torah alludes to the fact that God created reality in complementary pairs.[1]▶

According to Kabbalah, every pair is comprised of a male and a female component. The first man and woman were originally created "back-to-back,"[2] meaning that each of them had a separate consciousness, but, at this stage there was no communication between them.

Like the position and momentum of a particle, the masculine and feminine states of consciousness cannot be perceived simultaneously. Nonetheless, although each appears to be different, they remain as inseparable as two sides of the same coin. The sages teach us that the common soul-root of a married couple splits in two as their souls descend to this world.[3] Thus, from the Torah's perspective, every pair does have a single origin, as Bohr hypothesized. The union in marriage of two apparently opposite phenomena reveals their common source. The key to discovering this unity is effective communication (see Chapter 9), which is achieved when each partner in the marriage reveals compassion for his or her spouse. The result of this mutual sensitivity is that the couple turns around until they are face-to-face.▶▶

As such, they are able to bear the paradox of their unity. Indeed, In Hebrew, "marriage" (נְשׂוּאִין) is conjugate to "bearing paradox" (נְשִׂיאַת הַפָּכִים).

This "turning around" happens when the male partner reveals the

▶ As a rule, every verse in the Torah is divided into two parts by the *etnachta* cantillation mark, which serves as a punctuation mark. This is another manifestation of the creation of pairs in the Torah.

The first verse of the Torah divides into two at "In the beginning God created." The *etnachta* appears at the third word, God (אֱ-לֹהִים). The number of letters in each half of the verse is 14, which is the numerical value of "hand" (יַד). The first verse thus alludes to the two "hands" of God, with which He created the world.

▶▶ When we stand back-to-back, all I see is my own view of the world. When I turn aound to see you face-to-face, I can see your view of the world and I also see my view as it is reflected in your eyes.

feminine aspect that is inter-included in his masculine soul and the female partner reveals the masculine aspect inter-included in her feminine soul. The mysterious union of these two hidden facets is the revelation of their common soul-root as it existed before it descended into their two bodies.

This idea can offer us some insight into discovering the connection between complementary pairs in general. If, for instance, we discover the wave function of a particle or the particle aspect of the wave, we would be well on the way to discovering their common origin. Indeed, physicists have discovered that a wave function is present in larger than subatomic particles. In macroscopic objects this function is so small that it is negligible.

Day of Unity

Duality is apparent in the first pair of words of the Torah (בְּרֵאשִׁית בָּרָא), both of which begin with the three letters of the verb "create" (ב־ר־א).◄ This indicates the dual nature of creation in general, as explained in the Zohar.[4] The third word of creation, which follows this pair of words, is the Divine Name (אֱ־לֹהִים).◄◄ Grammatically speaking, this Name is in the plural form. Any plural in the Torah that is unqualified by another number, automatically implies two, the minimum plural,[5] thus the plural *Elokim* implies duality. Nonetheless, the verbs associated with this Name are always in the singular, unless the reference is to "other gods." Thus, *Elokim* implies that God is the one, single source of all complementary pairs. This idea is alluded to in the fact that *Elokim* is the third word in the Torah. The number 3 implies duality together with the source of unity from which it emanates. This is like two parents who are the representatives of the "third partner"[6] in creating a child, i.e., God, who is the source of their ability to bear children.

The fact that unity is the source of duality is apparent in the Biblical account of the first day of creation. "And God said, 'Let there be light!' and there was light ... And there was evening and there was morning, day one."[7] The expression "day one" is grammatically different from

▶ "In the beginning" (בְּרֵאשִׁית), is a complete manifestation, which corresponds to the male aspect of the pair. "He created" (בָּרָא) is "half" a manifestation of the same phenomenon, corresponding to the female aspect. One physiological illustration of this is in the genetic code in males and females. The male genetic code contains both X and Y genes (a complete manifestation), whereas the female genetic code contains only X genes (half a manifestation).

▶▶ The third word, *Elokim*, or "God" reflects the third letter of the Torah, the silent letter alef in the word *Bereishit* (בְּרֵאשִׁית). The numerical value of the *alef* is 1. It is the essence of unity. *Elokim* is the mysterious source of duality hidden within God's unity.

each of the other days of creation, which conclude, "second day," "third day," "fourth day," etc. The sages explain that on day one of creation, the Divine consciousness of absolute unity pervaded all of creation.

On the second day of creation, God created the angels. Only then did the dual nature of creation become apparent. Angels cannot perceive both aspects of a paradoxical situation at any given moment. An angel can only perform a task that is related to its one innate talent, and no other.[8]

Complementary Pairs

In the Torah verses that describe the first day of creation, there are five explicit references to complementary pairs: heaven and earth, chaos and void, darkness and light, day and night, evening and morning.

We have already seen that the numerical value of "chaos" (תֹהוּ) is 411, which equals 3 times 137.▸ Remarkably, the electromagnetic coupling constant—alpha—is reflected many more times in the numerical values of these pairs and their combinations.

- The sum of the numerical values of "the heaven" (הַשָּׁמַיִם, 395) and "the earth" (הָאָרֶץ, 296), and "day" (יוֹם, 56) and "night" (לַיְלָה, 75) equals 822, which equals 6 times 137.

- The sum of the numerical values of "chaos" (תֹהוּ, 411) and "void" (בֹּהוּ, 13), "darkness" (חֹשֶׁךְ, 328) and "light" (אוֹר, 207) is 959, which equals 7 times 137. Since each of these words has three letters, they can be written in the form below.▸▸

▸ The sum of the numerical values of "chaos" (תֹהוּ; 411) and "chaos" in the *atbash* transformation (אצפ; 171) is 582. The average value of these two words is 291, which is the numerical value of "earth" (אֶרֶץ). This reflects the original phrase in which "chaos" appears, "and the earth was chaos."[9] The average numerical value of each of the letters of these two words is 97, the numerical value of Me-heitavel (מְהֵיטַבְאֵל).

▸▸ The sum of the numerical values of the four final letters of these words (ו ו ר ר) is 232, which is the numerical value of the phrase "let there be light" (יְהִי אוֹר).[10]

				Numerical value of word
Chaos	ו	ה	ת	411 (= 3 times 137)
Void	ו	ה	ב	13
Darkness	ך	ש	ח	328
Light	ר	ו	א	207
Numerical value of column	**232**	**316**	**411**	**959 (= 7 times 137)**

The initial letters of the four words (in the right-hand column of letters: א ח ב ת) also have a numerical value of 411, meaning that the sum of the remaining letters is 548, 4 times 137. Unlike the word "chaos" (תהו) these four letters do not form a meaningful word. The division into 3 times 137 (411) and 4 times 137 (548), where 3 and 4 are the numerical values of the letters gimel (ג) and dalet (ד), respectively, which combine to form Gad (גָּד). This relates to the question of which number is more important, 3 or 4 (see Appendix B).

♦ Another explicit pair that appears later in creation is "good" (initially associated with the creation of light[11]) and "evil."[12] In the verse, "He forms light and creates darkness, makes peace and forms evil,"[13] "evil" parallels "darkness." The sum of the numerical values of the words for these two related pairs, "light" (207, אוֹר) and "darkness" (328, חֹשֶׁךְ), and "good" (17, טוֹב) and "evil" (270, רע) is 822, which once again is a multiple of 137. These two pairs have the same numerical value as "heaven" and "earth," and "day" and "night," as mentioned above. The heavens above and the earth below correspond to good and evil, as noted in *Sefer Yetzirah*: "There is no goodness above pleasure and no evil below affliction."[14] Similarly, day and night clearly correspond to light and darkness.▼

As mentioned above, the numerical value of "chaos" (תֹהוּ) is 411. This means that the numerical value of "chaos chaos" (תֹהוּ תֹהוּ) is also 822. This alludes to two states of chaos in creation, as taught in Kabbalah, "stable chaos" and "unstable chaos."

In the world of "stable chaos" many lights were contained in one vessel. In the world of "unstable chaos" there was an unsuccessful attempt to contain the lights in a multitude of immature vessels. This form of "unstable chaos" resulted in the shattering of the primordial World of Chaos. The ultimate purpose of creation, assigned to

▶ Another Torah verse that has many sets of complementary pairs also reveals a connection with the numbers 137 and 411. "So long as the earth exists, seedtime and harvest, cold and heat, summer and winter, and day and night shall not cease."[15] The numerical value of the four pairs, "seedtime, harvest, cold, heat, summer, winter, day, night" (זֶרַע וְקָצִיר קֹר וָחֹם קַיִץ וָחֹרֶף יוֹם וָלַיְלָה), is 1,644, which equals 12 times 137, or 4 times 411. This means that each pair of concepts has an average numerical value of 411, the numerical value of "chaos" (תֹהוּ). Between the words in this verse are seven letters *vav* (ו), which is also closely related to 137, as will be discussed in Chapter 9.

The ten units of these five pairs correspond to the ten *sefirot* in the following way: heavens and earth correspond to *tiferet* and *malchut* (the *sefirot* of beauty and kingdom, respectively), as explained in many works of Kabbalah. Light and darkness correspond to *chochmah* (the *sefirah* wisdom), which is the initial revelation of light in the conscious mind, and *binah* (the *sefirah* of understanding). *Chochmah* corresponds to the spiritual sense of sight. *Binah* is associated with the spiritual sense of hearing. Thus, *binah*, relative to *chochmah* is referred to as darkness. The Arizal explains that chaos corresponds to *keter* (the *sefirah* of crown). Void, the beginning of the rectification process, is *yesod* (the *sefirah* of foundation), as explained in the subsequent text. Morning and evening correspond to *chesed* and *gevurah* (the *sefirot* of loving-kindness and might) as represented by Abraham, of whom it is written on three separate occasions, "And Abraham rose early in the morning",[16] and Isaac, of whom it is written, "And Isaac went out to pray in the field towards evening".[17] Day and night, the final manifestations of light and darkness, correspond to *netzach* and *hod* (the *sefirot* of victory and acknowledgment). These two *sefirot* are the final extensions of *chochmah* and *binah*.

	keter (כֶּתֶר)
	chaos (תֹהוּ)
binah (בִּינָה)	*chochmah* (חָכְמָה)
darkness (חֹשֶׁךְ)	light (אוֹר)
gevurah (גְּבוּרָה)	*chesed* (חֶסֶד)
evening (עֶרֶב)	morning (בֹּקֶר)
	tiferet (תִּפְאֶרֶת)
	heaven (שָׁמַיִם)
hod (הוֹד)	*netzach* (נֶצַח)
night (לַיְלָה)	day (יוֹם)
	yesod (יְסוֹד)
	void (בֹהוּ)
	malchut (מַלְכוּת)
	earth (אֶרֶץ)

The sum of the numerical values of the names of all five pairs alone is 2,345, which equals 5 times 469, meaning that 469 is the average value of each pair. The number 469 is the number of words in the entire account of creation, as mentioned in Chapter 1, and is the 13th Shabbat (hexagonal) number, i.e. the Shabbat number of the numerical value of "one" (אֶחָד).

The numerical value of, "day and night" (יוֹם וָלַיְלָה) is exactly 137. This phrase is unique to this verse, but the complementary phrase, "night and day" (לַיְלָה וָיוֹם) has the same numerical value. This second phrase appears a total of three times in the Prophets and Writings.[18] The ordinal value of "day and night" (יוֹם וָלַיְלָה) is 74, i.e., the average ordinal value of each word is 37, the companion number to 137.

▶ When spelled in full, the numerical value of "day and night" (יוד וו מם מם למד יוד למד הא) is 298. When filled with its letters a second time (יוד וו דלת וו וו מם מם וו וו למד מם דלת יוד וו דלת למד מם דלת הא אלף), the value is 2,433. The total of these three values (137 plus 298 plus 2,433) is 2,868, which is the total numerical value of the names of the ten *sefirot*: keter, chochmah, binah, chesed, gevurah, tiferet, netzach, hod, yesod, malchut. (כֶּתֶר חָכְמָה בִּינָה חֶסֶד גְּבוּרָה תִּפְאֶרֶת נֶצַח הוֹד יְסוֹד מַלְכוּת).

A single word that is comprised of the same letters as "chaos, chaos" (תֹהוּ תֹהוּ) is "coming into being" (הִתְהַוּוּת), referring to the emergence of "something from nothing" (יֵשׁ מֵאַיִן), which is another phrase that equals 411, as mentioned above. This relates to the expression that recurs in Chassidic literature, "The condensation of the lights, produces the vessels" (מֵהִתְעַבּוּת הָאוֹרוֹת נִתְהַוּוּ הַכֵּלִים).[19]

mankind, is to create a world of many strong vessels that are capable of containing the intense energies of chaos, without shattering. ◄

We can take a step further in our analysis of the word pairs that appear until the creation of light by taking a look at the position of each word of every pair in the Torah text. Doing so reveals an interesting mathematical series. The first couple, "heaven" (שָׁמַיִם) and "earth" (אֶרֶץ), appear as the 5th and 7th words in the Torah, the sum of their placing is thus 12. The next pair is "chaos" (תֹהוּ) and "void" (בֹהוּ). In Chapter 1, we mentioned the significance of "chaos" as the 10th word in the Torah. Now, we see that its pair, "void" is the 11th word. The sum of the placing for this pair is therefore 21. The third pair is "darkness" (חֹשֶׁךְ) and "light" (אוֹר). Darkness is the 12th word, and light is the 25th word. The sum of the placing for this pair is thus 37, which is a complementary number of 137

$$10 \quad \mathbf{12} \quad \mathbf{21} \quad \mathbf{37}$$
$$2 \quad \mathbf{9} \quad \mathbf{16}$$
$$7 \quad \mathbf{7}$$

The number preceding these numbers in the series produced from them is 10. This alludes to another pair—the two prepositions in the phrase "the heavens and the earth" (אֵת הַשָּׁמַיִם וְאֵת הָאָרֶץ) in the first verse of the Torah, whose positions as 4th and 6th words add up to 10.

Stable Particles and Unstable Particles

three generations of matter
(fermions)

	I		II		III			
mass charge spin	≈2.4 MeV/c² 2/3 1/2	**u** up	≈1.275 GeV/c² 2/3 1/2	**c** charm	≈172.44 GeV/c² 2/3 1/2	**t** top	0 0 1	**g** gluon	≈125.09 GeV/c² 0 0	**H** Higgs

QUARKS: up, charm, top, gluon — SCALAR BOSONS: Higgs

| ≈4.8 MeV/c²
-1/3
1/2 | **d**
down | ≈95 MeV/c²
-1/3
1/2 | **s**
strange | ≈4.18 GeV/c²
-1/3
1/2 | **b**
bottom | 0
0
1 | **γ**
photon |

| ≈0.511 MeV/c²
-1
1/2 | **e**
electron | ≈105.67 MeV/c²
-1
1/2 | **μ**
muon | ≈1.7768 GeV/c²
-1
1/2 | **τ**
tau | ≈91.19 GeV/c²
0
1 | **Z**
Z boson |

LEPTONS | GAUGE BOSONS

| <2.2 eV/c²
0
1/2 | **νₑ**
electron
neutrino | <1.7 MeV/c²
0
1/2 | **νμ**
muon
neutrino | <15.5 MeV/c²
0
1/2 | **ντ**
tau
neutrino | ≈80.39 GeV/c²
±1
1 | **W**
W boson |

Particles in the Standard Model of QED

The standard model of quantum mechanics (see Chapter 4) includes various types of elementary particles (and their antimatter counterparts). There are two general groups of elementary particles, from which all other particles in the particle "zoo" are formed: matter particles (fermions) and force particles (bosons). Some of these, namely the electron and other first-generation fermions (appearing in the left column of the above chart), as well as the photon and the gluon, are stable particles that supposedly have infinite lifespans. Other particles, such as second and third-generation fermions, are unstable particles that have very short lifespans. Once they have been formed in interactions they quickly decay into more stable particle pairs of the next highest generation.

One unsolved question of modern physics is why there are exactly three generations of quarks and leptons, and no more.

We can explain this with the knowledge of Kabbalah that we have accumulated so far. The three generations of fermions correspond to two levels of chaos and one level of rectification. Third-generation fermions correspond to unstable chaos, second-generation fermions correspond to stable chaos, and first-generation fermions correspond to the World of Rectification. But, all this is only with reference to free fermions that are not a part of an atom. This is like the bachelor state mentioned above, with reference to the seven kings of chaos. The true state of rectification is when the particles are connected and work together in cooperation to form physical matter.

Bachelorhood is a flawed state in which one is not yet connected to another soul in marriage. This state is liable to deteriorate into permissiveness, which may eventually lead to a state of total immorality. This can be compared to a sharp knife with a very slight flaw that is imperceivable to the naked eye, but can be felt by testing the blade on one's finger. Such a knife is still functionable, but is disqualified by Jewish law as a slaughtering knife. An animal that was slaughtered with such a knife is not considered kosher. This corresponds to the flaw of bachelorhood. The second level of deterioration is a knife that is completely blunt. One can still perceive that it was once a knife, but it is no longer functional as such. This corresponds to the level of permissiveness, in which the individual's spiritual sensitivity becomes dulled. The third level is a knife that is completely broken and one can no longer recognize that it was once a knife. This level corresponds to total immorality and spiritual death, as described with reference to the seven kings of chaos.

Kabbalah teaches us that the only way to rectify these three stages of deterioration is by studying the deepest Kabbalistic secrets. The energies that drive the individual to return to their pre-flawed state are concealed in the spiritual light that emanates from the inner dimension of the Torah.

Chaotic, unstable particles "decay" into stable particles with an infinite lifespan. This process echoes revival from the spiritual "death" of immorality to living a moral, fruitful life with one's spouse in marital harmony. This revival process returns the individual's sensitivity to spirituality, thus healing his or her psyche.

The Rishon Model

Some theories suggest that the particles enumerated in the standard model are comprised of even smaller particles, generally referred to as "preons." Appropriately, the Rishon Model, the first preon model, developed by Israeli physicist Chaim Harari, proposes that all particles currently referred to as "elementary" are formed of t (tohu) and v (vohu) sub-particles. Tohu (תהו) and vohu (בהו) are the Hebrew terms for "chaos" and "void." No evidence has yet been found to uphold or dispute the preon theory.

Chaos and Order—a Complementary Pair

Above, we mentioned the seven kings of chaos. Even though they were all worthy of being kings, each king desired to reign by himself, without taking anyone else into consideration. Similarly, any system is chaotic if its components are of equal ranking, but none of them is prepared to relinquish its status to conform to a hierarchial system. Order is achieved when the elements of a system all take on a different task. When chaos rules, no productive action can take place and the system becomes impotent.

This relates to the second law of thermodynamics, which states that a closed system tends towards a state of maximum entropy. The measure of entropy is the macrostate in which the energy of the system tends towards its most probable distribution. Thus, for example, if an ice-cube is placed into a glass of warm water, we know intuitively that the ice will melt and the water will be cooled. As entropy increases, the overall temperature eventually becomes distributed evenly. The same process in reverse (the ice-cube growing larger and the water heating up) is totally inconceivable. Therefore, most authorities see the second law of thermodynamics as directly relating to the arrow of time.

The first law of thermodynamics, in keeping with the law of conservation of energy, states that the overall energy of a closed system remains static—never increasing and never decreasing. Thus, according to the laws of thermodynamics, in order to overcome the effects of entropy in a closed system, there must be an input of new energy. Once the energy source is removed, the system will inevitably return to its state of highest entropy (i.e., its lowest energy level).◀

Chaos Theory

Many academic theorists and scientists believe that when left to its own devices, given enough time, a system in total chaos will eventually transform spontaneously into a state of natural order. Researchers of Chaos Theory have discovered that random phenomena display organized patterns when repeated many times. For example, if each of a set of two simple rules is performed randomly more than about fifty times (e.g., tossing a coin and moving a point according to one rule for "heads" and a second rule for "tails"), certain patterns will begin to emerge.

Similarly, entropy usually results in disorder at the molecular level. However, there are some spontaneous processes that defy this rule. The formation of crystals, which have an exquisite internal molecular structure, is possible because the kinetic energy involved in their formation overrides the entropy of the system. Thus, in keeping with the second law of thermodynamics, the result of crystal formation is an increase in total entropy.

When you consider the type of chaos that is rampant in your children's bedroom, the radical idea of spontaneous order is counterintuitive. The situation does not resolve itself when they are left to do as they please! A better parable is the apparent chaotic state of a large group of individuals as they prepare for an imminent major event. Everyone is hustling and bustling around, each doing their

▶ The third law of thermodynamics states that the entropy of a system approaches a constant value as the temperature approaches absolute zero. Absolute zero is -273.15 degrees C. When rounded down to a whole number, this is -273.

The Celsius temperature scale is based on units that relate to a natural phenomenon (the freezing point and boiling point of water). Therefore this is a legitimate measurement system according to the Torah (see Chapter 13). The Kelvin scale sets this point as zero, while retaining the same ratio as the Celsius scale. The freezing point of water is thus 273 Kelvins.

The midpoint of 273 is 137.

own task. To the casual observer who is unaware of the reason for the commotion and the ultimate plan behind it, it looks as though the activity is random and chaotic. Yet, once the event begins, the order inherent in the chaos and the original plan of the organizer are revealed in all their glory.▶

Supporters of such self-organization claim that once accomplished, spontaneous order is preferable to any order that the human mind can devise. Chaos Theory claims that latent in chaos from its outset lies the seed of perfect order.

Five Strata of Spontaneous Order

Spontaneous order is classified by theorists into various categories. Here, we will discuss five general strata in which spontaneous order is observable:

1. The universe
2. Evolution
3. Natural ecosystems
4. The development of language
5. The free market system – capitalism

Chaos and Order in the Universe

The first verse of Genesis[20] describing creation corresponds to the Big Bang hypothesized by physicists.▶▶

The second verse of Genesis teaches us that after the initial act of creation, chaos ensued. But, how could a beautifully ordered world such as ours emerge from that chaos? Believers in the Big Bang theory have developed two complementary interpretations to solve this conundrum. Either a law is inherent in nature that eventually produces order from chaos (self-organization), or creation has intelligent design.

Let's use a common analogy to explain the difference: what happens when different colors of paint are splashed randomly on a canvas?

▶ The Hebrew phrase that describes such a phenomenon, with reference to Shabbat, appears in the *Lecha Dodi* song, traditionally sung on Friday evening before the Shabbat prayers, "The result of an action is [inherent] in the initial thought" (סוֹף מַעֲשֶׂה בְּמַחֲשָׁבָה תְּחִלָּה). Inbetween the initial thought and the final result, chaos is likely to ensue. The numerical value of this phrase is 1,371, which equals 10 times 137, plus an additional unit (the *kolel*; see Chapter 4).

▶▶ Just for fun, the initial letters of "Big Bang" (בִּיג בֶּנְג), are also the initial letters of the first two words of the Torah, "In the beginning [God] created" (בְּרֵאשִׁית בָּרָא). When the letters of "Big Bang" are spelled in full (בית יוד גימל בית נון גימל), the sum of their numerical values equals 1,116, the exact numerical value of these two words

According to the proponents of ordered chaos, in a counter-intuitive way, a beautiful picture will emerge (this obviously depends on the subjective imaginative powers of the onlooker. Indeed, certain forms of modern art rely on this theory!) For those who support intelligent design, a beautiful picture is unlikely to appear from this form of art unless a skilled artist applies the paint.

From Darkness to Light

Believers in Chaos Theory conjecture that the process of self-organization began with total darkness. Light, stars, galaxies and super-galaxies emerged from this state until the universe as we currently know it developed.

Proponents of intelligent design claim that after the Big Bang, conditions on earth were exactly suitable for life to materialize. This theory concludes that the order apparent in the universe proves that it is administrated by an Intelligent Designer.

One of the factors that uphold this theory is the precise value of the physical constants, particularly the cosmological constant, but also referring to the value of *alpha*, the fine-structure constant. As noted in Nature magazine:

> *"The possibility of life as we know it evolving in the Universe depends on the values of a few basic physical constants—and is in some respects remarkably sensitive to their numerical values."*[21]

If the values of the physical constants would deviate in the slightest, the universe as we know it would not exist. All known forms of life depend on carbon. Yet, minor variations to the physical constants would mean that there would be no carbon production in the stars and consequently, no life on earth.

Skeptics prefer not to accept the anthropic principle because it implies that there is an entity that governs the universe. For them, self-organization is a feasible alternative.

As mentioned above, complementarity supposes that multiple

properties of apparently contradictory phenomena all derive from a common origin. What then, can be the united source of these two polar opinions?

From a Torah perspective, the initial act of creation is a one-time event that happened, "In the beginning." Chassidic literature elaborates on this theme, teaching that this unique incident is repeated imperceptibly at every single moment in time. At every instant, the universe pulses in and out of existence. It returns to nothingness, but is immediately re-generated by the Creator.[22] This obviously supports intelligent design and the anthropic principle.

Yet, the Torah's account of the first day of creation describes a gradual process similar to that described by advocates of organized chaos. The second verse of Genesis describes the initial state of darkness. This is followed by the creation of light and the subsequent separation of light from darkness. This is the point of order that emerged from the primary state of chaos. In the following verse, light is then framed into a more orderly complementary set of day and night. The observer over a period of time perceives creation as a development from chaos to order.

Void—the Root of Order in Chaos

The second verse of Genesis states: "And the earth was chaos and void and darkness upon the face of the abyss" (וְהָאָרֶץ הָיְתָה תֹהוּ וָבֹהוּ וְחֹשֶׁךְ עַל פְּנֵי תְהוֹם). In Kabbalah, "void" (בֹהוּ) is interpereted to mean, "it is in it" (בּוֹ־הוּא).[24] ▶▶ This refers to an initial "something" inherent in the total "nothing" of chaos.[25] "Nothing" is the root of disorganization (chaos) and "something" (void, according to this interpretation) is the root of order within chaos. ▶▶

The sages explain that "the spirit of God" that "hovered over the waters" refers to the spirit of the Redeemer (*Mashiach*),[26] who will ultimately implement spontaneous order in the world. Thus, this expression also indicates the underlying order in chaos.

▶ God created the world with His 42-lettered Name.[23] This Name is encoded in the first 42 letters of the Torah, from the *bet* that begins the Torah until the 42nd letter, which is the *bet* that begins the word "and void" (וָבֹהוּ). A third letter *bet* is included in these 42 letters. It appears in the word "created" (בָּרָא). The sum of the numerical values of the first 42 letters in the Torah, when the three letters *bet* are not included (leaving 39 letters in all), is 3,836, which equals 28 times 137.

▶▶ As mentioned above, the numerical value of "something from nothing" (יֵשׁ מֵאַיִן) is 411 (3 times 137), which is the numerical value of "chaos" (תֹהוּ).

The sum of the numerical values of "chaos" (תֹהוּ), "void" (בֹהוּ), Bela (בֶּלַע, the first of the seven kings), Hadar (הֲדַר) and Meheitavel (מְהֵיטַבְאֵל) is 832, which is the numerical value of, "the Land of Israel" (אֶרֶץ יִשְׂרָאֵל). This number is the product of 32, [the numerical value of "honor" (כָּבוֹד)] and 26, the numerical value of God's essential Name, Havayah. Honoring God in the Land of Israel is the ultimate rectification of "the earth was chaos and void" and of all the Kings of Chaos.

Common Origin

When spelled in full, "chaos" (תו הא וו) contains three additional letters (ויאו). These three letters spell the name of the Hebrew letter *vav* (see Chapter 9). The sum of the numerical values of these extra letters is 13, which is also the numerical value of the complementary word "void" (בהו). The numerical value of "chaos" (תו הא וו) when spelled in full is 424, which is equal to the sum of the numerical values of the two words "chaos" (תהו) and "void" (בהו). The numerical value of "void" when spelled in full (בית הא וו) is 430, which is equal to the numerical value of the complete phrase, "chaos and void" (תהו ובהו).▼ These two numerical equalities indicate a form of cross-pollination between the two concepts. This mutual fertilization is referred to in Chassidic texts as *hit-kalelut*, or "inter-inclusion." The ability of two complementary qualities to inter-include one within the other is a manifestation of their common root, as mentioned above with reference to marriage. This idea is emphasized by the fact that the two Hebrew words are a rhyming pair—that is, they share the same vowel combination.◄◄ Their common sound represents their common origin.

In modern physics, the observer plays an important role in interpreting reality. Some go so far as to say that the observer even creates

▶▶ In the *tanta* (*ta'amim, nekudot, tagin, otiyot*) system, the *nekudot* (vowel points) emanate from a higher source than the *otiyot* (letters of the *alef-bet*).[27]

▶ The sum of the numerical values of "chaos" (תהו, 411) and "void" (בהו, 13)—without the connecting *vav*—is 424, which is the numerical value of the phrase, "Mashiach, son of David" (מָשִׁיחַ בֶּן דָּוִד).

Another method of calculating the value of two words when both have the same number of letters is by multiplying each letter of the first word by the corresponding letter in the second word. This is called the dot product. The dot product of "chaos" (תהו) and "void" (בהו) is 400·2 ⊥ 5·5 ⊥ 6·6; 861 (△41), which is the numerical value of "the Holy Temple" (בֵּית הַמִּקְדָּשׁ). The Temple is the abode of the Divine Presence on earth. It is the purpose of all creation and the ultimate rectification of chaos, as the verse in Isaiah states, "Not chaos did He create it, but to settle it did He form it."[28] Thus, the fact that Mashiach, the final redeemer, will achieve the ultimate objective of creation by building the Holy Temple, is alluded to in the chaos and void, which preceded the creation of light.

The name of the letter *bet* (בַּיִת) relates to a "house" (בַּיִת),. It also alludes to the Temple. The first two Temples in Jerusalem were destroyed and we await the construction of the third Temple, which will stand forever. The letter *bet* of "void" (בהו) is the third letter *bet* from the beginning of the Torah. Each of these three letters *bet* refers to one of the Temples in Jerusalem. Thus, the third *bet* corresponds to the eternal third Temple. The Third Temple represents the manifestation of the seed of order in the mundane world.

The initial letters of "chaos" (תהו) and "void" (בהו) are *tav* (ת) and *bet* (ב). Before creation, the letters approached God in reverse (i.e., chaotic) order. Each letter asked that God choose it to create the world. First came the *tav*, the final letter of the *alef-bet*, and lastly, the letter *alef*. God disqualified each of the letters, each for a different reason, until He chose the *bet*, which is the first letter of the word "blessing" (בְּרָכָה). He found this to be the most fitting letter with which to create the world.[29]

reality. A famous Chassidic dictum states, "Think good, and it will be good." This implies that our thoughts are able to redirect events to a positive outcome. The appropriate paraphrase in this case would be, "Observe good, and it will be good!" In order to find order, it is incumbent upon the conscious observer to choose to look for it.

The English translation, "chaos and void," corresponds to Unkelus' Aramaic translation of the Hebrew phrase. Targum Yonatan adds to this, "chaotic of men and void of animals." From here we infer that "chaos" refers to a lack of conscious humans, and "void" refers to emptiness of non-conscious animals. Chaos indicates a total lack of consciousness. By contrast, void represents a state of pre-consciousness that begins to notice the emptiness of conscious thought.

The literal explanation of "chaos" (תֹּהוּ) is the onlooker's subjective sense of astonishment as he or she observes the emptiness of the "void."[30] Human astonishment creates more chaos by wanting to step in to fill the gap. Returning to our parable of an imminent major event, the onlooker, seeing the chaotic hustle and bustle that precedes it, is shocked and wishes to intervene in order to "correct" the situation. However, any intervention on his part is liable to interfere with the order inherent in the commotion. Thus, the onlooker's intervention causes even more chaos. Paradoxically, if he withdraws his astonishment and watches the proceedings without involvement, he will see order emerge.

God created the world by contracting His Divine Light to make way for creation. Similarly, spontaneous order in the universe manifests when the prevailing consciousness retracts itself, and the something-ness of the ego returns to nothing. ▶ This explains why nullifying the human ego is such a central theme of Chassidic teaching. Order begins at the point that is empty even of chaos. We can activate our choice to find order and goodness in the world by first diminishing our inner chaos. Without this fundamental step, the chaos of our egos resonates so strongly with the chaos in the universe that we become an inseparable part of it.

▶ The numerical value of "Nothing from something" (אַיִן מִיֵּשׁ), like "something from nothing" (יֵשׁ מֵאַיִן), is 411.

The numerical value of the second filling of the letters of "chaos" (תו וו הא אלף וו וו) is 559, which is equal to half the numerical value of the quintessential proclamation of God's unity, "Hear O' Israel, *Havayah* is our God, *Havayah* is one" (שְׁמַע יִשְׂרָאֵל הוי' אֱ-לֹהֵינוּ, הוי' אֶחָד).

Once human consciousness humbly merges with the void (rather than identifying with chaos) the seed of Divine order that lies latent in the vacuum is released. Spontaneous order is thus the essence of creation ex-nihilo.

In Kabbalah, *keter* (the *sefirah* of the super-conscious crown) is the highest realm of the soul. The three inner powers of *keter*—will, pleasure and faith—are sometimes referred to as "zero" (אֶפֶס) "nothing" (אַיִן) and "chaos" (תֹּהוּ), respectively. The "chaos" (תֹּהוּ) associated with *keter* is holy "chaos." By annulling our egos, we can begin to resonate with the source of chaos at this level.

Two Divine Perspectives

Let's continue by explaining an important point relating to God's Names. In the Torah, God is referred to by a number of Hebrew Names. The two most common Names of God are *Havayah* and *Elokim*. The former is the essential Four-Lettered Name of God, which is so sacred that it is forbidden to articulate it.◀ The four letters of this Name are pronounced by rearranging the letters to read, *Havayah*, which represents the transcendental, or supernatural aspect of God.

The second Name of God is *Elokim*, mentioned previously with reference to its paradoxical plural form. This Name represents the immanent aspect of God that governs the laws of nature. This is reflected in the numerical value of *Elokim* (86, אֱ־לֹהִים), which is also the numerical value of "nature" (הַטֶּבַע).[31]◀◀

These two Names of God are two different perspectives on reality. From God's own point of view, as it were, He is *Havayah*—omnipotent in His ability to create something from nothing or perform miracles by overriding the laws of nature. From our human perspective, He is *Elokim*. The omnipotent, transcendental aspect of God is shrouded in darkness, and we are confronted by the natural phenomena of the universe. It seems that there is no way to reconcile these two paradoxical aspects of God. Man's knowledge of God begins from the moment of creation, when He introduced Himself to us as *Elokim*, "In

► The numerical value of "four letters" (אַרְבַּע אוֹתִיּוֹת) is 1,096, which equals 8 times 137. The numerical value of "four" (אַרְבַּע) is 273, the midpoint of which is 137. When the ten letters of this idiom are divided equally into two sets of five (אַרְבְּעָא וְתִיּוֹת), the numerical value of the first set of five letters is 274 (2 times 137) and the numerical value of the second set is 822 (6 times 137).

►► The numerical value of "heavens" (שָׁמַיִם) is 390, which equals 15 times 26, the numerical value of God's essential Name, *Havayah*. The numerical value of "earth" (אֶרֶץ) is 1,290 (when the numerical value of the letter *alef* is equal to 1,000), which equals 15 times 86, the numerical value of *Elokim* (אֱ־לֹהִים). Similarly, when filled, the numerical value of the Name *Elokim* (אלף למד הא יוד מם) is 291, which equals the numerical value of "earth" (אֶרֶץ).

the beginning, God [*Elokim*] created…" From that moment onwards, we became conscious observers. In order to discover God's Name *Havayah*, we need to relax our analytical observation and tune into a new world governed by our freedom to choose good.

Spontaneous order reflects the ascent of creation to the Creator, corresponding to *Elokim*. Intelligent design perceives the descent of the Creator to creation, corresponding to *Havayah*.

In the initial account of creation, only the Name *Elokim* appears, the Name that signifies nature, mankind's subjective perspective. Beginning from the second account of creation and throughout the Hebrew Bible, these two Names often appear together. Despite their apparent contrast, they form an idiom that is referred to by the sages as the "Complete Name." The numerical value of "Complete Name" (שֵׁם מָלֵא) is 411 (3 times 137), the numerical value of "chaos" (תֹהוּ). ▶ The "Complete Name" is thus alluded to in the three translations of the very first word of the Torah, as explained in Chapter 1. ▶▶

The revelation of the Divine Name *Havayah* within creation is like the revelation of the wave function of light that appears within the particle behavior.

Chaos and Order in Evolution

The second level at which chaos and order are apparent in creation relates to the development of life on earth.[33] According to evolutionists, this process took place gradually over a period of billions of years. Practically speaking, evolution is a function of time during which species develop from their predecessors. Creation, on the other hand, took place in a limited period of six days. From then on, life on earth has proceeded more or less unchanged. The only thing that has changed is the knowledge that mankind has accumulated about the laws of nature. So, evolution vs. creation is a complementary pair along the time scale: either evolution over relatively infinite time, or creation with limited duration. Just as in other complementary pairs (e.g., the position and momentum of a subatomic particle) only one

▶ The numerical value of "Complete Name" (שֵׁם מָלֵא) is 411. Its numerical value when written in full (שין מם מם למד אלף) is 705. The sum of 705 and 411 is 1,116. This is the numerical value of "In the beginning, He [God] created" (בְּרֵאשִׁית בָּרָא).

▶▶ We explained that there is an apparent paradox in the meaning of the Name *Elokim*, which implies the plural, where the minimum plural is two.[32] This means that there is a complementary pair inherent in this Name. The revelation of this duality becomes apparent in the full Name, *Havayah Elokim*, which is the principle complementary pair with relation to Divinity. A similar structure is apparent in the first complementary pair in the Torah, heaven and earth. The earth is subsequently described as "chaos and void." The second pair is the revelation in the lower worlds of the duality apparent in the first pair—i.e., chaos and void on the earth is like *Havayah Elokim* in *Elokim*.

of these qualities can be observed at any given moment. Thus, the transformation from chaos to order in creation depends either on belief in a Creator who created time, or in time as the creator of order. *Havayah* is God as He created time. *Elokim* is God as He ordained the laws of nature to govern evolution.

Mutation and Natural Selection

The two basic principles proposed by evolution are mutation and natural selection. If we contemplate these two carefully we find that they too correspond to spontaneous order (mutation) and intelligent design (natural selection). Mutation, which is governed by a type of unconscious "choice," corresponds to spontaneous order and to the Name *Elokim*. By contrast, natural selection is directed by Divine Providence, which corresponds to intelligent design and the Name *Havayah*.◄

The basic idea of gradual evolution from simple to complex species is not unfamiliar to Jewish sources. In his commentary on the verses of Genesis describing creation, Malbim, a 19th century Torah scholar,◄◄ definitely suggests that the stages of creation follow an evolutionary process. However, according to Malbim, this process was achieved during the six days of creation as they appear in the text of the Torah, and not over a period of billions of years. Thus, we see once again that the original Jewish approach synthesizes two complementary notions.

Chaos and Order in Ecology

There is one spark of truth in the legends that abound about lemmings—their population fluctuation is mysteriously chaotic. During a four-year cycle, the lemming population escalates chaotically before suddenly plummeting to near extinction. Fiction claims that every four years millions of lemmings jump off cliffs into the ocean to their certain death. But, there is no truth to the lemmings' alleged mass-suicidal tendencies. Neither is there any truth to the myth that these

▶ The first number that is divisible by 26 (the numerical value of *Havayah*) and 86 [the numerical value of *Elokim* (אֱ-לֹהִים)] is 1,118, which is the numerical value of the phrase "Hear O' Israel, *Havayah* is our God, *Havayah* is one" (שְׁמַע יִשְׂרָאֵל הוי׳ אֱ-לֹהֵינוּ, הוי׳ אֶחָד). This is a clear allusion to the innate union of these two Names at their ultimate source.

The sum of the numerical values of the four principle fillings of the Name *Havayah* (72, 63, 45, 52) is 232. The sum of the numerical values of the three possible values of the filled value of *Elokim* (300, 295, 291) is 886. The total of these seven numbers is also 1,118.

▶▶ Interestingly, Malbim was a contemporary of Darwin and their lives were more or less parallel. Malbim was born a month after Darwin and died a couple of years before him (Malbim: 7 March, 1809 – 18 September 1879, Darwin: 2 February 1809 – 19 April 1882).

rodents fall from the skies and die when the grass grows in spring. Perhaps their unexplained disappearance is due to the changes in predator populations.

Whatever the reason may be, these small rodents are a key factor in arctic ecosystems and an enthralling example of how ecology works throughout the world. Any existing ecosystem operates within a fascinating scheme of naturally occurring order. This involves all subsystems within it, including animal, vegetable and mineral. For this reason, ecology is one of the classic examples of nature's ability to arrange order out of chaos. If the stability of an eco-system is disrupted—for example, when a new species of insect is inadvertently introduced into a habitat—the entire system may become temporarily chaotic. Chaos reigns until equilibrium is eventually restored.

Edward Goldsmith, a British economist and philosopher, was one of the first to point out that the laws of thermodynamics do not necessarily apply to the dynamics of ecosystems. He formatted two "laws of eco-dynamics" that he perceived to govern ecological systems:

- "Living things seek to conserve their information, structure, and behavior."
- "Natural systems tend towards stability, not in the direction of entropy or disorder, but towards climax, which must correspond to ecological equilibrium, a point in which the system ceases to grow."

The ecological cycle from turmoil to equilibrium, chaos to order, is reminiscent of the waxing and waning of the moon. Indeed, the passage of time plays a major role in the changes in ecological balance.

Everything in its Right Time

There are twenty-eight "times" stated in what can be considered the most balanced and complementary chapter of the Hebrew Bible, which begins: "There is a time for everything and a season for every matter under the heavens. A time to give birth and a time to die…"[34]

Each "time" refers to sensing the appropriate moment to act. Every action, both the climax and the equilibrium, is necessary in order to achieve true status quo.

The final pair, "A time for war and a time for peace," refers to ecological chaos and order. This is the natural balance between humans and all living creatures. In the future, all beings will live in peace with one another, as Prophet Isaiah states, "the wolf shall dwell with the lamb and the leopard shall rest with the kid, etc."[35] In order to achieve this idyllic state, there must first be a chaotic stage of war—between animals and between nations. Just as death is the natural result of birth, so too, peace is the natural result of war.◄ In order to achieve order, it must be preceded by chaos. Without one, the other is inconceivable. Only when both members of the complementary pair are manifest is it possible to reach their united source.

This is also true of the various factions of the human psyche. In order to achieve true inner harmony, we must first battle and overcome our evil inclinations. This is the constant inner war waged by the intermediate individual (who is neither evil nor righteous) as defined by the classic Chassidic text, the *Tanya*. Under present circumstances, the intermediate level is the highest status we can hope to achieve.[36] This level is a state of constant inner conflict that can only be reconciled by special grace from God.

The sum of the numerical values of "war" (123, מִלְחָמָה) and "peace" (376, שָׁלוֹם) is 499, which is equal to the numerical value of the Divine Name *Tzevakot* (צְבָאוֹת). The literal meaning of this Name is the plural of "army" (צָבָא). This Name corresponds to the *sefirot* of *netzach* (victory) and *hod* (acknowledgment).[37] War and peace thus relate to an army that goes out to war, confident in their victory. They then return in peace, acknowledging that their success was through God. Peace is the vessel capable of containing God's blessing, but it can only be achieved by first defeating evil. As the verse in Psalms states: "God gives power to His nation," and only then, "God blesses His nation with peace."[38]

▶ "A time for war" corresponds to King David, whose reign was dominated by wars. "A time for peace" relates to King Solomon, whose Hebrew name, Shlomo (שְׁלֹמֹה), discloses the essence of peace—*shalom* (שָׁלוֹם)—in his nature. These two are the precursors to *Mashiach*, the harbinger of ultimate peace.

Maker of Peace

God Himself is the ultimate maker of peace.[39] He defined the movement of the heavenly bodies in a cyclic motion rather than in a straight line. This gives each entity the sense that it is the first in line, thus preventing rivalry between them, and creating peace. Similarly, He constantly makes peace between conflicting elements. "Heaven" (שָׁמַיִם) is made up of two words, "fire" (אֵשׁ) and "water" (מַיִם).[40] By nature, fire and water are diametrically opposed. But, God took the two of them and fused them together to create the heavens.

Not only is God the maker of peace, He is also the essence of peace, so much so that "Peace" (שָׁלוֹם) is one of His Divine Names.[41] In Isaiah, God states: "I form light and create darkness, I make peace and create evil."[42] While God is the ultimate source of peace, paradoxically, He is also the source of all conflict. In order to create true peace, it is incumbent upon us to fight against the evil that conceals God's unity.

By defeating evil, Pinchas merited God's covenant of peace.[43] This is why the High Priest in the Temple was regularly chosen from among his descendants.[44]

Peace between two opposites is possible only when both are in a state of selflessness. They look up at their Divine origin and recognize that just as they are one at their source, even now they are one. This is the secret of true harmony at all levels of creation. This is particularly true with regard to marital harmony.

Chaos and Order in Language

A man from Babylonia immigrated and married a woman in the Land of Israel. One day the man asked his wife to cook him "a couple" of lentils. This was a Babylonian idiom that meant that he wanted a small amount. His wife took him literally and did as he asked, cooking just two lentils, and the husband got annoyed. The next day, he asked her to cook him "a seiah" (a measure of approximately 8 liters), intending that she should cook him a good-sized portion, not just two lentils as she had done the day before. Once again, the wife took him literally and cooked a whole seiah. He asked her to bring him two watermelons, but she misinterpreted his dialect again and brought him two candles. The husband turned to her in anger and told her to break the candles at the top of the gatepost (baba, in Aramaic), but instead, she went and broke them on the head of Baba ben Buta, a sage who sat as a judge at the city gates. The happy ending to this somewhat chaotic misunderstanding came when Baba ben Buta inquired why she had done so. "My husband told me to do it," she answered, Baba ben Buta then replied, "Since you did as your husband wishes, you should be blessed with two sons like Baba ben Buta."[45]

According to the common perception of linguistics, language began as a collection of syllables and sounds articulated by primitive peoples. It developed in the human psyche from linguistic chaos into a classic example of spontaneous order. Language cannot effectively be designed and implemented to cause an entire nation to speak according to its rules. When linguists decide that a certain word should be used to describe something, there is no guarantee that the word will be accepted in common usage. People understand one another in their own language. Proponents of spontaneous order use this fact to uphold their view that self-organization is better than man-made order.

The Torah, however, relates a somewhat different story. Prior to the building of the Tower of Babel, there was one primordial language, common to all humans.[46] But, while they were building the Tower of Babel, God punished their attempt to make war with Him by confusing their language. Linguistic chaos ensued and that one primordial language shattered into the origins of the seventy fundamental languages on earth. Despite this chaotic outset to linguistics, ordered languages soon developed. The prophets have promised that in the future there will come an era of peace on earth in which all nations will speak one clear language:[47] Hebrew, the Holy Tongue.[48] Kabbalah teaches us that this is the language with which God created the world.

King James IV of Scotland once conducted an experiment. He wanted to determine the fundamental language of mankind. He sent a mute woman away to a lone island with two babies and all the necessary equipment and food that they would need. In this way, he would see which language the children would speak when they grew up. One Scottish historian of his times wrote that the two children grew up to speak fluent Hebrew![49]

Similarly, the Midrash teaches us that Abraham's first test was when he was born. In order to escape Nimrod's cruel decree to kill all male babies, Abraham's parents hid him in a cave, where he survived miraculously. When he exited the cave, he spoke in the Holy Tongue.[50]

Gates and Roots

Hebrew is the most organized natural language that exists. It is constructed from a logical development of gates and roots.

Each Hebrew letter contains a complete array of ideas derived from its form.[51] However, in general, none of the letters on its own bears any specific linguistic meaning. In order to generate meaningful units of language in Hebrew, there must be a minimum of two letters. One of the beauties of the Hebrew language is that the three (or four) letter root of every verb and noun is derived from a two-letter unit, called a "gate" (שַׁעַר).[52] Each letter in the unit is like a pillar on each side of a gateway, through which we can pass from either direction. In this way, we obtain two different permutations from each gate. Thus, each sub-root becomes a gateway to meaning and understanding. We are taught in Kabbalah that there are 231 gates. To calculate all possible permutations of two-letter units, we take each of the 22 letters and match it with every one of the 21 remaining letters.▶ This results in 462 ($22 \cdot 21$) possible matches. To calculate the exact number of gates, we divide 462 by the number of permutations available for the same two letters ($2!$) by which we arrive at the number 231.

▶ Two letter units of the same letter do exist in Hebrew. However, the rule in Hebrew is that a double letter reinforces its indivdual meaning and does not convey any new significance as a pair. This is why double letters are not considered.

The number of gates in the Hebrew language:

$$\frac{22 \cdot 21}{2!} = \frac{462}{2} = 231$$

The mathematical formula of a triangle number is $\triangle n = \frac{1}{2} n(n+1)$ (see Chapter 3). It is thus clear that 231 is the triangle of 21.

בת	גש	דר	הק	וצ	זפ	חע	טס	ין	כמ	**אל**
גת	דש	הר	וק	זצ	חפ	טע	יס	כן	למ	אב
דת	הש	ור	זק	חצ	טפ	יע	כס	לן	**במ**	אג
הת	וש	זר	חק	טצ	יפ	כע	לס	מן	בג	אד
ות	זש	חר	טק	יצ	כפ	לע	מס	**גנ**	בד	אה
זת	חש	טר	יק	כצ	לפ	מע	נס	גד	בה	או
חת	טש	יר	כק	לצ	מפ	נע	**דס**	גה	בו	אז
טת	יש	כר	לק	מצ	נפ	סע	דה	גו	בז	אח
ית	כש	לר	מק	נצ	ספ	**הע**	דו	גז	בח	אט
כת	לש	מר	נק	סצ	עפ	הו	דז	גח	בט	אי
לת	מש	נר	סק	עצ	**וף**	הז	דח	גט	בי	אכ
מת	נש	סר	עק	פצ	וז	הח	דט	גי	בכ	אל
נת	סש	ער	פק	**זצ**	וח	הט	די	גכ	בל	אמ
סת	עש	פר	צק	זח	וט	הי	דכ	גל	במ	אן
עת	פש	צר	**חק**	זט	וי	הכ	דל	גמ	בן	אס
פת	צש	קר	חט	זי	וכ	הל	דמ	גן	בס	אע
צת	קש	**טר**	חי	זכ	ול	המ	דן	גס	בע	אפ
קת	רש	טי	חכ	זל	ומ	הן	דס	גע	בפ	אצ
רת	**יש**	טכ	חל	זמ	ון	הס	דע	גפ	בצ	אק
שת	יכ	טל	חמ	זן	וס	הע	דפ	גצ	בק	אר
כת	יל	טמ	חן	זס	וע	הפ	דצ	גק	בר	אש
כל	ימ	טנ	חס	זע	ופ	הצ	דק	גר	בש	את

The 231 Gates of the Hebrew Language

The three-letter roots of the Hebrew language are developed from the two-letter sub-roots. Each three-letter combination has six (3!) permutations. We can thus calculate the number of possible three-letter root-combinations in the Hebrew language:

The number of root-combinations in the Hebrew language:▸

$$\frac{22 \cdot 21 \cdot 20}{3!} = \frac{9,240}{6} = 1,540$$

Just as we can describe numbers in the form of triangles and squares and other two-dimensional shapes, there are also forms for three dimensional numbers. One of these forms is a tetrahedron, i.e., a three-dimensional triangle. To get some idea of what this means physically, we could construct a tetrahedron by placing one storey of ten ($\triangle 4$) disposable cups in the form of a triangle, with six ($\triangle 3$) more cups in the form of a triangle on the second storey, three ($\triangle 2$) cups in the form of a triangle on the third storey, topped by one ($\triangle 1$) more cup. The sum of all the cups is 20, the tetrahedron of 4.

The mathematical formula for tetrahedron numbers is $Te_n = \frac{1}{6}n(n \perp 1)(n \perp 2)$. Thus, 1,540 is the tetrahedron of 22. But, it is also the triangle of 55 ($\triangle 55$)! There is a total of only five such numbers that are both triangles and tetrahedrons. Thus, 1,540 (the third of this series) is a rare number indeed.▸▸

The only numbers that are simultaneously tetrahedral and triangular are:

$$Te1 = \triangle 1 = 1$$
$$Te3 = \triangle 4 = 10$$
$$Te8 = \triangle 15 = 120$$
$$Te20 = \triangle 55 = 1,540$$
$$Te34 = \triangle 119 = 7,140$$

Number under examination (a)	1	10	120	1,540
a is the triangle of (n)	1	4	15	55
a is the tetrahedron of (m)	1	3	8	20
Product of n and m	1	12	120	1,100

The sum of the products of the triangular and tetrahedral roots (n and m) of the first four numbers in this series is 1,233, which equals 3 times 411, the numerical value of "chaos" (תֹהוּ). Thus, the Hebrew

▸ Not all of the 1,540 different possible root-combinations have specific meaning. Similarly, not all of the six permutations of each three-letter combination is meaningful. In the future, one of the innovations of the Mashiach will be to teach the meanings of the "missing" roots.

▸▸ 1,540 is equal to 10 times 154, which is the numerical value of the Divine Name *Akvah* (א־הוה) when spelled in full (אלף הי ואו הי). The filling alone of this spelling is equal to 137. The average numerical value of the four letters when spelled in this way is 77, which is the numerical value of "might" (עז), which refers to God's ability to "articulate" two words at once (Chapter 15).

language, as a manifestation of order in linguistic chaos, is related to the number 1,233, which also equals 9 times 137.

Anyone familiar with the mathematics of group theory will immediately note that the formula for calculating the number of gates and roots is identical to the method used in permutation theory, which is the basis of group theory. Today, the ultimate use of group theory is to devise a unified theory of everything that will explain the source of all subatomic forces and particles.

Chaos and Order in the Competitive Market

The fifth level of spontaneous order is free economy and the competitive market. Upholders of the free market theory believe that the best and most efficient economic situation emerges when no government legislations control the market. The theory claims that the current government-controlled system is not optimal. If the market was allowed to work itself out freely, it would fall into a far more efficient state of economic equilibrium.

Initially, the free market appears to be chaotic because every individual has his or her own interests at heart. Yet, the belief is that without government subsidies, taxation or coercion, people will only pay as much as they feel the goods are worth to them. The claim is that, at an intrinsic level, there is a common social interest that relies on a built-in supply and demand mechanism. This eventually brings about economic equilibrium.

As mentioned, scientists who believe in spontaneous order theorize that the seed of order is latent in any chaotic state from its outset. This theory is aligned with those Torah commentaries that teach how the initial state of chaos and void contained a hidden allusion to perfect order. Some of these commentaries[53] see "chaos" as formless matter, and "void" as the seed of form as it connects with matter, making it tangible to our senses. From the first moment of time these two were inseparably bound together so that we can neither experience matter that has no form, nor form that does not shape matter.[54]

▸ The sum of the numerical values of "matter" (חֹמֶר) and "form" (צוּרָה) is 549, which equals 4 (representing the definition of matter in the form of four fundamental elements: fire, water, earth and air) times 137 plus an additional unit (the *kolel*; see Chapter 4) that unifies the two concepts. The additional unit in this case alludes to the fact that matter and form are only perceptible to human consciousness when they are together.

Void is the intangible root of order in chaos. It corresponds to stable chaos. Thus, in effect, pure matter and pure energy are both types of chaos.

"Chaos" (תֹהוּ) appears twenty times in the Bible. By contrast, the root of "order" (ס־ד־ר) appears only twice.▶ The most literal appearance of the root is in Job: "A land of gloom like darkness itself; the shadow of death without order, and where radiance is as darkness."[55] "Without order" is a synonym for "chaos."▼▼ Rashi explains that it refers to a lack of man-made order. This corresponds to the economic lifestyle recommended by the free market theory.▶▶▶

Another verse in which the root of "order" (ס־ד־ר) appears is in Judges: "And Ehud exited to the portico and he closed the doors of the attic behind him and locked [them]."[56] "To the portico" (הַמִּסְדְּרוֹנָה) refers to a hall set with organized rows of benches and clearly connotes man-made order.[57]

Anarchy and Freedom

Anarchy is a belief held—sometimes by the most intelligent minds—that society would be much better without any form of government. The principle of anarchy is that if everyone would do what they think is right, society would spontaneously settle into a natural order. This order would be far better than the current situation when governments make all the decisions regarding the rights and wrongs of society. One example offered of "order" that has been created through such anarchistic freedom is the Internet.

The theory that there is a seed of order within chaos contains a

▶ The dot product of "chaos" (תֹהוּ) and "order" (סֵדֶר) is 25,220, which equals 97 [the numerical value of "time" (זְמַן) or Meheitavel (מְהֵיטַבְאֵל) who represents the origin of order] times 260, which is the numerical value of the phrase, "all existing [matter] disintegrates" (כָּל הֹוֶה נִפְסָד). This alludes to the ultimate rectification of the law of entropy through time.

▶▶▶ Ibn Ezra teaches that this refers to the order of the stars, i.e., a state in which even the natural order of time is void.

▶▶ The numerical value of "without order" (לֹא סְדָרִים) is 345, which is the numerical value of Moses (מֹשֶׁה). Moses' soul-root is derived from the level of chaos that preceded order. This level is represented by the Divine Name pronounced *Kel Shakai* (אֵיל שַׁדַּי). The numerical value of this Name is also 345. The sum of the numerical values of "orders" (סְדָרִים) and "the portico" (הַמִּסְדְּרוֹנָה) equals 684, which is the numerical value of the idiom "a set table" (שֻׁלְחָן עָרוּךְ). Adding an additional unit to 684 brings its value to 685, which equals 5 times 137. The root of "set" (עָרוּךְ) relates to "array" (מַעֲרָכָה), a word that appears a number of times in the Hebrew Bible. The Aramaic translation of this word is generally related to the root of "order" (ס־ד־ר).

The sum of all four words relating to two types of chaos and two types of order (הֹמֶר צוּרָה סְדָרִים הַמִּסְדְּרוֹנָה) equals 1233; 9 times 137. The numerical value of "the portico" (הַמִּסְדְּרוֹנָה) is 370, which equals 10 times 37, the companion number to 137, as we have seen.

▶ The numerical value of the entire verse is 5 times 441, which is the numerical value of "truth" (אֱמֶת).

certain element of truth. However, it cannot be accepted in its entirety in every area of life. Political anarchy is one example that the Torah does not promote in the current era. The *mishnah* states explicitly that we must, "Pray for the welfare of the government for if it was not for fear of it, people would swallow one another alive."[58] Indeed, chaos ensued during the time of the Judges, when "There was no king in Israel and each man did as he deemed right in his own eyes."[59]◀

One of the most noted advocators of anarchy stated that freedom is a prerequisite for spontaneous order. In his words, "freedom is the 'mother' and not the 'daughter' of order." In Kabbalah, *binah* (the *sefirah* of understanding), the mother figure, is called the "World of Freedom." She "gives birth" to her "sons," the six lower *sefirot*. But, these six emotive *sefirot* are far from being orderly. They are the vessels of the World of Chaos that shattered. It is there where order must be reinstated. Rectified government—i.e., a Divinely appointed king—accomplishes the necessary inter-inclusion of all the *sefirot*. This is achieved through the power of the king's judgment. *Malchut* (the *sefirah* of kingdom) is the rectified government for which we pray; the seventh child, the "daughter" of *binah*. The rectified state of ordered government is in constant need of support through our prayers. The free-fall state of *malchut* is indicated in the name of Mashiach, *bar-naflei*, "The Fallen One."[60]

The king's primary task is to rid the land of criminals. Eventually, he receives a more exalted status of sovereignty to reveal the softer, more moderate side of judgment. This is the new Torah that will be revealed by *Mashiach*, which will be wondrous to all who perceive it. The Torah of *Mashiach* contains no harsh judgment at all. This will result in the ideal state of redemption when the "earth will be full of knowledge of God";[61] an eternal state of refined anarchy. Only then will there no longer be a need for a police force. Ultimately, *Mashiach* returns his monarchy to God, the first and final King of the Universe. Then, human government will become obsolete. "And God will become King over the entire earth; on that day, God will be one and His Name one."[62] This is the ultimate manifestation of rectified chaos.

Chaos and Order in the Redemption Process

In Kabbalah, the king corresponds to the mouth,[63] as we are taught, "There is one spokesman [i.e., leader] to the generation."[64] In Hebrew, Passover is called *Pesach* (פֶּסַח), which the Arizal[65] interprets to mean, "The mouth speaks" (פֶּה־סָח). Indeed, Passover is the most appropriate time to generate order. It is called, "The Time of our Freedom." This is because it is a holiday that commemorates the Exodus, the moment the Jewish People became acquainted with the concept of freedom. On the first night of Passover there is a *mitzvah* to retell, by word of mouth, the miracles that occurred during the Exodus. This is carried out through a ritual, set out in the text of the *Haggadah*, that follows an order of fifteen stages.▶ Appropriately, this night is called the *Seder* Night (לֵיל הַסֵּדֶר), which means "the Night of Order." In the concluding phrase of the very first passage of the *Haggadah*, we pray that "this coming year [we will be] in Jerusalem… this coming year as free men."▶▶

Jewish exile in foreign countries is one aspect of chaos, and the destruction of the Temple is another.▶▶▶ The seed of order in every chaotic phenomenon lends even greater hope that a glorious orderly redemption will sprout from the chaos of exile and a more beautiful Temple will be constructed. Rabbi Akiva was one sage who recognized that seed of order within the chaos.[66]

> *Rabban Gamliel, Rabbi Elazar ben Azaria, Rabbi Joshua and Rabbi Akiva went up to Jerusalem [after the destruction of the Temple]. When they reached Mt. Scopus [from which the Temple Mount can be seen], they tore their garments [in mourning]. When they reached the Temple Mount itself, they saw a fox emerging from the place of the Holy of Holies. The others started weeping, but Rabbi Akiva laughed.*

▶ The sum of the numerical values of the first five stages of the *Seder*, up to and including the stage of retelling the story of the Exodus in the *Haggadah*: (קַדֵּשׁ וּרְחַץ כַּרְפַּס יַחַץ מַגִּיד) is 1,233, which equals 9 times 137.

▶▶ The numerical value of "free [men]" (חוֹרִין) is 274, which equals 2 times 137 (see Chapter 5).

▶▶▶ There were 137 years between the exile of the last of the ten tribes and the destruction of Jerusalem (Abarbanel, Hosea 1:5).

"Why are you laughing?" they asked him.

"Why are you weeping?" he retorted.

They replied, "This place was once so holy that it was said of it, 'the stranger that approaches it shall die,'[67] and now foxes traverse it. Should we not weep?"

Said he to them: "That is why I laugh. For it is written [in Isaiah], 'I shall have faithful witnesses testify for Me—Uriah the Priest and Zachariah the son of Jeberechiah."[68] Now what is the connection between Uriah and Zachariah? Uriah lived in the time of the First Temple, and Zachariah lived in the time of the Second Temple! But, the Torah makes Zachariah's prophecy dependent upon Uriah's prophecy. Of Uriah's prophecy, it is written: 'Therefore, because of you, Zion shall be plowed as a field; [Jerusalem shall become heaps, and the Temple Mount like the high places of a forest.]'[69] Of Zachariah's prophecy, it is written, 'Old men and women shall yet sit in the streets of Jerusalem.'[70]

"As long as Uriah's prophecy had not been fulfilled, I feared that Zachariah's prophecy may not be fulfilled either. However, now that Uriah's prophecy has been fulfilled, Zachariah's prophecy will certainly be fulfilled."

With these words they replied to him: "Akiva, you have consoled us! Akiva, you have consoled us!"

Rabbi Yosef Ruzhin, the Rogachover Gaon, was one of the greatest rabbinic sages of the previous generation. He taught[71] that, by definition, the Temple must be eternal. Since the two Temples proved to be temporal, we have never yet fulfilled the commandment of building a Temple for God. He uses similar logic to explain that since King David's dynasty has not yet manifested eternally and *Mashiach* has not yet materialized, we have never really fulfilled the commandment to appoint a king.

In Jewish law, the only justification to demolish a synagogue is to rebuild a better version. The Temple had to be destroyed twice and returned to chaos so that eternal order will result when the third Temple is built.

Tisha B'Av—the ninth day of the Hebrew month of Av—is the day on which we mourn the destruction of the two Temples. On the Shabbat following this day of mourning we read the chapter of Isaiah that begins with the verse, "Be consoled, be consoled, My people, says your God."[72] When Isaiah received his prophecy, only one Temple had been destroyed. Nonetheless, the word "be consoled" (נַחֲמוּ) appears twice.▶

This can be compared to the revelation of a fine-structure within the projected ray of consolation. Similarly, the following verse mentions that the sin was atoned "twofold." The final consolation of the Jewish People will come with the building of the third, eternal Temple in Jerusalem. Then the creative purpose of the destruction of both Temples will become apparent.

The ultimate consolation of the Jewish People will be the rekindling of the golden menorah in the Temple. The olive oil used to light the menorah had to be "pure olive oil, crushed for the luminary" (שֶׁמֶן זַיִת זָךְ כָּתִית לַמָּאוֹר).[73] The duration of both Temples is alluded to in the numerical value of "crushed" (כָּתִית), which is 830. The word divides exactly into 420 (כת), followed by 410 (ית). Lighting the golden menorah with pure olive oil thus alludes to integrating the chaotic lights of destruction in rectified vessels when the Temple is rebuilt.▶▶

Light Touches Matter

In a Torah scroll, the passages of the Torah are divided into paragraphs. Each paragraph is defined by a space before and after the paragraph. If the passages are written without these spaces, the Torah scroll is disqualified for use. Similarly, the handwritten letters of the Torah scroll must be discrete. If one letter touches another, the Torah scroll is also disqualified. The fact that the paragraphs and letters in the Torah are so clearly defined, encourages us to consider the number of words and letters in each paragraph.

▶ The numerical value of the entire verse, "Be consoled, be consoled, My people, says your God" (נַחֲמוּ נַחֲמוּ עַמִּי יֹאמַר אֱ־לֹהֵיכֶם) is 685, which equals 5 times 137. Since there are five words in the verse, the average value of each word is 137.

▶▶ Rabbi Ya'akov ben Asher, *Ba'al HaTurim*, alludes to a direct connection between the phrase "and the earth was chaos and void" and the destruction of the first and second Temples. He explains that the numerical value of "[And the earth] was [chaos...]" (הָיְתָה) is 420, which is the number of years that the second Temple stood. We are already familiar with the next word in the verse, "chaos" (תהו, 411). Its numerical value indicates the 410 years (plus the *kolel*, the additional unit) that the first Temple stood.

The portion in which God told Moses to command the Jews to light the menorah in the Temple with pure olive oil is the only portion in the entire Torah that contains exactly 137 letters. The number of words in the passage is 37, the companion number to 137.

And you shall command the Children of Israel, and they shall bring to you pure olive oil, crushed for lighting, to kindle the lamps continually.	וְאַתָּה תְּצַוֶּה אֶת בְּנֵי יִשְׂרָאֵל וְיִקְחוּ אֵלֶיךָ שֶׁמֶן זַיִת זָךְ כָּתִית לַמָּאוֹר לְהַעֲלֹת נֵר תָּמִיד:
In the Tent of Meeting, outside the dividing curtain that is in front of the testimony, Aaron and his sons shall set it up before Havayah from evening to morning; [this shall be] an everlasting statute for their generations, from the Children of Israel.	בְּאֹהֶל מוֹעֵד מִחוּץ לַפָּרֹכֶת אֲשֶׁר עַל הָעֵדֻת יַעֲרֹךְ אֹתוֹ אַהֲרֹן וּבָנָיו מֵעֶרֶב עַד בֹּקֶר לִפְנֵי הוי' חֻקַּת עוֹלָם לְדֹרֹתָם מֵאֵת בְּנֵי יִשְׂרָאֵל:

This commandment in particular symbolizes the ultimate goal of the Jewish People to be a spiritual light to all nations. Their light will shine out from the Temple in Jerusalem to illuminate the physical world. The Temple is the location that unites all peoples in prayer, as the verse states, "For My House is a house of prayer, for all peoples."[74]

Yet, this passage is not the only passage in the Torah that relates to oil and the menorah, there are two more. The first contains 45 words:[75]

And God spoke to Moses, saying,	וַיְדַבֵּר הוי' אֶל מֹשֶׁה לֵּאמֹר:
Command the Children of Israel, and they shall bring to you pure olive oil, crushed for lighting, to kindle the lamps continually.	צַו אֶת בְּנֵי יִשְׂרָאֵל וְיִקְחוּ אֵלֶיךָ שֶׁמֶן זַיִת זָךְ כָּתִית לַמָּאוֹר לְהַעֲלֹת נֵר תָּמִיד:

Outside the dividing curtain of the testimony in the Tent of Meeting, Aaron shall set it up before God from evening to morning continually. [This shall be] an eternal statute for your generations.	מִחוּץ לְפָרֹכֶת הָעֵדֻת בְּאֹהֶל מוֹעֵד יַעֲרֹךְ אֹתוֹ אַהֲרֹן מֵעֶרֶב עַד בֹּקֶר לִפְנֵי הוי' תָּמִיד חֻקַּת עוֹלָם לְדֹרֹתֵיכֶם:
Upon the pure menorah, he shall set up the lamps, before God, continually.	עַל הַמְּנֹרָה הַטְּהֹרָה יַעֲרֹךְ אֶת הַנֵּרוֹת לִפְנֵי הוי' תָּמִיד:

And the second contains 55 words:[76]▶

And Havayah spoke to Moses, saying:	וַיְדַבֵּר הוי' אֶל מֹשֶׁה לֵּאמֹר:
Speak to Aaron and say to him: "When you light the lamps, the seven lamps shall cast their light toward the center of the menorah."	דַּבֵּר אֶל אַהֲרֹן וְאָמַרְתָּ אֵלָיו בְּהַעֲלֹתְךָ אֶת הַנֵּרֹת אֶל מוּל פְּנֵי הַמְּנוֹרָה יָאִירוּ שִׁבְעַת הַנֵּרוֹת:
Aaron did so; he lit the lamps toward the center of the menorah, as God had commanded Moses.	וַיַּעַשׂ כֵּן אַהֲרֹן אֶל מוּל פְּנֵי הַמְּנוֹרָה הֶעֱלָה נֵרֹתֶיהָ כַּאֲשֶׁר צִוָּה הוי' אֶת מֹשֶׁה:
This was the form of the menorah: hammered work of gold, from its base to its flower it was hammered work; according to the form that God had shown Moses, so did he construct the menorah.	וְזֶה מַעֲשֵׂה הַמְּנֹרָה מִקְשָׁה זָהָב עַד יְרֵכָהּ עַד פִּרְחָהּ מִקְשָׁה הִוא כַּמַּרְאֶה אֲשֶׁר הֶרְאָה הוי' אֶת מֹשֶׁה כֵּן עָשָׂה אֶת הַמְּנֹרָה:

▶ The two numbers 45 and 55 are the triangles of 9 and 10 respectively. Their sum equals 10^2 [according to the rule that $\triangle n + \triangle(n+1) = (n+1)^2$].

The total number of words in these three passages is 137 (37 + 45 + 55), i.e., the number of letters in the first passage reflects the total number of words in all three passages that relate to the menorah.▼

Chaos and Entropy

In his interpretation of the first verse of the Torah, Malbim states that within the context of human consciousness, all heavenly bodies—and all species on earth—remain intact and do not die. Only individual forms die and do not have a fixed existence.[79] This alludes to the constant energy of the universe that is inherent in the first law of thermodynamics. According to this law, energy can neither be created nor destroyed.

The second verse of the Torah, describes the effects of the second law of thermodynamics, the law of entropy: "and the earth was chaos and void and darkness upon the face of the abyss."

In living creatures in general and human beings in particular, death is the effect of a maximal entropic state. God warned Adam that he should expect death as the punishment for sinning, "On the day that you will eat from it [the Tree of Knowledge of Good and Evil], you will surely die."[80] Indeed, one explanation is that Adam would then become subject to the universal law of entropy.[81] The first "time" is "a time to give birth and a time to die."[82] From the moment a baby is born, although its body is growing and developing, the ultimate expectation is death.

▶ One important function that is applied in discrete mathematics is logical sequences of numbers. From any given sequence of numbers, the next number in the series can be calculated. In this way, we can discover the function that produces the sequence. One way of achieving this is by calculating the finite differences between the numbers.

This method is one of the methods that have been adopted in Kabbalistic mathematics to discover the connections between various numbers that appear in the Torah. The sequence of numbers that begins 37, 45, 55 can be continued by taking the finite differences between the numbers. Upon doing so, we find that the series begins with 25, the numerical value of the word, "Let there be" (יְהִי). The 14th number in the series is 207, the numerical value of "light" (אוֹר).

25	27	31	**37**	**45**	**55**	67	81	97	115	135	157	181	207
	2	4	6	**8**	**10**	12	14	16	18	20	22	24	26
		2	2	**2**	2	2	2	2	2	2	2	2	2

Thus, these three passages, all of which relate to lighting the menorah in the Temple, allude to the initial phrase relating to the creation of light, "Let there be light!"[77] The sum of the first 26 numbers in the series is 6,500, which equals 100 times 65, the numerical value of the Divine Name *Adni* (אי־דני). It also equals 250 [the numerical value of "candle" (נר)] times 26 (the numerical value of *Havayah*, the essential Name of God). This is a clear allusion to the verse in Proverbs "The candle of God is the soul of man" (נר הוי' נשמת אדם).[78]

The Hebrew idiom coined by medieval Jewish philosophers for the obvious phenomenon of entropy is, "all existing [matter] disintegrates."[83]▶

The ancient Egyptian priests attempted to overcome the limiting order of the natural world. Their goal was to preserve the chaotic potency of the ego. As a result of this creed, the Egyptians sank to the fiftieth gate of impurity. At this stage, they apathetically refrained from action. The fiftieth gate of impurity is a state of total spiritual death, achieved through belief that taking action is futile, because nothing will ever come of man's deeds. Then, the only enterprise with any significance is preservation, i.e. building mausoleums, mummification etc. Thus, Egyptian philosophy came to worship the preservation of physical matter.▶▶

The Exodus of the Jews from Egypt was liberation from despair and apathy. The Exodus initiated revitalization accompanied by the recognition that the Jewish People is governed by laws that transcend the finite laws of nature.

When God destroys something, His ultimate goal is to achieve true spontaneous order from the apparent chaos that ensues. Chaos appears in the second verse of the Torah and was not initially present in creation. Kabbalah teaches us that chaos was the result of the vessels shattering. Having created the world in an initially ordered state, God broke it. Fragments flew in all directions, making it appear as if the world was in total chaos.▶▶▶ Yet the Ba'al Shem Tov explains that each splinter of reality was pre-programmed to reach an exact address. This pre-programming manifests as Divine Providence in every earthly phenomenon. In terms of modern science, the Big Bang was not a haphazard explosion that sent elementary particles shooting randomly in every direction. It was a controlled detonation with precise information encoded into each and every atom.

True freedom is thus freedom from the law of entropy and the ability to act, create, renew, and elevate ourselves above the regularity of the cold nature of reality, to create a new order. As mentioned above

▶ The numerical value of "all existing [matter] disintegrates" (כָּל הֹוֶה נִפְסָד) in primordial numbering is 1,096, which equals 8 times 137. The initial letters of this phrase spell the word, kohen (כֹּהֵן) meaning "priest"; this refers in particular to the ancient Egyptian priests of On. The numerical value of this phrase is 260 (10 times the numerical value of God's essential Name), which equals the sum of the numerical values of the lifespans of Aaron, the High Priest (123), and his father, Amram (137 years). The rectification of the priests of On is by Aaron the High Priest, who "loves peace and pursues peace".

▶▶ All egotism is destined to return to "nothing," to the "void" that is the source of all existence. Note that the Hebrew word for "nothing" (אַיִן) is a permutation of the letters of "ego" (אֲנִי). As the Mishnah[84] states, "Know from where (מֵאַיִן) you have come and to where you are going." An alternate reading is, "Know that you have come from nothing…"

▶▶▶ Breaking vessels in one's anger, with no creative reasoning behind the act, is like serving idolatry.[85] Such an act is considered an act of vandalism and is prohibited.[86]

with relation to crystals, this involves active involvement in the form of "kinetic energy."

Chaos and Time

If we choose to observe light as a wave of energy, it can no longer be measured accurately as a particle. So too, accessing the lights of chaos immediately shatters the orderly vessels that should contain them. This is why chaos and order appear to be essentially paradoxical. However, we can experience two complementary states simultaneously in the context of God's service (see Chapter 12). Chaos and order manifest simultaneously in the paradoxical phenomenon of spontaneous order.

In one of his most powerful discourses, the Lubavitcher Rebbe, Rabbi Menachem Mendel Schneerson, said that in order to bring *Mashiach*, we must energize the chaotic lights, but contain them in rectified vessels. It is our responsibility to reach up to harness those infinite energies. We must contain them in a system of mature order of the highest degree. *Mashiach*'s unique task is to tap into the lights of chaos and draw them into the world without shattering the vessels.

As mentioned, the factor that determines whether or not chaos will transform into order is time. The more time that is available for the process, the more likely it is that the system will regulate into order. Kabbalah teaches us that the higher we rise in the spiritual worlds of consciousness, the more condensed and chaotic time becomes.[87] The highest lights of chaos are thus rooted in the essential point at which time does not exist. If we could reach up to that moment, we could access the lights of chaos. However, retaining those energies as we traverse the worlds back into our human time-frame is a formidable task. The Ba'al Shem Tov taught his disciples that the secret of retaining one's consciousness of the higher worlds while simultaneously concentrating on physical actions in this world is hidden in the power of the Goodly Name, *Akvah* (אכוה), mentioned in Chapter 1. ◄ This Name bears the secret of rectifying the shattered vessels of the World

▶ This Divine Name does not appear explicitly in the Torah. It is derived in particular from the initial letters of the phrase "The heavens and the earth" (אֵת הַשָּׁמַיִם וְאֵת הָאָרֶץ) at the beginning of creation.

of Chaos. When this Name is spelled in full (אלף הי ואו הי), the numerical value of the filling is 137.

The original vessels of chaos that shattered during creation are associated in Kabbalah with the seven primordial kings, who "ruled and died."[88] Each of these royal "vessels" shattered because of an exaggerated sense of ego. To rectify the breakage and the chaos that ensued, we need to incorporate a sense of humility into our lives. We must relax our belief in our own ability to achieve, and rely instead on Divine Providence. The seed of order that lies in the void can only flourish when human intervention is absent. This idea is expressed in the verse, "Cast your burden on *Havayah*, and He will bear you."[89] ▶

The archetypal soul who successfully contained chaotic lights in rectified vessels was the Prophet Elisha. He achieved this through the power of his prayer. One miracle that particularly illustrates this power was the way Elisha assisted the impoverished widow of another prophet. She had nothing left in the house but a vial of oil. ▼▼ Elisha told her to borrow vessels from her neighbors and to begin pouring oil from the flask. The oil kept flowing until all the vessels were filled. By selling the oil, the woman supported herself and her two sons for the remainder of their lives.[90] On another occasion, Elisha successfully revived a dead child. He did this once again through the power of his prayers.[91] This too was a manifestation of his ability to contain the lights of chaos within rectified vessels.

▶ The total numerical value of the first six words of the verse "Cast your burden on *Havayah*, and He will sustain you" (הַשְׁלֵךְ עַל הוי' יְהָבְךָ וְהוּא יְכַלְכְּלֶךָ) is 666 (see Appendix B). The average value of each word is 111, which is the numerical value of *alef* (אָלֶף), or "wonder" (פֶּלֶא), which is also the sum of the numerical values of the 6 initial letters of the phrase (ה עייוי). 111 is equal to 3 times 37 (the companion of 137). The numerical value of the first three words (הַשְׁלֵךְ עַל הוי') is 481, which equals 13 times 37; the numerical value of the next word, "your burden" (יְהָבְךָ) is 37; and the numerical value of the final pair of words, "and He will sustain you" (וְהוּא יְכַלְכְּלֶךָ), is 148, which equals 4 times 37. Here, we see that the numerical value of the entire phrase divides into three parts, echoing perfectly the syntax of the verse.

▶▶ All the woman had was "only a vial of oil" (כִּי אִם אָסוּךְ שָׁמֶן), which alludes to the simple essence of faith in God that is in every Jewish individual. The numerical value of this phrase is 548, which is equal to 4 times 137, the numerical value of Kabbalah (קַבָּלָה) and the number associated with the fine-structure constant in physics. Since there are four words in the phrase, 137 is the average numerical value of each word. The vessels that eventually contained this chaotic light are mentioned four times in the story, "vessels from outside" (כֵּלִים מִן הַחוּץ), "empty vessels" (כֵּלִים רֵקִים), "the vessels" (הַכֵּלִים) and once again, "the vessels" (הַכֵּלִים). The numerical value of these seven words is 959, which equals 7 times 137, which is once again the average of each word. This clearly suggests that the vessels were from the same source as the light. The numerical value of Elisha (אֱלִישָׁע), the prophet who performed this miracle, is 411, the numerical value of "chaos" (תֹּהוּ), which equals 3 times 137. This means that the sum of the miracle worker (3 times 137) and the miracle light (4 times 137) is 959, 7 times 137, which is the numerical value of the vessels that contain the light. The soul of a righteous person, such as the Prophet Elisha, is strong enough to contain the chaotic lights without shattering.

Another figure with the same name was Elisha ben Avuya, who was not able to contain the chaotic energies with which he was born. He was one of those who entered the orchard with Rabbi Akiva and two other contemporaries, but he emerged as a heretic

REFERENCE NOTES FOR CHAPTER 2

1. See *Bereishit Rabah* 11:8; 17:4.
2. *Berachot* 61a.
3. See *Eiruvin* 18a; Rashi on Genesis 5:2.
4. *Sifra Detzniuta* Ch. 5.
5. See *Midrash Aggadah Bereishit* 28.
6. *Kidushin* 30b.
7. Genesis 1:3-5.
8. See *The Wondering Jew* p. 36.
9. Genesis 1:2.
10. Ibid 1:3.
11. Ibid 1:4.
12. Ibid 2:9.
13. Isaiah 45:7.
14. *Sefer Yetzirah* 2:4.
15. Genesis 8:22.
16. Ibid 19:27; 21:14; 22:3.
17. Ibid 24:63.
18. I Kings 8:29; Isaiah 27:3; Esther 4:16.
19. See for example, *Torat Menachem* 29:8.
20. Genesis 1:1.
21. B. J. Carr & M. J. Rees, "The anthropic principle and the structure of the physical world," *Nature*, vol. 278 (12 April 1979), p. 612.
22. See *Tanya, Sha'ar Hayichud Ve'haemunah*.
23. *Tosfot, Chagigah* 11b
24. *Zohar Bereishit* 16a, etc.
25. *Sefer Habahir* 2.
26. *Zohar Bereishit* 31b; see also *Ba'al Haturim* Genesis 1:3.
27. See our book in Hebrew, *Mavo Lekabbalat Ha'ari* p. 244-245.
28. Isaiah 45:18.
29. *Yalkut Shimoni* 1.
30. Rashi on Genesis 1:2.
31. For a detailed definition of the Names of God, see our book, *What You Need to Know about Kabbalah*, part III.
32. See *Midrash Aggadah Bereishit* 28.
33. See our book on evolution, *The Breath of Life*.
34. Ecclesiastes 3.
35. Isaiah 11:6.
36. *Tanya* 27: "Would it be that he be an intermediate."
37. See our book *What You Need to Know about Kabbalah*, part III, "Names of God."
38. Psalms 29:7.
39. Job 25:2.
40. Rashi on Genesis 1:8; *Chagigah* 12a.
41. Judges 6:24.
42. Isaiah 45:7.
43. Numbers Ch. 25.
44. *Tosfot, Zevachim* 101b.
45. *Nedarim* 66b.
46. Genesis 11:1.
47. Zephaniah 3:9.
48. *Metzudat David*, ibid ad loc.

49. Robert Lindsay, *The Cronicles of Scotland*, vol. 1, p. 249-250.

50. *Midrash Pirkei Derabi Eliezer* 26:1 (Warsaw version, 5612); see also, *Otzar Hamidrashim*, p. 2.

51. See our book, *The Hebrew Letters*.

52. For a detailed account of many of the attributes of the Hebrew language, see our article, "Hebrew as the Ideal Language for Information Processing and Management," accessible online at: http://torahscience.org/communicat/hebrew_language.html.

53. See for example, Nachmanides, Rabeinu Bachyei and Sforno, Genesis 1:2.

54. Nachmanides, Genesis 1:2 as explained by the *Gaon*, Rabbi Eliyahu of Vilna, *Aderet Eliyahu*.

55. Job 10:22.

56. Judges 3:23.

57. *Metzudat Tzion*, Judges 3:23.

58. *Avot* 3:2.

59. Judges 17:6; 26:5.

60. *Sanhedrin* 96b.

61. Isaiah 11:9.

62. Zachariah 14:9.

63. *Tikunei Zohar, Patach Eliyahu*.

64. *Sanhedrin* 8a.

65. *Pri Etz Chayim, Chag Hamatzot* 1:4, etc.

66. *Makot* 24b.

67. Numbers 3:10, etc.

68. Isaiah 8:2.

69. Jeremiah 26:18.

70. Zachariah 8:4.

71. *Mefa'aneach Tzfunot* 5.

72. Isaiah 40:1.

73. Exodus 27:20; Leviticus 24:2.

74. Isaiah 56:7.

75. Leviticus 24:1-4.

76. Numbers 8:1-4.

77. Genesis 1:3.

78. Proverbs 20:27.

79. *Malbim*, Genesis 1:1.

80. Genesis 2:17.

81. *Or Hachayim*, Genesis 3:19.

82. Ecclesiastes 3:2.

83. See also our book in Hebrew, *Mudaut Tiveet*, p. 82. *Avot* 1:12.

84. *Avot* 3:1.

85. *Shabbat* 105b.

86. Maimonides, *Hilchot Melachim* ch. 6:10.

87. See our book on evolution, *The Breath of Life*, Chapter 1.

88. Genesis 32:36-39.

89. Psalms 55:23.

90. II Kings 4:1-3.

91. Ibid, verses 32-35.

Maximal Product, Minimal Investment and Moral Purity

Euler's Number (e) and Euler's Constant (γ)

Leonhard Euler, an 18th century mathematician and physicist, was one of the greatest mathematicians of all time. Two important numbers are named in his honor, Euler's Number and Euler's Constant. Like *pi*, Euler's Number (*e*) is an irrational, transcendental number.▶ The value of *e* to 3 decimal places is 2.718. It is the base of natural logarithms and is indirectly related to *alpha*, the electromagnetic coupling constant.[1] Some of the properties of *e* may be difficult for most people to comprehend; however, there is one simple property that can be understood by everyone. When a number is divided into smaller segments, the closer the segments are to 2.718, the larger their product.

For example, 10 can be divided in many different ways. If it is divided into ten 1s, the product is 1 (1^{10}), which is the smallest possible product. If 10 is divided into five 2s, the result is 2^5, which equals 32; much larger than 1, but not yet the maximal product. The maximal product of 10 is produced when 10 is divided into two 3s and two 2s (or one 4), and the product that results is 3^2 times 2^2, which equals 36. This is the maximal possible product when dividing 10 into whole integers. The maximal product of any number can be achieved by dividing it into as many 3s as possible (for multiples of 3 – all 3s) with one or two remaining 2s (never leaving over a 1). When we allow the use of fractions, dividing 10 into four 2.5s, $(2.5)^4$ produces a result of 39.0625, which is significantly greater than 36, because 2.5 is closer to *e* (2.718…) than 3.▶▶ The same is true for any number: when divided

▶ A rational number is a whole number, or any number that can be expressed as a fraction of whole numbers. Such numbers have either a finite number of figures after the decimal point, or a sequence of figures that repeats itself infinitely.

An irrational number is a number that can only be expressed by a decimal number that has an infinite number of figures after the decimal point with no repeating order to them (usually truncated to a short number of three or four decimal places, as in the case of *e*, 2.718).

Irrational numbers are either algebraic or transcendental. If the number is a root of a non-constant polynomial equation with rational coefficients then it is algebraic, if not, it is transcendental. The two most important transcendental numbers are π and *e*. Significantly, π is 3.14, a little more than 3 (the sages often use the number 3 to approximate π) while *e* is 2.718, a little less than 3. Both round off to the whole number 3.

▶▶ There are 39 categories of work forbidden on Shabbat. The Talmud[2] derives this number from the numerical value of "these" (אֵלֶּה), which is 36 (the largest product generated by dividing 10 into whole numbers). The additional three categories are derived from the next word in the verse, "the things" (הַדְּבָרִים), bringing the total to 39 (the maximal product generated by dividing 10 into fractions). Thus, the 39 categories of work represent the maximal product of a division of 10 (the 10 *sefirot*).

▶ Another significant example of this quality relates to one of the secrets of the Jewish calendar, which is based on a 19 year cycle. Every 19 year cycle includes 7 leap years that appear 5 times with 3 year intervals and 2 times with 2 year intervals (see Chapter 14). Dividing 19 into whole integers (5·3 ⊥ 2·2) yields a maximal product of $3^5 \cdot 2^2$, which equals 972.

Dividing 19 by 7 results in 2.7143 (to 4 decimal places; close to Euler's number). 2.7143^7 is 1,085.4… and may be rounded off to 1,086 (often in the Torah a fraction/part is considered as a whole, e.g., a part of a day is considered a whole day). 19 is the numerical value of Eve (חַוָּה, the first woman, relating to the lunar cycle, the basis of the Jewish calendar). The sum of the numerical values of the names of the four Matriarchs (שָׂרָה רִבְקָה רָחֵל לֵאָה), who rectified the soul of Eve, is 1,086 (see Chapter 8).

The difference between the product of the whole segments of 19 (972, when dividing into whole integers) and the product when using the fraction (1,086) is 114, which equals 6 times 19. This difference corresponds to the full array of 6 permutations of the three-letter name Eve (חַוָּה).

▶▶ The discrete value of γ, 577, is the sum of the numerical values of Israel (יִשְׂרָאֵל) and Leah (לֵאָה).

into m times n, the closer n is to e the closer the product will be to its maximum. ◀

10 divided into	product
1, 9	9
2, 8	16
3, 7	21
4, 6	24
5, 5	25
3, 3, 3, 1	27
2, 2, 2, 2, 2	32
3, 3, 2, 2	36
2.5, 2.5, 2.5, 2.5	39.0625
$e^{3.686}$	39.6000

The Euler-Mascheroni constant (γ, *gamma*) recurs in analysis and in number theory. It is the companion number of e and is defined as the limiting difference between the harmonic series (see Chapter 15) and the natural logarithm.[3] Its value is 0.577.

$$\gamma = \lim_{n \to \infty}\left(\sum_{k=1}^{n}\frac{1}{k} - \ln(n)\right) = \int_{1}^{\infty}\left(\frac{1}{[x]} - \frac{1}{x}\right)dx$$

The γ constant has neither been proven algebraic nor transcendental. In fact, it has not yet been established whether or not γ is irrational. ◀◀

Euler calculated the value of e^{γ} as 0.17810. As mentioned in the Introduction, the Torah takes into consideration only whole, positive integers. The discrete value of e^{γ} is 1,781, which is exactly 13 times 137. e^{γ} has proven valuable in calculating prime numbers. Both 13 and 137 are primes.

A Blessing of Maximal Abundance

God commanded the Jewish priests, the *kohanim*, to bless the Jewish

People with the Priestly Blessing specified in the Torah.[4] This is also the blessing with which many Jewish fathers customarily bless their children on the eve of Yom Kippur, the holiest day of the year; others have the custom to bless their children with this blessing every Shabbat eve.

May God bless you and protect you.	יְבָרֶכְךָ הוי' וְיִשְׁמְרֶךָ:
May God shine His countenance upon you and be gracious to you.	יָאֵר הוי' פָּנָיו אֵלֶיךָ וִיחֻנֶּךָ:
May God raise His countenance towards you and grant you peace.	יִשָּׂא הוי' פָּנָיו אֵלֶיךָ וְיָשֵׂם לְךָ שָׁלוֹם:

The rule is that every first appearance of a word in the Torah testifies to its essential significance. The first time that the root of the verb "to bless" (ב-ר-ך) appears in the Torah is in God's blessing to "be fruitful and multiply."[5] This suggests that the Priestly Blessing affects a blessing of maximal multiplication in the Jewish People. Appropriately, in each of the three verses that comprise the Priestly Blessing there is a progressive increase in the number of words and letters.[6] The number of words in each verse is 3, 5 and 7, respectively, an increase of 2 words in each verse. The verses contain 15, 20 and 25 letters, respectively, which is an increase of 5 letters in each verse.

Although the numerical values of the verses do not illustrate the same progressive increase, the total numerical value of the entire Priestly Blessing is 2,718, the discrete value of *e*, the number that produces the maximal product.▸ The blessing is comprised of three complete verses, which demonstrates an explicit relationship between *e* and 3, the whole number segment that generates a maximal product.

The three verses of the Priestly Blessing correspond to the three *sefirot*, *chesed* (the *sefirah* of loving-kindness), *gevurah* (the *sefirah* of might) and *tiferet* (the *sefirah* of beauty), the first three "legs" of the Divine Chariot. The fourth leg is *malchut* (the *sefirah* of kingdom).

▸ The numerical value of the entire Priestly Blessing when the first word "[God] shall bless you" (יְבָרֶכְךָ) is excluded, is 2,466, which equals 18 [the numerical value of "life" (חַי)] times 137.

▶ The first verse of the *Shema* is the climactic concluding verse of of the *malchuyot* verses recited in the *Musaf* prayer of Rosh Hashanah, the day of God's coronation as King of kings. This is one indication of its correspondence to *malchut* (the *sefirah* of kingdom).

The Torah verse that corresponds to *malchut* and completes the Divine Chariot is the opening verse of the *Shema*.◀ The 25 letters of the third verse of the Priestly Blessing allude to the *Shema*, which also contains 25 letters. The numerical value of the entire *Shema*, "Hear O Israel, *Havayah* is our God, *Havayah* is one" (שְׁמַע יִשְׂרָאֵל הוי' אֱ-לֹהֵינוּ הוי' אֶחָד) is 1,118. As mentioned, the numerical values of the entire Priestly Blessing is 2,718. The sum of the numerical values of these four verses relating to the Divine Chariot is thus 3,836, which equals 28 times 137.

The three verses of the Priestly Blessing are summarized in the 28 letters of the Torah verse that follows it: "And they shall place My Name upon the Children of Israel, and I will bless them" (וְשָׂמוּ אֶת שְׁמִי עַל בְּנֵי יִשְׂרָאֵל וַאֲנִי אֲבָרְכֵם). The numerical value of this verse is 2,136. Adding this number to the sum of the three preceding verses, 2,718, results in 4,854. This number equals 3 times the discrete version of the golden ratio, *phi* (ϕ), 1,618. The Priestly Blessing thus offers maximal abundance in the most aesthetic proportions.

Here we have seen interesting relationships between some of the transcendental numbers significant in mathematics and modern physics. These relationships are only revealed by meditating on the numerical mysteries hidden in the verses of the Torah.

Figurate numbers

Before we go any further, let's explain another essential idea that connects Kabbalah to mathematics.

If you were to ask a child to arrange a number of coins into a symmetrical shape, they might begin by forming a square or a hexagon. Both of these shapes are created from the most basic geometrical form—a triangle.[7]

The general mathematical formula for an equilateral triangle is $\triangle n = n(n \perp 1)/2$. That is, the triangle of *n* is *n* times its mid-point $(n \perp 1)/2$.

This function produces the series:

1 3 6 10 15 21 etc.

Below, we see how the geometric shapes formed from these numbers are indeed triangles.▶

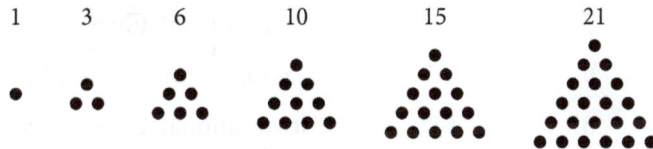

A third type of figurate number is inspirational numbers. They are defined as the sum of two consecutive square numbers. The mathematical formula for an inspirational number is $\boxdot n = n^2 + (n-1)^2$.

The first five inspirational numbers are 1, 5, 13, 25 and 41.

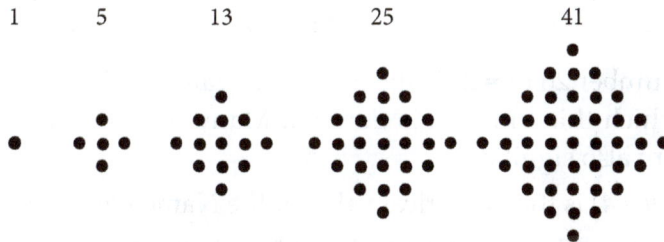

The Mathematical Series for God's Name

The first and foremost source for Kabbalistic mathematical meditation is in the essential Name of God, *Havayah*. This Name appears once in each of the three verses of the Priestly Blessing.

Below is a table of each letter of the Name, its numerical value and its expression as a figurate number.

▶ The fourth number in the series of triangular numbers is 10. In the Zohar,[8] the classic text of Kabbalah, the ten *sefirot* are derived from the four letters of the essential Name of God, *Havayah*. The first letter of the Name, *yud* (י), corresponds to the *sefirah* of *keter* (the super-conscious crown). It is referred to as "a single song." The second letter of the Name, *hei* (ה), corresponds to the next two *sefirot*—*chochmah* (wisdom) and *binah* (understanding)—which are "a double song." The third letter of the Name, *vav* (ו), corresponds to the next three *sefirot*, *chesed* (loving-kindness), *gevurah* (might) and *tiferet* (beauty), "a triple song." The fourth and last letter of the Name, a second *hei* (ה), corresponds to the last four *sefirot*: *netzach* (victory), *hod* (acknowledgement), *yesod* (foundation) and *malchut* (kingdom), which are "a quadruple song."

In general, the *yud* of God's essential Name, *Havayah*, corresponds to *chochmah* (the *sefirah* of wisdom), the hei corresponds to *binah* (the *sefirah* of understanding), the *vav* corresponds to the six *sefirot* of the heart, and the final *hei* corresponds to *malchut* (the *sefirah* of kingdom). Here, although the interpretation is apparently different, there is an explanation that solves the contradiction.

When presented graphically, this development of the ten *sefirot* from the four letters produces an equilateral triangle with each side measuring four units.

Yud	•	single song
Hei	• •	double song
Vav	• • •	triple song
Hei	• • • •	quadruple song

Letter of God's Name	Numerical value	Expression as figurate number	General formula for figurate number
Yud	10	△4	Triangle: $\triangle n = n(n + 1)/2$
Hei	5	□2	Inspirational: $\square n = n^2 + (n - 1)^2$
Vav	6	△3	Triangle: $\triangle n = n(n + 1)/2$
Hei	5	□2	Inspirational: $\square n = n^2 + (n - 1)^2$

The generalized additive function of the letters of God's Name is thus, $\triangle(n + 2) + \square n + \triangle(n + 1) + \square n$ (where $n = 2$). Neatly simplified, this function is: $f(n) = 5n^2 + 6$. ◄

n	-4	-3	-2	-1	0	1	2	3	4	5	6	7	8	9	10	11	12	13
$f(n)$	86	51	26	11	6	11	26	51	86	131	186	251	326	411	506	611	726	851

The number 26 ($n = 2$) is the numerical value of the Divine Name from which this function is derived. Many other numbers in this series are also significant:

- 86 ($n = 4$) is the numerical value of the Name *Elokim* (אֱ-לֹהִים).

- 186 ($n = 6$) is the value of *Makom* (מָקוֹם), another common reference to God. *Makom* means "space," referring to God's omnipresence. 186 equals 6 times 31, the numerical value of the Name *Kel* (אֵ-ל), which also means "to" (אֶל) alluding to six vector forces of three-dimensional space.

- 411 ($n = 9$) is the numerical value of "chaos" (תֹּהוּ), which we have seen is a holy word. 411 equals 3 times 137.

In the verse that concludes the Priestly Blessing, "And they shall place My Name upon the Children of Israel, and I will bless them" (וְשָׂמוּ אֶת שְׁמִי עַל בְּנֵי יִשְׂרָאֵל וַאֲנִי אֲבָרְכֵם), there are two references to God, "My Name" (שְׁמִי) and "I" (אֲנִי). When we contemplate these two references carefully, we see that they are a permutation of the phrase, "something from nothing" (יֵשׁ מֵאַיִן). Above, we saw that this phrase has a numerical value of 411. These two

This series is symmetric. The midpoint of the series is $n = 0$, where $f(n) = 6$ (i.e., the series has "masculine" symmetry). The rule is that the average number of the seven numbers surrounding the midpoint of any quadratic series with masculine symmetry (where the midpoint is a single number, as opposed to a series with two identical numbers in the middle, which has feminine symmetry) is the second number from either side. In this case, the sum of the seven numbers (51 26 11 6 11 26 51) is 182, the numerical value of Jacob (יַעֲקֹב), which equals 7 times 26 (the numerical value of the essential Name of God, from which the function of the series is derived).

Taking the finite differences, we discover that the base of this series is 10.

6   11   26   51   86   131   186   251

  5   15   25   35   45   55   65

    10   10   10   10   10   10

This refers to the first letter of the essential Name and also alludes to the ten *sefirot*.

references appear together in another verse, "I am *Havayah*, that is My Name" (אֲנִי הוי' הוּא שְׁמִי)

- 506 (*n* = 10). 10 is the numerical value of the first letter of God's Name, the *yud* (י). 506 is the diamond form [$n(n \perp 1)$] of 22, the number of letters in the *alef-bet*. (In Chapter 4, the number 506 is discussed further with reference to another important mathematical series that relates to the appointment of a king. God is the King of Kings.)

- 611 (*n* = 11) is the numerical value of Torah (תּוֹרָה).

- 851 (*n* = 13) is the product of 23 [(the numerical value of *chayah* (חַיָּה)] times 37 [the numerical value of *yechidah* (יְחִידָה)]. These two words are the names of the two highest levels of the soul. (See Chapter 12 for more about the relationship between 23 and 37.)

Euler's Constant and the Age of the Patriarchs

Having generated this series directly from the letters of God's Name, we can discover more of its secrets by meditating on it further. Adding every pair of consecutive numbers in the series (beginning with *n* = -1 and *n* = 0; *n* = 1 and *n* = 2, etc.) results in a new series:▶

17   37   137   317   577

This series, derived from the essential Name of God, directly links the number 137 (the third number in the new series) to 577 (the fifth number), the discrete value of γ.

Each of the three verses of the Priestly Blessing corresponds to one of the three Patriarchs, Abraham, Isaac and Jacob, respectively. The new series relates to the ages of the Patriarchs at points of major significance in their lives.

- The number 17 relates to Joseph, who was 17 when he experienced his dreams and was subsequently sold into slavery in Egypt. It also relates to Jacob's seventeen final years, spent in Egypt.▶▶

▶ As mentioned in Chapter 1, the numerical value of the "Goodly Name" (אהרה) is 17. The numerical value of the filling of the Name *Sag* is 37. The numerical value of the filling of the Goodly Name when filled with the same letters is 137.

▶▶ When Joseph revealed his identity to his brothers in Egypt, "He raised his voice and wept… [and Joseph said to his brothers], 'I am Joseph. Is my father still alive?'" (וַיִּתֵּן אֶת קֹלוֹ בִּבְכִי... [וַיֹּאמֶר יוֹסֵף אֶל אֶחָיו] אֲנִי יוֹסֵף הַעוֹד אָבִי חָי) (Genesis 45:2-3). The sum of the numerical values of these two expressions is 1,370, which equals 10 times 137.

The numerical value of "I am Joseph. Is my father still alive?" (אֲנִי יוֹסֵף הַעוֹד אָבִי חָי) is 333, which equals 9 times 37. For more about Joseph and 137, see Chapter 9.

- Prior to his descent to Egypt, Jacob spent 20 years in Charan. Thus, the total number of years that he spent outside the Holy Land is 37. Isaac was 37 years old when God commanded Abraham to bind him and offer him as a sacrifice on the altar.

- At the time of this final and most difficult test—known as the *Akeidah*, or the Binding of Isaac—Abraham was 137 years old (see Chapter 7).[9] The next number in the series, 317, a permutation of the same integers as 137, is the numerical value of the common expression, *Avraham Avinu* (our Patriarch Abraham, אַבְרָהָם אָבִינוּ).

This sequence of numbers can also be identified in Psalm 15, which we shall examine in the next section.

Psalm 15: Refining the Imagination

What do scientists do when they come across a query in their quest for knowledge? Having no authentic solutions at hand, they dream up a hypothesis with "creative imagination," or "educated guesswork." The scientist then attempts to prove his assumption using inductive logic, as mentioned in the Introduction. Although guessing doesn't seem to be the most scientific way of relating to science, this is how quantum physicist, Richard Feynman, described the process:

> In general we look for a new law by the following process. First we guess it. Then we compute the consequences of the guess to see what would be implied if this law that we guessed is right. Then we compare the result of the computation to nature, with experiment or experience, compare it directly with observation, to see if it works. If it disagrees with experiment, it is wrong. In that simple statement is the key to science. It does not make any difference how beautiful your guess is. It does not make any difference how smart you are, who made the guess, or what his name is—if it disagrees with experiment it is wrong. That is all there is to it.[10]

The quality of our imagination, and our ability to "guess," is greatly affected by our associative conceptions.[11] These associations depend upon our beliefs and culture. A person's logic and power of deduction are conditioned by their own opinions and desires.[12]

Rectifying our imagination nurtures the development of authentic scientific theories. The more our associations are in tune with the Torah, the more likely we are to come up with genuine ideas. The first step we can take toward refining our minds and imaginations is to refine our behavior.[13] Before embarking on a journey to the root of all existence, it is a good idea to assess our actions and optimize them.

Psalm 15, which is in tune with the above-mentioned mathematical series, helps us accomplish this refining process by condensing the Torah's 613 commandments▸ into eleven essential principles.

The Talmud advocates such condensation to allow the Torah to penetrate deeper into the heart and consciousness. The more condensed the principles, the more potent they become. At the same time, they become more universal, becoming progressively more relevant to Jew and non-Jew alike. Beginning from one fundamental principle is something that each of us can easily handle. Gradually expanding and deepening our knowledge will eventually give us the ability to keep all of God's commandments, which is the ultimate refinement of character.

King David identified eleven principles. The Prophet Isaiah further condensed them into six principles, the Prophet Micah into three, then came Isaiah again and condensed them into two principles, until the Prophet Amos condensed them into one all-inclusive verse. Finally, the Prophet Habakkuk condensed them all into one single principle.[14] Psalm 15 contains the first such effort at condensation undertaken by King David, as follows:

▸ The sages arrive at this conclusion from the verse, "Moses commanded us the Torah." They base their calculation on the numerical value of "Torah" (תּוֹרָה), which is 611, plus the first two of the Ten Commandments, which the Jewish people heard directly from God at Mt. Sinai.

The Talmud states that there are 613 commandments in the Torah, however, there is nowhere in the Talmud where they are all enumerated. As a result of this, there are a number of different systems for enumerating the 613. The most generally accepted system is that of Maimonides.

Verse 1	Song of David: God, who may reside in Your tent? Who may dwell on Your holy mountain?	מִזְמוֹר לְדָוִד הוי' מִי יָגוּר בְּאָהֳלֶךָ מִי יִשְׁכֹּן בְּהַר קָדְשֶׁךָ:
Verses 2-5	He who walks in sincerity and acts with justice and speaks truth in his heart. He who withholds his tongue from gossip nor harms his neighbor, nor brings disgrace upon one close to him. He who finds a base person contemptible in his eyes but who honors those who fear God. He who stands by his oath without wavering. He who does not lend money on interest nor takes bribes against the innocent.	הוֹלֵךְ תָּמִים וּפֹעֵל צֶדֶק וְדֹבֵר אֱמֶת בִּלְבָבוֹ: לֹא רָגַל עַל לְשֹׁנוֹ לֹא עָשָׂה לְרֵעֵהוּ רָעָה וְחֶרְפָּה לֹא נָשָׂא עַל קְרֹבוֹ: נִבְזֶה בְּעֵינָיו נִמְאָס וְאֶת יִרְאֵי הוי' יְכַבֵּד נִשְׁבַּע לְהָרַע וְלֹא יָמִר: כַּסְפּוֹ לֹא נָתַן בְּנֶשֶׁךְ וְשֹׁחַד עַל נָקִי לֹא לָקָח
Verse 6	He who does these things will never falter.	עֹשֵׂה אֵלֶּה לֹא יִמּוֹט לְעוֹלָם:

As can be seen in the table above, Psalm 15 is composed of three distinct segments:

- The first segment is the introductory phrase in which King David asks: "Who may reside in Your tent, who may dwell on Your holy mountain?"◄
- The second segment is a list of eleven desirable principles of behavior.
- The concluding segment states that "he who does all of these things will never falter."

The first verse has 37 letters; the eleven principles of behavior together have exactly 137 letters; and the concluding phrase contains 17 letters. Thus, the ages of Abraham and Isaac at the moment of the Binding of Isaac are reflected in this psalm, together with the number 17, the number of years that Jacob spent in Egypt at the end of his life. This

► The sum of the numerical values of "Your tent" (אָהֳלֶךָ) and "Your holy mountain" (הַר קָדְשֶׁךָ) is 685, which equals 5 times 137. The sum of the numerical values of the words when filled (אלף־הא־למד־כף האי ריש־קוף־דלת שין־כף) is 1,887, which equals 51 (3 times 17) times 37.

The value of the filling letters alone (לף־א י מד־ף א יש־וף־לת־ין ף) is 1,202, which is the numerical value of "In the beginning, God created" (בְּרֵאשִׁית בָּרָא).

Thus, these two words alone relate to the first three numbers in our series, 17, 37 and 137. The ordinal value of "Your tent" (אָהֳלֶךָ) is 29, and the ordinal value of "Your holy mountain" (הַר קָדְשֶׁךָ) is 80. The reduced numerical value of "Your tent" is 11, and the reduced numerical value of "Your holy mountain" (הַר קָדְשֶׁךָ) is 17. The sum of these four numbers is 137.

refers us to the sequence of numbers shown above: 17 (letters in final phrase) 37 (letters in introductory phrase) 137 (letters in body of the eleven traits).

These three numbers reflect the ages at which the Patriarchs passed their most taxing trials in their service of God. This psalm should therefore be meditated upon when undergoing trials.

Eleven Principles of Education

The sages quote this entire psalm and then restate each of the eleven principles with its interpretation, including role models for us to emulate.

Each of the eleven principles contains one main perspective that divides into finer aspects. Each principle can thus be compared to a spectral line that needs to be examined for a fine-structure.

Be Sincere

The first principle is "he who walks in sincerity" (הוֹלֵךְ תָּמִים). This is the most all-inclusive principle, representing the absolute, uncountable crown of the soul.▶ Of all the eleven principles enumerated, this is the only one that refers directly to the relationship between man and God.[15]▼▼

▶ In Kabbalah, 11 relates to the unit that transcends the 10 sefirot. This is the super-conscious, whose apex is the "unknowable head," from which the ten sefirot develop. It is called the "one that is not considered in the account" (חַד וְלֹא בְחֻשְׁבָּן). The numerical value of this phrase is 411, which equals 3 times 137, and is the numerical value of "chaos" (תֹּהוּ).

The Lubavitcher Rebbe, Rabbi Menachem Mendel Schneerson, explained that in the terminology of Kabbalah, the super-conscious crown has two aspects: an external aspect and an internal aspect. The external aspect that is represented by the initial unit is "countable." The eleventh is the absolute one, which is "uncountable." In the terminology of Kabbalah, the countable one, the first of the ten, is Arich, the source of will in the soul. The uncountable level is Atik, which represents the chaos of "nothingness" that preceded creation ex-nihilo.

▶▶ There are two general categories of commandments: those between man and God (בֵּין אָדָם לַמָּקוֹם), and those between man with his fellowmen (בֵּין אָדָם לַחֲבֵרוֹ). The sum of the numerical values of these two phrases is 676, which is equal to 26^2. The numerical value of God's essential Name, Havayah, is 26, thus the union of these two phrases indicates that all of our actions, both those that relate to man and those that relate to God, all refer directly to God.

These two aspects are reflected in the Name Havayah. The first two letters of the Name—yud-hei (יה)—are alluded to at the beginning of the verse, "The hidden things are for Havayah, our God,"[16] which refers to our relationship with God. The second two letters of the Name—vav-hei (וה)—refer to the second half of the phrase, "and the revealed things are for us and for our children," which relates to our relationship with our fellowmen.

When asked which is greater, loving God or loving Israel? Rebbe Schneur Zalman of Liadi, the founder of the Chabad Movement, would reply that loving what one's beloved loves is the ultimate expression of true love. Loving the Jewish People is therefore greater than loving God, because God loves the Jewish People. In this sense, the origin of the last two letters of the Divine Name is higher than our relationship to God Himself. This idea appears repeatedly in various Chassidic texts.

From here we learn that in order to relate properly to one's fellowman, one must first have a profound love for God.

The role model associated with this quality is Abraham.[17] When God commanded Abraham to circumcise himself, He told him, "Walk before Me and be sincere" (הִתְהַלֵּךְ לְפָנַי וֶהְיֵה תָמִים).[18]

The word "sincere" (תָּמִים) has various connotations, each of which represents a different level of this quality. Some commentaries explain that the phrase in this verse refers to the all-inclusive verse, "Be sincere with *Havayah*, your God."[19] Rashi explains that we should live our lives with simplicity, without occupying ourselves with what the future holds in store, and without turning to superstitious rituals to predict it. Rather, we should trust in God and accept with equanimity all that He sends our way. We can rest assured that He will take responsibility for the optimal outcome.

God's injunction to Abraham to "walk before Me and be sincere" was in the context of Abraham's circumcision. The process of circumcision is the removal of the foreskin from the male reproductive organ. So, this commandment in particular represents matrimonial fidelity.

The heart also has a spiritual foreskin, and it is incumbent upon us to circumcise our hearts by cleansing our subconscious desires. To accomplish this, we must constantly guard our minds to prevent unwanted thoughts from entering our consciousness. The rectification of subconscious desires is the manifestation of true sincerity in the soul.

Act with Justice

The second principle mentioned in Psalm 15 is "[he] acts with justice" (פֹּעַל צֶדֶק). "Justice" (צֶדֶק) is related to "charity" (צְדָקָה). The Talmud explains that to act with justice means to help others through acts of kindness. The role model the Talmud[20] proposes is the sage, Abba Chilkiah. Elsewhere in the Talmud,[21] we find a story that explains how Abba Chilkiah earned this distinction:

> *Once, during a heavy drought in the Land of Israel, the leading sages sent two colleagues to ask Abba Chilkiah to pray for rain.*

▶ Nachmanides includes this precept as one of the 613 commandments. However, Maimonides does not include it in his enumeration. In this he follows the rule, which he mentions in the introduction to his book, that all-inclusive commandments are not included in the count.

The two scholars first came to his house, but he was not there. So they went out and found him digging a field. They greeted him respectfully, but he paid them no attention. At the end of the day, when he had completed his work, they followed him home.

After the family had eaten, Abba Chilkiah called his wife aside and said to her: "I know that these scholars came to ask me to pray for rain. Let us go up to the roof and pray; perhaps God will accept our prayer and send rain, so they won't need to ask us for anything. Then they won't have to feel grateful to us."

Abba Chilkiah and his wife went up to the roof. He stood in one corner and his wife stood in another, and they prayed for rain. It was not long before their prayers were answered and heavy rain clouds covered the sky. The visitors noticed that the clouds first began accumulating from the side where Abba Chilkiah's wife had been praying.

Abba Chilkiah returned to his visitors and asked them, "Why have you come to visit me, my respected teachers?"

"The sages sent us to ask you to pray for rain," they answered,

"Blessed is the Master of the Universe who spared you from having to rely on Abba Chilkiah," replied he.

However, the scholars had been following Abba Chilkiah's every move since they had arrived, to see what they could learn from the actions of this pious man. They replied, "We know that the rain came because of your prayers. However, we would like you to please explain some of your strange actions to us. Please tell us why you did not acknowledge our greeting in the field?"

Abba Chilkiah replied, "I had hired myself out as a day laborer, and I had no right to use my time to interrupt my work by greeting you."

"And why did the clouds first appear from the side where your wife prayed?" asked the messengers.

"My wife's prayers were heard first, because she spends her day at home. When a poor man comes to the door, she gives him ready-to-eat food. This saves him from trudging farther to buy something to eat. When a pauper asks me to help him, I can only give him a coin, and he still has to trouble himself to purchase food. Or, perhaps it is because when there were once villains in my street, I prayed that God do away with them, but my wife prayed for them that they should repent!"

These are just some of the good deeds that are mentioned regarding Abba Chilkiah's exemplary conduct.

Behave with Integrity

The third principle identifies the one "who speaks truth in his heart" (וְדֹבֵר אֱמֶת בִּלְבָבוֹ). The literal interpretation of this phrase is that we should never disguise our true thoughts in words that express something else.[22] Yet, the Talmudic model[23] for this quality expresses a more far-reaching type of integrity than simply saying what one means:

Once someone approached Rav Safra wishing to purchase an item from him. He offered Rav Safra a price. But, Rav Safra was reciting the Shema and was thus unable to reply. The customer presumed that Rav Safra's silence was because the price was not high enough, so he made a much higher bid. When Rav Safra finished saying the Shema, he sold the item to the customer for the initial price mentioned. Detecting the customer's surprise, Rav Safra explained that he had heard when the customer mentioned the first price and had already agreed in his heart to accept the lower price. Taking the higher price would have indicated that he had been speaking falsely in his heart.

According to Jewish law, planning to do a good deed without verbally expressing our intention does not obligate us to do the deed. Should we later decide against doing it, we are not required to act on our original plans.

According to one opinion,[24] in the context of sanctity, we are obliged to fulfill what we think of doing. Only regarding secular matters are we are not obliged to fulfill an action unless we verbally expressed our intention to do so. But, by implementing our good thoughts as if we had spoken them, as illustrated by Rav Safra's act, we refine our behavior as if we are constantly within God's "tent," i.e., the Temple. This equates our mundane deeds with our religious life. It sanctifies our secular life.

Along similar lines, the Tzemech Tzedek, said (Yiddish), *Mach do Eretz Yisroel*. Every Jew should sanctify the place where he lives as if he was living in the Holy Land of Israel.

One of the greatest challenges of Jewish education is to incorporate the secular world into the context of holiness. In particular, this means integrating science into the framework of Torah. This phrase in Psalm 15 teaches us that the proper attitude begins by actualizing our good thoughts in the secular realm as we are obliged to do in the realm of sanctity.

Hold Your Tongue

The next principle specified is, "he who withholds his tongue from gossip" (לֹא רָגַל עַל לְשֹׁנוֹ) referring to someone who is careful not to talk about other people. "Gossip" here shares its root with the word for "spy" (ר-ג-ל). Thus, this quality refers to not spying on other people's actions with the intention of relating them to a third party. Even neutral gossip that contains no slander often has negative implications and leads to undesirable results.

The sages add another dimension to this "spectral line," basing themselves on the behavior of Jacob. His mother, Rebecca, told him to receive his father's blessings in his brother's place. But, Jacob hesitated; it was not his nature to deceive. He was honest and sincere. Yet, he knew that if Esau received the blessings, it would result in tremendous harm for the whole of humanity. Moreover, for Jacob and his descendants, receiving the blessings was crucial.

Even knowing that he would be preventing great harm and receiving great benefit, Jacob was reluctant to practice any form of deception. He agreed to do so only when Rebecca reassured him that her request had been ordered by God through prophecy. From here the sages teach us that the fine-structure of this particular phrase, "withholds his tongue from gossip" refers to someone who refuses to resort to any form of corruption for his personal advantage, or even for public benefit.

Cause No Harm

The fifth principle identifies the one "who does not harm his neighbor" (לֹא עָשָׂה לְרֵעֵהוּ רָעָה). While the meaning is unambiguous, once again, the sages offer a more precise interpretation, this time without presenting a role model. They describe acting over-competitively in business.

Modern society is based on a competitive free market. Nonetheless, we must realize that not all forms of competition are good. Some are not considered at all ethical by Torah standards. For example, opening a store selling the same wares as a nearby store is forbidden by Torah law. One is not allowed to invade another's business territory in a way that jeopardizes his livelihood.

In contrast, in the context of Torah education, there is no forbidden competition. It is permitted to open a school on the same street as another school, as we are taught, "rivalry among scholars increases wisdom."[25]

Don't Shame Others

The next principle speaks of a person who "does not bring disgrace upon one close to him" (וְחֶרְפָּה לֹא נָשָׂא עַל קְרֹבוֹ). "One close to him," could be referring to a blood relative, a friend, or even a business partner. This quality too, has a simple explanation and a finer structure. The simple explanation is that this refers to someone who is sensitive to others and careful never to embarrass them.

The fine-structure of this quality is the Maharsha's interpretation of the Talmud. He explains that this refers to someone who takes care to reprimand those close to him—his business associates for example—should they behave inappropriately. He does so because if he would ignore such behavior, he would ultimately bring embarrassment and shame upon them.

The third interpretation complements the previous one. It relates to not covering up for a close relative or associate whose actions are detrimental to others, or to society as a whole. If it becomes necessary to divulge the details of the damaging behavior to the authorities, he does so in order that his associate will refrain from his activities and receive just punishment for his deeds.

Be Humble and Despise Evil

The next quality featured is humility. The Hebrew phrase used in the verse is most frequently translated as "He who finds a base person contemptible in his eyes" (נִבְזֶה בְּעֵינָיו נִמְאָס). This refers to one's self—he who finds himself contemptible in his own eyes. Even if others praise him as a good and righteous person, a truly humble individual discredits the good deeds that he has accomplished. He knows that all he has done is unworthy of consideration when compared to what he should be doing for God. So, no matter how much I am praised, I should regard myself as lowly and unworthy of praise. One might think that such a low state of self-esteem would paralyze a person. Indeed, this is what happens to someone whose ego is master over him. However, in a truly humble individual, low self-esteem becomes a motivator to do more, because he is never satisfied that he has achieved enough. ▶

Another interpretation is that the righteous individual treats someone who behaves in a base way with contempt, appropriate to his evil ways. The sages illustrate this idea by describing King Hezekiah's treatment of his father, Ahaz, upon his death. King Ahaz was indeed "a base person" who brought idols into the Temple. By contrast, his

▶ In material matters, the highest level is to be satisfied with our portion. By contrast, in matters of spirituality, we must always strive and aspire to achieve more.[26]

son Hezekiah was a righteous king. When his father passed away, Hezekiah took his bones and publicly dragged them through the street in the most dreadful manner. The sages commended Hezekiah for doing so.

A number of commentaries explain why Hezekiah did this. Rashi states that his aim was to discipline the people. He showed them that an evil person must receive the punishment he deserves, even if that person is the king's father and was once king himself.

It would appear that dragging the bones of one's father through the street is the antithesis of the commandment to honor one's parents. However, by doing so, Hezekiah affected rectification of his father's soul.[27] Rectifying the soul of his deceased father was an act of greatest respect.

Respect Wisdom

The next quality contrasts the humility described above. It describes how one interacts with other righteous individuals, He "who honors those who fear God" (וְאֶת יִרְאֵי הוי' יְכַבֵּד). These two are a complementary pair.

The previous quality relates to our left eye, which looks inward critically at our own conduct. The right eye looks outward and sees the positive deeds that others do. Whenever we see someone doing good, we should be positively influenced by their behavior and act respectfully towards them.

The sages relate this quality to King Yehoshafat, who ruled some time before King Hezekiah. As a king, he was worthy of the greatest respect. Yet, whenever a Torah scholar or any God-fearing Jew would enter his presence, he would stand up and greet them respectfully, saying, "My father, my father, my teacher, my teacher, my master, my master!"[28]

Be Steadfast

The ninth principle describes one "who stands by his oath without wavering" (נִשְׁבַּע לְהָרַע וְלֹא יָמִר). This refers to someone who has decided to take upon himself a fast. The model is Rabbi Yochanan, one of the greatest sages of his time, who would sometimes vow not to eat until he arrived at his own home. He did this so that he would always have a polite excuse not to eat at somebody else's house, even if the food was absolutely kosher. In this way, he avoided accepting gifts, as stated in Proverbs, "He who hates gifts, will live."[29] He learned this from the Prophet Samuel, who would never accept gifts of food. ▸

In general, this is a form of modesty. Righteous individuals dislike accepting gifts to the extent that they are prepared to forfeit their own physical needs.

Even though we have vowed to deprive ourselves, we should not waver or change our mind. We should have the courage and the willpower to persevere in our self-inflicted duty.

▸ The Talmud states that in this context there are two complementary qualities that are both good when used with the correct intention. In contrast to the Prophet Samuel and Rabbi Yochanan, the Prophet Elisha always agreed to be a guest when he was on the road. By accepting gifts of food and hospitality, he intended it as a rectification of his host's soul.

Charge No Interest

As mentioned, the tenth in a series is holy. The tenth principle describes one "who does not lend money on interest" (כַּסְפּוֹ לֹא נָתַן בְּנֶשֶׁךְ). The injunction prohibiting loaning money to a fellow Jew on interest is mentioned explicitly in the Torah. The sages explain that this principle goes beyond the letter of the law. It refers to not taking interest even from a non-Jew. According to Jewish law, one is permitted, or even required, to take interest from a non-Jew. In some European countries, where Jews were restricted from many trades, this was one way that they could earn a living. Rashi explains that the person who wishes to dwell in God's house refrains from taking interest even from non-Jews. Ultimately, this self-restriction inhibits the temptation to ever take interest from a Jew.

In contrast to Rashi's interpretation, Radak explains that this means refraining from taking interest from non-Jews who are good to the

▶ The numerical value of "interest" (נֶשֶׁךְ) is 370, which equals 10 times 37, the companion number to 137. This relates to the abovementioned series, the second integer of which is 37. The original discovery of Euler's number was made when Jacob Bernoulli studied the effects of compound interest.

Jews, as a gesture of mutual fidelity. He perceives it as an ethical principle that goes beyond the letter of the law, to reward the non-Jew's positive attitude towards Jews. In this particular respect, the best way to reward a non-Jew is to treat him like a Jew.

Furthermore, Radak explains that not taking interest is better than giving charity. It is human nature to be ashamed to receive charity. But, even prosperous people borrow money from time to time, so people should not be embarrassed to take a loan. ◀

Be Objective

The eleventh and final principle listed in Psalm 15 describes one "who does not take bribes against the innocent" (וְשֹׁחַד עַל נָקִי לֹא לָקָח). This quality refers to an ingrained aversion to any form of bribery or flattery that is liable to mar one's objective consciousness of reality. The prohibition against accepting bribery is also one that is mentioned explicitly in the Torah. The innovation here is that a judge or a leader is so sensitive to accepting bribery that he refuses to accept gifts even when they are rightfully his.

The sages illustrate this quality with the story of Rabbi Yehoshua. His tenant worked his fields and would pay Rabbi Yehoshua regularly by delivering him a set portion of the produce. On one occasion, the tenant had a court case and Rabbi Yehoshua was one of the judges. At that time, the worker wanted to pre-pay Rabbi Yehoshua the sum that he owed him, because he was concerned that after the case he might not be able to do so. But, Rabbi Yehoshua refused to accept pre-payment, even though it was rightly his. He knew that the pre-payment might be considered bribery and could pervert his objective scrutiny of the facts.

In a rectified world, such "moral objectivity" should be a guiding light for all judges and politicians. They should constantly be aware that a bias—whether conscious or subconscious—causes their ability to render objective judgment to "falter." This brings us to the concluding phrase of the entire psalm.

Do Not Falter

The final phrase in the chapter is, "He who does these things will never falter" (עֹשֵׂה אֵלֶּה לֹא יִמּוֹט לְעוֹלָם).

In the Torah, the phrase, "He who does these" (עֹשֵׂה אֵלֶּה) appears three times, always referring to negative actions. The first refers to superstitious acts, the opposite of "walking sincerely"[30]; the second refers to wearing suggestive clothing[31]; and the third to deceitful business practices.[32] These are the antitheses of many of the principles enumerated in this psalm. The numerical value of "he who does these" (עֹשֵׂה אֵלֶּה) is 411 the numerical value of "chaos" (תֹהוּ). This numerical equivalence teaches us that the impeccable behavior outlined in this chapter is the ultimate rectification of the warped beliefs, immorality, and impropriety that created havoc in the World of Chaos.

REFERENCE NOTES FOR CHAPTER 3

1. James G. Gilson, *Strong Quantum Coupling and Relativity*, 2002.
2. *Shabbat* 97b, Rashi.
3. See also our article: http://www.inner.org/torah_and_science/mathematics/euler-equation.pdf
4. Numbers 6:23-27.
5. Genesis 1:22, 28.
6. See also, our book, *The Hebrew Letters*, p.39.
7. See the introduction to our book *913: The Secret Wisdom of Genesis*, and (in Hebrew), *Einayich Breichot Becheshbon*.
8. *Zohar Bamidbar* 119a.
9. See *Emek Hamelech, Parashat Chayei Sarah*. For more about 137 as a lifespan, see Chapter 7. See also our book, *The Twinkle in Your Eye*, Chapter 13.
10. Richard Feynman in a lecture given in 1964 at Cornell University.
11. The subject of rectifying imagination is discussed in detail in our book, *The Mystery of Marriage*, Chapter 2. See also, our book, *The Breath of Life*, Chapter 2.
12. See Rabbi Dr. Abraham Twersky, *Addictive Thinking*.
13. See *Sefer Hachinuch* 16.
14. See *Makot* 24a.
15. For example, see Malbim's commentary on this phrase.
16. Deutereonomy 29:28.
17. *Makot* 24a.
18. Genesis 17:1.
19. Deuteronomy 18:13.
20. *Makot* 24a.
21. Abridged story from *Ta'anit* 23a-b.
22. Rashi on Psalm 15:2.
23. *Rashi, Makot* 24a; ; see also *Baba Batra* 88a.
24. *Maharsha* on *Makot* 23a in the name of the Mordechai.
25. *Baba Batra* 21a.
26. See *Hayom Yom*, 30 Sivan.
27. See *Maharshah* on *Makot* 24a.
28. *Ketubot* 103b; *Makot* 24a.
29. Proverbs 15:27.
30. Deuteronomy 18:12.
31. Ibid 22:5.
32. Ibid 25:16.

THE ORIGIN OF UNITY

From Four to Unity

The behavior of objects is defined by the forces that act upon them. A force is an influence that causes an object to undergo a change in speed, direction, or shape. There are four fundamental natural forces: gravitation, electromagnetism, the weak nuclear force and the strong nuclear force.

We are all familiar with the law of gravity. Simply put, it states that "what goes up must come down." The acceleration of objects that fall under its influence equals the strength of the gravitational field. The standard average measurement of the gravitational force (g) at the earth's surface is 9.81 m/s^2. Quantum physicists postulate the existence of an elementary particle, the graviton, which mediates the gravitational force.

As its name suggests, electromagnetism is responsible for electricity and magnetism. Initially, these were thought to be two separate phenomena, until Maxwell discovered the unity between them. The basic law for electromagnetism is "opposite forces attract; similar forces repel." Like minuscule magnets, subatomic particles exert forces on each other. For example, positively charged protons hold negatively charged electrons in orbit around the nucleus of an atom, and electrons of one atom attract protons of neighboring atoms to form a residual electromagnetic force. This holds atoms and molecules together to form larger constellations of particles.

Electromagnetic radiation, produced by an accelerating electric charge or a changing magnetic field, is classified by the frequency of its wave. These frequencies include radio waves, microwaves, infrared radiation, the visible color spectrum (light), ultraviolet radiation, X-rays and gamma rays. The elementary force particle associated with electromagnetism is the photon. It is the basic quantum of light and all other forms of electromagnetic radiation. The coupling constant that determines the strength of the electromagnetic interaction between electrons and photons is *alpha*, which is closely approximate to $\frac{1}{137}$.

The weak nuclear force is responsible for the radioactive decay of subatomic particles, which affects all known matter particles (fermions). Physicists hypothesize that the weak interaction is caused by the exchange (i.e., emission or absorption) of two force particles (i.e., W and Z bosons). This force is termed *weak* because its typical field strength is several orders of magnitude less than that of both electromagnetism and the strong nuclear force. Weak interactions are most noticeable when particles undergo beta decay, which is the variable measured in carbon dating.

All protons have positive electric charges. Thus, in any atom that contains two or more protons, we would expect them to repel one another. Yet, the atom remains in one piece. We deduce from this that there must be a stronger force that holds the nucleus together, preventing the protons from repelling one another. This force is the strong nuclear force. The force particles responsible for the strong interaction are called gluons.

At regular energy levels, each force acts independently of the others. At high energies, the electromagnetic and weak forces become stronger, while the strong nuclear force becomes weaker. Under conditions of extreme temperature, the particles that convey the weak nuclear force apparently become identical to those that convey the electromagnetic force. The distinction between photons (electromagnetism) and W and Z bosons (the weak force) disappears, thus unifying the two forces into one, called the electroweak force. This phenomenon is the basis of the belief that at the level of grand unification energy, all four forces unite.

Following the discovery of the electroweak force, physicists were encouraged to continue their attempts at unifying the forces of nature.

A number of theories endeavor to incorporate the strong force together with the electroweak interaction, thus uniting three of the four forces. These theories are referred to as grand unified theories (GUT). They include M Theory, String Theory and Superstring theory. *Alpha* is one of the key numbers that must be involved in such a union. As physicist Leon Lederman stated, "[*alpha*] contains the crux of electromagnetism (the electron), relativity (the velocity of light), and quantum theory (Planck's constant)." However, no association between 137 and the gravitational force has yet been established.▸

According to the most recent GUTs, all four forces originated in one force. These theories conjecture that for a split second after the Big Bang, the temperature of the universe was so infernally hot that everything was unified into one perfectly symmetrical force. As the universe began to cool and expand, gravitation first broke away from the other three forces, which remained unified. The next symmetry break occurred between the strong nuclear force and the electro-weak force. The final symmetry break severed the weak force from electromagnetism. According to these theories, most of the cooling process took place before the first second of time had elapsed. Once the universe had cooled sufficiently for the electro-weak force to break apart from electromagnetism, the four forces as we know them today came into play.

▸ The numerical value of "gravitation" (מְשִׁיכָה) is 375, adding 137 brings the total to 512, which equals 2^9. This is the relationship between the sum of the numerical values of the 137th letters of the five books of the Torah and their filling letters, as we shall see later in this chapter. The gravitational force corresponds to *malchut* (the *sefirah* of kingdom), which further corresponds to the Matriarch Rachel. The numerical value of Rachel (רָחֵל) is 238, which is the difference between 137 and 375. All this refers to Eddington's theory and the passages relating to appointing a king, as we shall see.

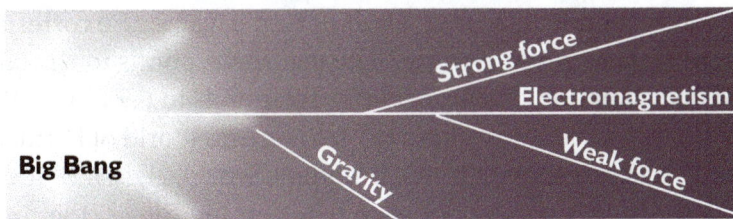

Unified Force Breaks into Four Forces after Big Bang

From Four Forces to Four Letters

Just as there are four forces of nature, so there are four letters to the essential Name of God, *Havayah*. The four letters of God's Name divide into two pairs. The first two letters, *yud-hei*, represent the Divine, hidden aspects of creation. The last two letters, *vav-hei*, correspond to the revealed aspects.[1]◄

Each letter of God's Name corresponds to one of the four forces of nature. The very essence of the strong and the weak forces is that we cannot see them, since they are hidden inside the atom. They correspond to the first two letters of God's Name.

The letter *vav* is often considered the most significant letter of the essential Divine Name. It represents the force that connects the hidden aspects of creation to the more tangible aspects. Electromagnetism therefore corresponds to the *vav* of God's Name. The two aspects of electromagnetism, electricity and magnetism, correspond to the *chasadim* and *gevurot* of *da'at*. The letter *vav* connects these two together. The *vav* begins in *da'at* (the *sefirah* of knowledge), the origin of electricity, and extends to *yesod* (the *sefirah* of foundation), the seat of magnetism.

The final *hei* of God's Name corresponds to *malchut* (the *sefirah* of kingdom), which relates to the earth in general. This is why the final *hei* corresponds to the gravitational force.

In Kabbalah, each of the four letters of God's Name also corresponds to one of four spiritual Worlds:[3] The *yud* corresponds to *Atzilut* (the World of Emanation); the upper *hei* corresponds to *Beriah* (the World of Creation); the *vav* corresponds to *Yetzirah* (the World of Formation). The final *hei* corresponds to the mundane world, *Asiyah* (the World of Action). In this correspondence between the forces, the final *hei* represents gravity, which we sense as the gravitational pull of our physical world, the earth.

The hypothesis that all four fundamental forces of nature began as one unified force is upheld in the Kabbalistic teaching that all four

► This is alluded to in the verse, "The hidden things are for *Havayah*, our God."[2] The numerical value of this phrase is 19 [the numerical value of Eve (חַוָּה)] times 67 [the numerical value of "understanding" (בִּינָה)].

spiritual Worlds emanated from a fifth level that corresponds to the cusp of the *yud* of God's Name. It is called *Adam Kadmon* (Primordial Man).▸

To summarize:

Letter of Havayah	Spiritual World	Physical Force
Cusp of *yud*	*Adam Kadmon* (Primordial Man)	Unified force
Yud	*Atzilut* (World of Emanation)	Strong nuclear force
Hei	*Beriah* (World of Creation)	Weak nuclear force
Vav	*Yetzirah* (World of Formation)	Electromagnetism
Hei	*Asiyah* (World of Action)	Gravitation

▸ This division of 5 into 1 and 4 relates to the Kabbalistic mystery of the letters *alef* (א, with a numerical value of 1) and *dalet* (with a numerical value of 4), which form the word "mist" (אד) in the verse "and a mist arose from the earth" (וְאֵד יַעֲלֶה מִן הָאָרֶץ).[4]
The numerical value of this phrase is 512, which equals 2^9, which is the numerical value of "paradox" (נוֹשֵׂא הַפָּכִים).

The verse in the Torah that alludes to the four Worlds is, "Everything that is called by My Name [*Atzilut*] and for My honor [*malchut* of *Atzilut*], I have created it (*Beriah*), formed it (*Yetzirah*), and even have I made it (*Asiyah*)."[5]

The references to *Atzilut*, *Beriah* and *Yetzirah* (the three upper Worlds: Emanation, Creation and Formation), are divided from *Asiyah* (the World of Action) by "even" (אַף) which has a variant meaning of "wrath." *Asiyah* is the most easily separated from the other three Worlds. This directly corresponds to the first symmetry break, when gravitation split away from the other forces. The three remaining forces were still in a state of symmetry in relation to themselves until the second split occurred. In our Kabbalistic scheme,

▸ The sum of these four numbers is 64, which is the numerical value of Adam-Eve (אָדָם חַוָּה). Thus, this progression, which relates to the creation of light (electromagnetism), is also connected to the creation of man.

▸▸ 16 is the numerical value of the last three letters of *Havayah*, *hei-vav-hei* (ה־ו־ה). In this context, it refers to uniting the weak nuclear force and electromagnetism with gravitation. 16 is also the numerical value of *yud-vav* (י־ו), which would mean uniting the strong nuclear force with the electromagnetic force. These are ideas that scientists should investigate on the basis of these Kabbalistic insights.

after the lower *hei* (corresponding to gravitation) split away from the others, the first letter, *yud* (corresponding to the strong force), separated from the remaining letters, the upper *hei* and the *vav* (the weak force and electromagnetism). These two remained together like a mother and her infant until it is weaned. Finally, the universe cooled down sufficiently to allow the weak nuclear force, represented by the upper *hei*, to divide from the electromagnetic force, the *vav*, each letter thus becoming separate. This corresponds to the stage at which the child grows up and becomes independent.

Translating this progression into numbers, we begin with the numerical value of the complete Tetragrammaton, 26. After the first "symmetry break" the lower *hei* is removed and the sum of the remaining letters (*yud-hei-vav*) is 21. After the *yud* is removed the sum of the remaining letters (*hei-vav*) is 11. Lastly, the *vav* remains isolated with its numerical value of 6.◂

$$26 \quad 21 \quad 11 \quad 6$$

The average value of each of these four numbers is $^{64}/_4$, which equals 16.◂◂ If we expand this series of four numbers by calculating the finite differences between them and creating a cubic progression, we discover that the series is not symmetric. However, the number on both sides of the original sequence is 16, the average value of each of the four numbers in the original sequence. This is an important finding with reference to Eddington's theory (see below), which relies on a scheme constructed on the basis of 16^2.

16	**26**	**21**	**11**	**6**	16
	10	**-5**	**-10**	**-5**	10
		-15	**-5**	**5**	15
		10	**10**	10	

Let's expand the series to fourteen places:▼

1	2	3	4	5	6	7	8	9	10	11	12	13	14
26	**21**	**11**	**6**	16	51	121	236	406	641	951	1346	1836	2431
	-5	**-10**	**-5**	10	35	70	115	170	235	310	395	490	515
		-5	**5**	15	25	35	45	55	65	75	85	95	810
			10	10	10	10	10	10	10	10	10	10	

From GUTs to TOEs

The system by which scientists currently describe the forces of nature is referred to as the standard model of Quantum Electrodynamics (QED). This quantum model offers a sophisticated understanding of the subatomic world. However, it is cumbersome to use. It incorporates dozens of constants that have to be experimentally fine-tuned to reproduce our physical world, including the constant of gravity, the speed of light, and our current favorite, the fine-structure constant. Physicists would much prefer a theory that has a less complex array of constants. Moreover, many queries in this model continue to be

▸ The sum of the first (26) and third (11) numbers in the series equals 37 (the companion of 137).

The sum of the fifth (16 = 4^2) and seventh (121 = 11^2) numbers equals 137. 16 is an even number, it therefore has no midpoint. Yet, it is the final midpoint of 121 (61 is the midpoint of 121, 31 is the midpoint of 61 and finally, 16 is the midpoint of 31).

The sum of the 5 odd numbered elements that follow the original sequence (16, 121, 406, 951, 1836) is 3,330, which equals 90 times 37, or 5 times 666 (see Appendix B).

The sum of the first three even-numbered elements from the beginning of the original series (21, 6, 51) is 78 (an average of 26, the numerical value of the essential Name of God, from which the series was derived)

The sum of the first seven even-numbered elements (21, 6, 51, 236, 641, 1,346, 2,431) is 4,732, which equals 7 times 26^2, i.e., the average of the seven numbers is 26^2.

The sum of the squares of the first four numbers in the series (26, 21, 11 and 6) is 1,274, which equals 49 (7^2) times 26 (the first number in the series and the numerical value of *Havayah*).

The eighth number in the series, 236 (multiplied by 1 million), is the "measurement of the height of the Creator"[6] The next number is 406 (the triangle of 28), which equals 7 times 58, the numerical value of "grace" or "symmetry" (חֵן).

The thirteenth number is 1,836, which is the proton/electron mass ratio.

The sum of the 13 [the numerical value of "one" (אֶחָד)] numbers from 21 to 2,431 is 8,073, which equals 39 [the numerical value of "*Havayah* is one" (הוי׳ אֶחָד), and also the numerical value of "dew" (טַל)] times 207 [the numerical value of "light" (אוֹר)]. This alludes to the expression "a dew of lights."[7]

The sum of the first seven numbers of the series (26 ± 11 ± 16 ± 121 ± 406 ± 951 ± 1,836) is 3,367, which is equal to 91 times 37, the companion of 137, which was the sum of the first pair. The sum of the remaining numbers is 90 times 37, or 5 times 666. The average value of each of the seven numbers is 481, which equals 13 times 37.

veiled in mystery. In addition, although this model satisfactorily unifies three of the four forces, it does not include gravity. These are some of the reasons why physicists aspire to develop a unified "Theory of Everything" (TOE). Such a theory would incorporate answers to all modern scientific enigmas and present a unified way to describe the physical universe. In the words of the late quantum physicist Stephen Hawking:[8]

> *The real reason we are seeking a complete theory is that we want to understand the universe ... The standard model is clearly unsatisfactory in this respect. First of all, it is ugly and ad hoc. The particles are grouped in an apparently arbitrary way, and the standard model depends on 24 numbers whose values cannot be deduced from first principles, but which have to be chosen to fit the observations. What understanding is there in that? Can it be Nature's last word?*

> *The second failing of the standard model is that it does not include gravity. Instead, gravity has to be described by Einstein's General Theory of Relativity. General relativity is not a quantum theory, unlike the laws that govern everything else in the universe. ... All the structures in the universe, including ourselves, can be traced back to quantum effects in the very early stages. It is therefore essential to have a fully consistent quantum theory of gravity, if we are to understand the universe.*

So, the quest for unification is driven by practical, philosophical and aesthetic considerations. When scientists are successful in merging theories, it clarifies their understanding of the universe and leads them to discover new insights they might otherwise never have suspected. In addition to predicting new physical effects, a unified theory would provide a more aesthetically satisfying picture of how the universe operates.

God, the Source of all unity, Who created the world, revealed all the secrets of nature in the Torah. Thus, the ultimate TOE must be present in the Torah and Kabbalistic writings. It is our duty to decipher the teachings of the sages throughout the generations. By doing so, we may uncover the original unified theory that encompasses all of creation.

As a precursor to discovering the ultimate TOE in the Torah, we can analyze some of the theories that scientists have already proposed through the prism of the Torah, and extract the sparks of truth that lie dormant in them.

Adding One

One of the greatest British astrophysicists, Sir Arthur Eddington, was a staunch follower of Einstein, so much so that Einstein himself said that Eddington was the only one who understood the general theory of relativity. Indeed, in the famous eclipse of 1919, Eddington was the one to establish proof of the theory.

Early measurements of the inverse of the fine-structure constant had inaccurately assessed it to be 136.▸

Eddington attempted to derive the number 136 mathematically by investigating it from a numerological point of view using sophisticated mathematical methods. He argued that according to relativity, particles cannot be considered in isolation. They are only significant in their interactions with one another. Therefore, any theory of the electron has to deal with at least two electrons. By applying his own mathematical calculations to this theory, Eddington concluded that there were 16 different ways to define an electron. Multiplying 16 by 16 gave a total of 256 (16^2) different ways in which electrons could combine with each other. Eddington hypothesized that 256 was the "perfect" number of physics.▸▸

Any square number (n^2) can be divided into two triangular numbers, a large triangle and a smaller one. The large triangle is equal to the triangle of n ($\triangle n$) and the smaller triangle is the triangle of the previous number [$\triangle(n - 1)$]. Thus, $n^2 = \triangle n \perp \triangle(n - 1)$. Eddington showed that 120 ($\triangle 15$) of the 256 electron combinations were not viable. This

▸ As mentioned above, current advances in quantum electrodynamics (QED) hold that, depending on the conditions at which it is measured, the fine-structure constant can vary between $1/137$ and $1/128$. Thus, $1/136$ remains a viable proposition under certain circumstances. 136 is the numerical value of each of the three words "voice" (קול), "fast" (צום) and "money" (ממון). These three represent prayer, fasting and giving money to charity, respectively. They are the three most potent actions that sustain *teshuvah* (repentance). When "voice" (קול) is spelled in full (קוף ואו למד) the numerical value of the filling letters alone is 137.

▸▸ As mentioned in Chapter 1, *alef-yud-hei* (איה) is the only combination of three letters in the Torah that when spelled in full (אלף יוד הא) have a numerical value of 137. In this order, these three letters correspond to the three *sefirot* of *keter*, *chochmah* and *binah*. The numerical value of the complete phrase from the *Kedushah* prayer that contains this combination of letters as a word—"where is the place of His Honor?" (איה מקום כבודו)—is 240, which is 15 [the numerical value of the two letters *yud-hei* (יה)] times 16 [the numerical value of the complete word "where" (איה)]. Thus, the phrase alludes to "where" (איה) when the two end letters are read again from the beginning (היאיה). An integral part of Eddington's calculations of 256 was based on the sum of the triangles of 15 and 16 (120 and 136, respectively).

121

left exactly 136 ($\triangle 16$) available possibilities. Eddington saw this fact as an important basis for the fine-structure constant.[9]◄

When the fine-structure constant was later observed to be closer to $1/137$, Eddington altered his calculations by adding an extra unit. It was then that his peers began to regard his occupation with the fine-structure constant as an obsession. They viewed his mathematical meanderings as superstitious numerology, which was unscientific in their minds. As such, his associates nicknamed him "Sir Arthur Adding-One" and disregarded most of his work.

Eddington was perhaps the greatest astrophysicist of his time and the logic of his calculations is based on sophisticated mathematics. Simply neglecting his theory is not so simple. Eddington believed he had identified an algebraic basis for fundamental physics, which he termed "E-frames." Similar algebraic notions underlie many modern attempts to develop a GUT. Moreover, Eddington's emphasis on the values of the fundamental constants, as well as the dimensionless numbers derived from them, is nowadays a central concern of physics.

Almost a century has passed since Eddington made his calculations. Yet, 137 still remains the grand enigma to all. Perhaps there was some truth in his work.◄◄

▶ Eddington predicted the number of hydrogen atoms in the universe as $136 \cdot 2^{256}$, or half of the total number of particles. When equalized with the non-dark energy equivalent number of hydrogen atoms $(3/10) \cdot Rc^2/GmH$, this corresponds to a Universe radius: $R = 13.8$ Giga light years. This value is predicted from universal constants using atomic-cosmic symmetry, and compatible with c-times the Universe age 13.80(4) Gyr, as determined by the Planck mission (March 2003).

▶▶ The integers 2, 5, 6, which form the number 256, are the numerical values of the letters bet (ב), hei (ה), and vav (ו), which spell "void" (בהו), discussed in Chapter 2.

Remove, Add and Explain

When Eddington added 1 to the initial triangle of 136, he inadvertently tapped into a profound Kabbalistic principle, referred to as "the secret of amputation [lit.: sawing]." The concept originates in the creation of the first woman. Eve was originally joined to Adam back-to-back, like a Siamese twin.[10] God put Adam to sleep and amputated one of his ribs. He then constructed Eve from the amputated limb. Similarly, but at an abstract level, there is a Talmudic principle that states, "We remove, and add, and explain."[11] This refers to removing a letter or a unit from one word and adding it to another word, in order to interpret the meaning of a phrase.

In a similar way, Eddington took a unit from the triangle of 120 and added it to the triangle of 136 to create a new pair of numbers, 119 and 137.▶

The All-Inclusive Value

In addition to its letters, words, and numerical value, etc., any text in the Torah possesses an extra unit value, the *kolel* (literally, "all-inclusive" value). This is a manifestation of the unitary, all-encompassing light that connects the individual elements into one textual unit. So, from a Torah perspective "adding one" is acceptable practice under certain conditions.

The source for this principle is in the statement, "There are ten [*sefirot*] and not nine, ten and not eleven."[14] The Arizal explains that this is true in the rigid World of Chaos. However, in the World of Rectification, which is more flexible, there are situations in which nine or eleven *sefirot* are viable options (either by adding *da'at* or excluding *keter*, i.e., adding or subtracting 1).

In further appreciation of Eddington's work, in rabbinic literature, there is an expression that often appears referring to two things that divide or become four. Similarly, 4 divides into 16, and 16 divides into 256.[15]▶▶ In this case, Eddington's construction of a 16-by-16 square is a valid hypothesis. Three times has credible legal status in Jewish law.[16]

The numbers 4, 16, 64 and 256 are particularly significant in Ezekiel's esoteric vision of the Divine Chariot. The chariot had four animals, each of which had four personas (a total of sixteen personas). Each persona had four faces (a total of sixty-four faces). Each face had four wings bringing the number of wings to a total of 256. This fact becomes even more significant when we recall that the word for "wheel" (אוֹפָן) in Ezekiel's vision has a numerical value of 137. As mentioned above, the wheels of the chariot are related to the electron, with which Eddington began his calculations.▶▶▶

▶ The regular numerical value of our suggested word for "electron" (אוֹפָן) is 137. When this word is written in the *albam* lettering system (לפוג) its numerical value is 119. The *albam* system is the letter substitution system associated with *chochmah* (the *sefirah* of wisdom).[12] This relates directly to the creation of the world, as the verse states, "All of them have You made with wisdom."[13]

▶▶ Furthermore, the number 16, the inter-inclusion of the number 4, relates to the inter-inclusion of the four principal values of the essential Name of God, *Havayah*.

	ב"ן	מ"ה	ס"ג	ע"ב
ע"ב	ע"ב דב"ן	ע"ב דמ"ה	ע"ב דס"ג	ע"ב דע"ב
ס"ג	ס"ג דב"ן	ס"ג דמ"ה	ס"ג דס"ג	ס"ג דע"ב
מ"ה	מ"ה דב"ן	מ"ה דמ"ה	מ"ה דס"ג	מ"ה דע"ב
ב"ן	ב"ן דב"ן	ב"ן דמ"ה	ב"ן דס"ג	ב"ן דע"ב

▶▶▶ "Wheel" (אוֹפָן) begins with the letter *alef* (א), which has a numerical value of 1. This alludes to the original calculation of the fine-structure constant of $1/136$ and the later correction to $1/(136 \pm 1)$.

123

Three Communal Commandments

In Chapter 2 we saw that anarchy is not upheld by the Torah; law and order is in the hands of the king. As mentioned above, the passage that relates to lighting the menorah is the only one in the Torah scroll that has 137 letters. So too, the only passage that contains 137 words is the commandment to appoint a king.[17]

When you come to the land that Havayah, your God, is giving you, and you possess it and live therein, and you say, "I will set a king over myself, like all the nations around me,"

כִּי תָבֹא אֶל הָאָרֶץ אֲשֶׁר הוי' אֱלֹהֶיךָ נֹתֵן לָךְ וִירִשְׁתָּהּ וְיָשַׁבְתָּה בָּהּ וְאָמַרְתָּ אָשִׂימָה עָלַי מֶלֶךְ כְּכָל הַגּוֹיִם אֲשֶׁר סְבִיבֹתָי:

you shall surely appoint a king over you, whom Havayah, your God, chooses. From among your brothers, you shall set a king over yourself; you shall not appoint a foreigner over yourself, one who is not your brother.

שׂוֹם תָּשִׂים עָלֶיךָ מֶלֶךְ אֲשֶׁר יִבְחַר הוי' אֱלֹהֶיךָ בּוֹ מִקֶּרֶב אַחֶיךָ תָּשִׂים עָלֶיךָ מֶלֶךְ לֹא תוּכַל לָתֵת עָלֶיךָ אִישׁ נָכְרִי אֲשֶׁר לֹא אָחִיךָ הוּא:

However, he may not acquire many horses for himself, so that he will not bring the people back to Egypt in order to acquire many horses, for Havayah said to you, "You shall not return that way anymore."

רַק לֹא יַרְבֶּה לּוֹ סוּסִים וְלֹא יָשִׁיב אֶת הָעָם מִצְרַיְמָה לְמַעַן הַרְבּוֹת סוּס וַהוי' אָמַר לָכֶם לֹא תֹסִפוּן לָשׁוּב בַּדֶּרֶךְ הַזֶּה עוֹד:

And he shall not take many wives for himself, as his heart must not turn away, and he shall not acquire much silver and gold for himself.

וְלֹא יַרְבֶּה לּוֹ נָשִׁים וְלֹא יָסוּר לְבָבוֹ וְכֶסֶף וְזָהָב לֹא יַרְבֶּה לּוֹ מְאֹד:

And it will be, when he sits upon his royal throne, that he shall write for himself this Torah on a scroll from [that Torah which is] with the priests from the [tribe of the] Levites.

וְהָיָה כְשִׁבְתּוֹ עַל כִּסֵּא מַמְלַכְתּוֹ וְכָתַב לוֹ אֶת מִשְׁנֵה הַתּוֹרָה הַזֹּאת עַל סֵפֶר מִלִּפְנֵי הַכֹּהֲנִים הַלְוִיִּם:

And it shall be with him, and he shall read it all the days of his life, so that he may learn to fear Havayah, his God, to keep all the words of this Torah and these statutes, to fulfill them.

So that his heart will not be haughty [while ruling] over his brothers, and so that he will not turn away from the commandments, neither to the right nor to the left, in order that he may prolong [his] days in his kingdom, he and his sons, among Israel.

וְהָיְתָה עִמּוֹ וְקָרָא בוֹ כָּל יְמֵי
חַיָּיו לְמַעַן יִלְמַד לְיִרְאָה אֶת
הוי' אֱ-לֹהָיו לִשְׁמֹר אֶת כָּל
דִּבְרֵי הַתּוֹרָה הַזֹּאת וְאֶת
הַחֻקִּים הָאֵלֶּה לַעֲשֹׂתָם:
לְבִלְתִּי רוּם לְבָבוֹ מֵאֶחָיו
וּלְבִלְתִּי סוּר מִן הַמִּצְוָה
יָמִין וּשְׂמֹאול לְמַעַן יַאֲרִיךְ
יָמִים עַל מַמְלַכְתּוֹ הוּא וּבָנָיו
בְּקֶרֶב יִשְׂרָאֵל:

In the Book of Samuel[18] there is another passage that relates to appointing a king. It contains 119 words. ▶ The sum of the 119 words in the passage of the king in Samuel and the 137 words in the Torah gives a total of 256, or 16^2. Thus, the number that Eddington believed to be the source of creation is reflected in the same ratio in the two passages that relate to appointing a king.

Mashiach, the definitive King of Israel, will ultimately transfer his own kingdom on earth back to God, "The Master of the Universe, who ruled before any creature was created."[19]▶▶ Then God's purpose for creating the universe will be revealed to all and His status as King of Kings will become manifest, as all nations of the earth acknowledge that He is the Creator of the world. "Once all His desire has come to be, then King His Name will be."

According to Maimonides, the commandment to appoint a king is only applicable to the Jewish People when they live in the Land of Israel. Thus, this commandment is the first and foremost reason why the nation is obliged to live in its homeland.▶▶▶ The Land of Israel is the only place in the world where the Jewish People are defined as a congregate entity. There they acquire a collective consciousness. The king reflects God's sovereignty to the nation, and he reflects the collective consciousness of the nation back to God.

▶ The number of words (137 and 119) in the two passages relating to the king perfectly reflect the numerical value of "wheel" (אוֹפָן) and its spelling in the *albam* letter substitution system (לפוג), as seen above. 119 equals 7 times 17 [the numerical value of "good" (טוֹב)]. This reflects the fact that "good" appears 7 times in the account of creation.

119 is half the numerical value of Rachel (רָחֵל), and 137 is half the sum of the numerical values of Leah (לֵאָה) and Rachel (רָחֵל), both of whom are related to royalty and *malchut* (the *sefirah* of kingdom), as we shall discuss in detail in Chapter 8.

▶▶ The reduced numerical value of *Mashiach* (מָשִׁיחַ), the ultimate king of Israel, is 16. Appropriately, the sixteenth word in the passage in the Torah is "king" (מֶלֶךְ).

The numerical value of *Adon Olam*, "Master of the universe" (אֲדוֹן עוֹלָם) is 207, which is the numerical value of "light" (אוֹר). The filled numerical value of "Master of the universe" (אלף דלת וו נון עין וו למד מם) is 959, which equals 7 times 137. Abraham was the first to refer to God by the Name *Adni*, meaning "Master."[20] The primordial value of Abraham (אברהם; 1 ⊥ 3 ⊥ 795 ⊥ 145 ⊥ 15) is also 959.

▶▶▶ Maimonides does not state explicitly that there is a commandment to live in the Land of Israel. However, Nachmanides[21] states that one of the 613 commandments is that every Jew should live in the Land of Israel. There are many reasons to explain why Maimonides does not explicitly mention this commandment.

Appointing a king is the first of three commandments that the Jewish People must perform when they return to function as a nation upon re-entering the Land of Israel. The second commandment, under the leadership of their king, is to fight against the enemies who rise up against them. The third commandment, once the king has been victorious in his wars, is to rebuild the Temple in Jerusalem. The latter two depend on the first commandment to appoint a king. Before all else, there has to be a Torah-defined leadership. Without correct leadership, there can be no absolute victories over enemies. Building the Temple, the epitome of the process, cannot be fulfilled until the first commandment has been accomplished.

The two Hebrew words that define the entire commandment of appointing a king are *som tasim* (שׂוֹם תָּשִׂים; lit. "Appoint, you shall appoint [a king over you]").▼[22] In this phrase, the verb is repeated, first in the source form (שׂוֹם) and then in the future tense of the verb (תָּשִׂים). The source of a verb is a tense-less form that is above time definition; it is the origin of all other tenses. The unity at the origin manifests in many forms as it descends into the world of physical reality.

▶ The numerical value of the phrase "You shall surely appoint" שׂוֹם (תָּשִׂים) is 1,096, which equals 8 times 137. This is also the value of the phrase, "an intelligent woman" (אִשָּׁה מַשְׂכָּלֶת),[23] a phrase discussed in greater detail in Chapter 8. The feminine intelligence is particularly receptive to this commandment. She is the one who most appreciates the need to appoint a king above and below. This is apparent in the meeting between Abigail and David. Abigail recognized David as king, but her husband, Naval, did not accept his leadership.[24]

Subsequently, Bathsheba became King David's wife and mother of his heir, King Solomon. She, too, was particularly sensitive to David's royalty. Upon approaching David, she said "Long live my master King David"[25] (יְחִי אֲדֹנִי הַמֶּלֶךְ דָּוִד). The part of the primordial value (*chelek kidmi*) of this phrase is also 1,096. The numerical value of, "Long live ... King David," (יְחִי... הַמֶּלֶךְ דָּוִד) is 137. The numerical value of "Long live... David" (יְחִי...דָּוִד) is 42 [3 times 14, the numerical value of David (דָּוִד)]. The numerical value of "[the] King" (הַמֶּלֶךְ) is 95. This division of 137 into 95 and 42 relates to the 95 times that the root of the Hebrew word for "fear" (יר׳א), appears in the Torah and the 42 times that the root of the Hebrew word for "love" (אה׳ב), appears in the Torah (see Chapter 12). This is particularly appropriate for this phrase, since it relates to fear of the King and love of David himself.

The filled numerical value of "You shall surely appoint" (שׂי׳ן ו׳ו מ׳ם ת׳שׂין ׳יוד מ׳ם) is 1,318 (1,096 + an additional 222). The number 222 is equal to 6 times 37. This is another example of the appearance of 37 and 137 as a mathematical idiom.

The verse following "You shall surely appoint..." begins "Only he should not accumulate" (רַק לֹא יַרְבֶּה).[26] The numerical value of this phrase is 548, which equals 4 times 137. The ratio between the two phrases is 2:1, which is defined as "a whole and one half." The total numerical value of the two phrases is 1,644, which equals 12 times 137. This is the numerical value of the verse, "And you shall love Havayah your God with all your heart and with all your soul and with all your might" (וְאָהַבְתָּ אֵת הוי׳ אֱ-לֹהֶיךָ בְּכָל לְבָבְךָ וּבְכָל נַפְשְׁךָ וּבְכָל מְאֹדֶךָ).[27]

In this chapter, "not" (לֹא) appears 8 times. There are 34 words before the first appearance of the word and 61 words after the final appearance. The total number of words before and after is 95. Between the first and last time the word appears in the chapter there are 42 words. Once again, this refers to the division of the 137 words in the chapter into 42 and 95, as mentioned.

With reference to this phrase, the Zohar states, "Appoint—above; you shall appoint—below."[28] First, God must be appointed as King above, at the origin of the soul. By doing so, we touch upon the ultimate soul root of the human king to be appointed. This brings down the soul of the king from its origin.

The king corresponds to the *sefirah* of *malchut* (kingdom), which is a feminine *sefirah*. The first figure associated with royalty in Kabbalah is Rachel. King Saul, the first king of Israel, was one of Rachel's descendants, as was Queen Esther. After King Saul, the monarchy was transferred to King David, a descendant of the royal tribe descending from Judah, who was born to Leah. *Mashiach*, the ultimate redeemer, will incorporate the royal aspects of Rachel together with his lineage from Leah.▸

▸ The sum of the numerical values of Rachel (רָחֵל) and Leah (לֵאָה) is 274, which equals 2 times 137.

A Royal Series

As mentioned, the number of words in the passage referring to the commandment to appoint a king is 137. The number of letters in the passage is 506. We discover this number following 137 in a very important mathematical series that begins with the five most important numbers in Kabbalah:

1   4   10   26   137

The entire study of Kabbalah is related to the secret of God's essence and the secret of His Name. God's essence is obviously His absolute unity, therefore 1 is the first number in the series. Next, God's unique, essential Name is defined in the Talmud as "the Four Lettered Name," therefore the number 4 is a most significant number in Kabbalah. The four letters of God's Name manifest as four different spiritual worlds, as mentioned above with reference to the four fundamental forces of nature. Four is also the basic number of inter-inclusion, by which two entities become four, as mentioned above with reference to Ezekiel's vision.

The number 10 in this series is the sum of all numbers from 1 to 4 ($\triangle 4$); it represents the ten *sefirot*, as mentioned in Chapter 3 with reference to the single, double, triple and quadruple songs. Moreover, God's Name of four letters begins with the letter *yud*, which has a numerical value of 10. The next number in the series is 26, which is the basic numerical value of the Name.

Thus, the progression from 1 to 4 to 10 to 26 is a series that relates to Kabbalah.

The next number in the series is 137, the numerical value of "Kabbalah" (קַבָּלָה). One allusion to the relationship between 137 and God's Name, which further justifies 137 as the next number in this progression, is the fact that in all of the five books of Moses there are exactly 26 different letter combinations that form words with a numerical value of 137. Similarly, in the Prophets there are exactly 26 other words that have a numerical value of 137.◀

By calculating the finite differences between the members until we reach the base number of the series, we can expand the series further to discover the next number in the series. It is 506, the exact number of letters in the paragraph referring to the appointment of a king. But, as we saw in Chapter 3, this is also the tenth number in the generalized additive function of the letters of God's Name [$f(n) = 5n^2 \perp 6$]. In that series, 506 follows the number 411, which equals 3 times 137. In our present series, it follows directly after 137.

▶ In the Writings of the Bible, there are 18 other words with a numerical value of 137, bringing the total number of words with this numerical value in the entire Bible to 70, the numerical value of "secret" (סוד), alluding to the phrase, "God's secret is for those who fear Him."[29]

1	4	10	26	137	506	1374
	3	6	16	111	369	868
		3	10	95	258	499
			7	85	163	241
			78	78	78	

▶▶ One mathematical rule relating to any series of this kind is that after p steps—where p is a prime number—the sum of all p members of the series will be a multiple of p. In this case, the sum of the first 13 numbers is 106,496, which equals 13 times 8,192, i.e., 13 times 2^{13}. Thus, the average value of each of the 13 members is 2^{13}.

The first number in the series is 1. "One" (אֶחָד) has a numerical value of 13, a fact that is particularly significant in this case.

The sum of the first seven numbers in this series is 2,058.◀◀ This number divides exactly by 7, resulting in an average of 294, the numerical value of the phrase "there is no king without a people" (אֵין מֶלֶךְ בְּלֹא עָם).

In fact, 2,058 equals 6 times 7^3. Similarly, 506 is equal to 6 times 7^2. This emphasizes the relationship between this series and the number 7, which represents the seventh of the lower *sefirot*, *malchut* (the *sefirah* of kingdom).

The base of the series, the number that defines the highest power of *n*, is 78, which equals 3 times 26, the value of the essential Name of God. This alludes to the Priestly Blessing, discussed in the previous chapter, in which the essential Name is mentioned three times. This relates to the collaboration between the king and the High Priest in governing the nation, as we shall see.

The fourth, fifth and sixth numbers in this series (26, 137 and 506) together with the base value (78) all allude to a verse in the Torah: "Not by bread alone does man live, but by all that originates from the mouth of *Havayah* does man live."[30]▶ The base, 78, is the numerical of "bread" (לֶחֶם). All the other three numbers relate to the key phrase, "originates from the mouth of *Havayah*." The numerical value of *Havayah* is 26; the numerical value of the word "originates from" (מוֹצָא) is 137 (see Chapter 10). The numerical value of the entire phrase is 253, which is exactly one half of 506. The number of letters in the passage relating to the appointment of a king is thus equal to twice the numerical value of the phrase "originates from the mouth of *Havayah*" (מוֹצָא פִּי הוי).

In Kabbalah, *malchut* (the *sefirah* of kingdom) relates to the mouth and the power of speech.[31] This is also clear from the Talmudic dictum, "There is one spokesman (i.e. king) for every generation."[32] The number 506 [twice the value of the phrase "originates from the mouth of *Havayah*" (מוֹצָא פִּי הוי)] thus alludes to two aspects of the Divine "mouth." These are the higher mouth of *tevunah* (intuitive intelligence), and the lower mouth of *malchut* (kingdom).▶▶

One of the most important contexts in which "origin" (מוֹצָא) is used is in the idiom, "the five origins of the mouth" (חֲמִשָּׁה מוֹצָאוֹת הַפֶּה) from which speech is formed: the throat, the palate, the tongue, the teeth and the lips.[33] The 22 letters divide into these groups according to their phonetic pronunciation.

▶ This phrase states that, ultimately, God is the origin and the sustenance of every created being (see also, Chapter 4). The numerical value of, "originates from the mouth of God" (מוֹצָא פִּי) is 253, which equals △22—i.e. the sum of all numbers from 1 to 22. This alludes to the twenty-two letters of the Holy Tongue with which God created the world. Similarly, we are taught that the lifeforce of the food we eat emanates from the twenty-two letters of the Hebrew *alef-bet*. This is further alluded to in the numerical value of "wheat" (חִטָה, 22), from which bread is produced.

The first ten numbers that contain the integers of 137 (excluding the numbers that include the same integer twice) are 1, 3, 7, 13, 17, 31, 37, 71, 73 and 137, all of which are prime numbers. Amazingly, the first nine of these numbers total 253, the numerical value of "originates from the mouth of God" (מוֹצָא פִּי), while the tenth number is 137, which is the numerical value of "origin" (מוֹצָא). The sum of all ten numbers is 390. The average value is 39, the numerical value of "*Havayah* is one" (הוי אֶחָד).

▶▶ *Malchut* is the seventh of the lower *sefirot*. The sum of the numerical values of *tevunah* (תְּבוּנָה) and *malchut* (מַלְכוּת) is 959, which equals 7 times 137. *Tevunah* and *malchut* correspond to Leah and Rachel, respectively. The average numerical value of their two names is also 137.

Counting Jewels

In the Torah, the king and the High Priest act together to govern the people.◀ Moses' brother, Aaron, was the first to hold the position of High Priest in the *Mishkan*, the mobile Tabernacle that served the Children of Israel in the wilderness. The numerical value of "Aaron" (אַהֲרֹן) is 256, the number that Eddington attempted to link to 137 mathematically. As mentioned, the two passages in the Hebrew Bible that relate to a king have a total of 256 words, thus numerically relating the king to Aaron, the original High Priest, who blesses the congregation. The High Priest is the only one permitted to articulate the sacred Name of God.

Another example of a connection between 137 and 256 spans the Five Books of the Torah, which correspond to the above-mentioned "five origins of the mouth."

Counting 137 letters from the beginning of each of the five books, we come to the letters ה, י, י, נ, ש.◀◀

We can contemplate these letters at a deeper dimension by including the filling letters of their names (הא, יוד, יוד, נון, שין).◀◀◀ The sum of the numerical values of the five letters is 375 [the numerical value of "gravitation" (מְשִׁיכָה)]. The sum of the numerical values of the letters when spelled in full is 512. As we see in the table below, the total numerical value of the pregnant letters is 137.

▶ During the Second Temple period, after the miraculous salvation that we celebrate on Chanukah, Mattathias the High Priest and his descendants even became the ruling power in the country.[34]

▶▶ The numerical value of the first 137 letters in Genesis is 8,228, which equals 17 [the numerical value of "good" (טוב)] times 22^2 (or 68 times 11^2). This alludes to the creation of the world through the ultimate inter-inclusion of the 22 letters of the Hebrew language. It also alludes to the recurring phrase of creation, "it was good" (כִּי טוֹב). The ultimate property of creation is goodness.

[The numerical value of the first 137 words in Exodus is 8,661, which equals 3 times 2,887; in Leviticus 7,868, 28 times 281; in Numbers 11,669; in Deuteronomy 10,132; for a total of 36,058 = 298 times 11^2.]

▶▶▶ The filling letters are called "pregnant letters." Pregnant letters are implicit in the names of individual letters. But, while the letters are combined into words, the pregnant letters are concealed. The potential of these letters manifests when the letters are written in full, "at birth."

130

Book of the Torah	137th letter in book	Numerical value of letter	Pregnant letters	Numerical value of pregnant letters	Letter spelled in full	Numerical value of letters spelled in full
Genesis	*Hei* (ה)	5	א	1	הא	6
Exodus	*Yud* (י)	10	וד	10	יוד	20
Leviticus	*Yud* (י)	10	וד	10	יוד	20
Numbers	*Nun* (נ)	50	ון	56	נון	106
Deuteronomy	*Shin* (ש)	300	ין	60	שין	360
Total		**375**		**137**		**512**

The average value of the revealed part of the letters and their concealed part is $512/2$, which equals 256 (16^2), the number that Eddington proposed as the foundation of all creation.

By this simple calculation, we have discovered that the 137th letter of each of the five books of the Torah reflects the relationship that Eddington was looking for between 137 (the numerical value of the concealed letters) and 256 (the average of the revealed letters and the concealed letters).

One TOE, 248 Particles

Many physicists share an intuition that, at the deepest level, all physical phenomena match the patterns of a beautiful mathematical structure. One such structure is the E8 Lie (pronounced: "Lee") Group. There is one TOE (Theory of Everything) that proposes that all the subatomic particles discovered to date are part of this group. The E8 Lie Group has 248 points that hypothetically correspond to 248 sub-atomic particles. These particles include the particle responsible for mass, the recently confirmed Higg's boson. It also includes other particles that scientists expect to discover, such as the graviton, thought to be responsible for the gravitational force.

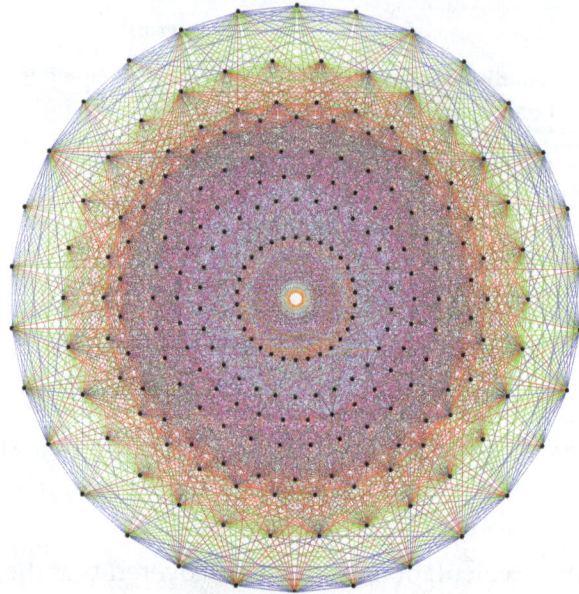

The E8 Lie Group

From a Torah perspective, the number 248 bears significance: it is the numerical value of Abraham (אַבְרָהָם); it is the number of limbs in the male human body; it is the number of positive commandments in the Torah; and most significantly, it is the numerical value of "matter" (חֹמֶר). Moreover, the numerical value of the word selected by the Hebrew language committee for "particle" (חֶלְקִיק)◄ is 248. Clearly, they did not choose this word based on its numerical value. Thus, this is a beautiful example of Divine Providence (or, synchronicity; see Chapter 6) acting in the world.[35]

From our current perspective through the prism of the fine-structure constant, 137 links 248 directly to the numerical value of the essential Name of God (26) in a linear progression with a constant difference of 111.

$$26 \quad 137 \quad 248$$
$$111 \quad 111$$

► "Particle" (חֶלְקִיק) is a derivative of the word "part" (חֵלֶק). The two-letter suffix yud-kuf (יק) is added to indicate extreme smallness. The numerical value of "part" (חֵלֶק) is 138, one more than 137.

The difference of 111 equals 3 times 37, the numerical companion of 137. Furthermore, 111 is the numerical value of the letter *alef* when spelled in full (אָלֶף). The *alef*, as the first letter of the *alef-bet*, refers us to the unity of God. Its name also alludes to God as the Chief of the World (אַלּוּפוֹ שֶׁל עוֹלָם).[36] The sum of 137 and 248 is 385, which is the numerical value of *Shechinah* (שְׁכִינָה, "the Divine Presence"). The *Shechinah* is the feminine, royal aspect of Divinity. It derives from the male aspect of Divinity represented by the Name *Havayah*, which has a numerical value of 26, the first number of the series. The sum of the three numbers (26, 137 and 248) is 411, which equals 3 times 137 and is the numerical value of "chaos" (תֹּהוּ). This offers us a new perspective on the rectified aspect of chaos, which alludes to the ultimate union of male and female—*Havayah* (26) and the Divine Presence (שְׁכִינָה, 385).▼

▶ Expanding this series to 13 places, we find some more interesting allusions:

26 137 248 359 470 581 692 803 914 1025 1136 1247 1358

111 111 111 111 111 111 111 111 111 111 111 111

The fourth number in the series is 359, which is the numerical value of Satan (שָׂטָן), the minister of death and darkness, whom Jacob defeated in battle at the Yabok Passage.[37] Rashi states, "He [Jacob] forgot small vessels there and returned to retrieve them." The numerical value of the phrase, "small vessels" (פַּכִּים קְטַנִּים) is 359, the value of the word "Satan" (שָׂטָן). Similarly, the numerical value of "the son of Rebecca" (בֶּן רִבְקָה), referring to Jacob, is also 359. Thus, by returning to retrieve the small vessels, Jacob instigated the battle with Satan. His mother's merit enabled him to overcome him.

Jacob's defeat of Satan is commemorated in the new name—Israel—which God gave him after the battle.[38] The numerical value of Israel (יִשְׂרָאֵל, 541) is the sum of Satan (שָׂטָן, 359) and Jacob (יַעֲקֹב, 182).

359 appears in this series before the fifth number, 470, which is the numerical value of "'Let there be light,' and there was light" (יְהִי אוֹר וַיְהִי אוֹר), as we saw in Chapter 1. Another idiom with the same numerical value is, "Temple of God" (מִקְדָּשׁ הוי).

The eleventh number in the series is 1,136. In Hebrew lettering, thousands are represented by large letters. In particular, one thousand is represented by a large letter *alef* (א). The name of this letter (אָלֶף) can be read alternatively as "one thousand" (אֶלֶף). The number 1,136 is thus written as 136 with a large *alef* (אקל״ו). When the *alef* is reduced to its normative value of 1, this number equals 137.

The thirteenth number in the series is 1,358 (26 + 12·111), which is the numerical value of the phrase, "Blessed be the Name of the glory of His kingdom forever and ever" (בָּרוּךְ שֵׁם כְּבוֹד מַלְכוּתוֹ לְעוֹלָם וָעֶד). This phrase refers to the transcendental aspect of God's Divinity, *above* time and space, as it manifests *within* time and space. This is the union of Torah and science that we aspire to reach.

The sum of the first 13 numbers in the series from 26 to 1,358 is 8,996, which equals 26 (the numerical value of *Havayah*) times 346 [the numerical value of "His Name" (שְׁמוֹ)], alluding to the phrase, "God is one and His Name is one." Alternatively, 8,996 equals 52 times 173 [the numerical value of "I am *Havayah* your God" (אָנֹכִי הוי אֱלֹהֶיךָ) and the numerical value of "raise my eyes" (גֵּל עֵינַי) etc.].

The next number in the series is 1,469, which is the sum of the first 13 inspirational numbers, i.e., the octahedron of 13.

133

REFERENCE NOTES FOR CHAPTER 4

1. See *Zohar Chadash* II.

2. Deuteronomy 29:28.

3. For a detailed description of the correspondence between the letters of God's Name and the four Worlds, and the meaning of each World, see our book, *"What You Need to Know about Kabbalah,"* p. 133 ff.

4. Genesis 2:6.

5. Isaiah 43:7.

6. See *Pardes Rimonim* Gate 4, Chapter 1; *Etz Chayim* Gate 44, Chapter 7 etc.

7. Isaiah 26:19.

8. Stephen Hawking, *Gödel and the End of Physics*, http://www.hawking.org.uk/godel-and-the-end-of-physics.html.

9. See Francis M. Sanchez, Valery Kotov, *Remarkable Properties of the Eddington Number 137 and Electric Parameter 137.036 excluding the Multiverse Hypothesis*, 2015.

10. *Berachot* 61a.

11. *Yoma* 48a; *Baba Batra* 111b; *Zevachin* 25a; *Bechorot* 44b.

12. See *The Mystery of Marriage*, p. 419.

13. Psalms 104:24.

14. *Sefer Yetzirah* 1:2.

15. See *Rashi*, Ezekiel 1:6.

16. See *Baba Metzia* 106b; *Sanhedrin* 81b.

17. Deuteronomy 17:14-20.

18. I Samuel 8:11-21.

19. The *Adon Olam* liturgical poem.

20. Genesis 15:2.

21. See Nachmanides' Additions to Maimonides' *Sefer Hamitzvot*, *Mitzvah* 4.

22. Deuteronomy 17:15.

23. Proverbs 19:14.

24. I Samuel 25.

25. I Kings 1:31.

26. Deuteronomy 17:16.

27. Ibid 6:5.

28. *Zohar Bamidbar* 275b.

29. Psalms 25:14.

30. Deuteronomy 8:3.

31. *Tikunei Zohar, Patach Eliyahu*.

32. *Sanhedrin* 8a.

33. *Sefer Yetzirah* 2:3.

34. For more on the connection between the king and the High Priest, see our book (in Hebrew) *Shloshah Ketarim*.

35. For more on the number 248 in Torah and science, see http://www.inner.org/torah_and_science/physics/E68-0422.php.

36. See our book, *The Hebrew Letters*, p. 24 ff.

37. Genesis 32:25.

38. Ibid 32:29.

From Binary Logic to Counter-intuition

2 to the Power of *n* (2^n)

The Greek philosopher Aristotle founded a system of logic based on two propositions: either an expression is true, or it is false. This led to the development of three fundamental laws of logic:

- the Law of Identity (B is A);
- the Law of Non-contradiction (the two statements, B is A and B is not A are mutually exclusive); and
- the Law of the Excluded Middle (if B = -A then either A or B is true).

This type of binary logic is the basic intuitive logic of human common-sense.

The English mathematician George Boole (1815-1864) sought to give symbolic form to Aristotle's system of logic. Boole codified several rules of relationship between mathematical quantities limited to one of two possible values: true or false, 1 or 0. Boole's theoretical work became known as Boolean algebra and was eventually put to use in a way Boole never could have imagined. It became a powerful mathematical tool for designing and analyzing digital circuits. All computers currently in use rely on this type of binary logic, which utilizes the base-two numbering system in place of the regular decimal base of ten. Base-two incorporates only the two numbers 0 and 1 to express values in binary code. The powers of 2 (2^n) are thus particularly important in computing. One illustration of this is in the measurements of memory in computer hardware.▸

▸ The mathematical series 2^n is also significant in other fields of science besides computing, including cellular division in biology. A healthy human cell can divide up to 40 (2^{40}) or 60 (2^{60}) times.

2ⁿ in Creation

In Kabbalah, the entire array of ten *sefirot* relates to the decimal system of counting. The first of the ten *sefirot* is *keter* (the *sefirah* of crown). The second is *chochmah* (the *sefirah* of wisdom). In the psyche, *keter* represents the pre-conscious, or super conscious power of the soul. From the perspective of human consciousness, *keter* reflects the nothingness that exists before something is created. Numerically, *keter* relates to zero.

Chochmah is the *sefirah* with which God created the world.[1] In the psyche, it is the initial flash of insight in the conscious mind. It relates to the number 1. Together, these two *sefirot* (*keter* and *chochmah*) relate to binary code, which incorporates only 0 and 1.

Binary code and the function 2^n are alluded to in the large letter *bet* (בּ) that begins the Book of Genesis.◄ It also refers to the complementary pairs of creation, as explained in Chapter 2.

The first three elements in the series produced by the function 2^n are 1,2 and 4. The sum of these three numbers is 7, alluding to the seven days of the week.

There are three phrases God spoke at creation that begin with "Let there be" (יְהִי). The three phrases appear on the first, second and fourth days of creation. These three days allude to the first three elements in the series 2^n. The next elements in the series are 8, 16, 32 and 64.

$$a = 2^n = 1, 2, 4, 8, 16, 32, 64\ldots$$

We have already discovered that this series relates to creation in various ways. Let us see what happens when we limit it by constraining it to modulus 7, to expose its relationship with the seven days of the week.

$$a = 2^n = 1, 2, 4, 8 \ (= 7 \perp 1), 16 \ (= 2 \text{ times } 7 \perp 2), 32 \ (4 \text{ times } 7 \perp 4), 64$$
$$(= 9 \text{ times } 7 \perp 1)$$
$$a \equiv 2^n \ (\text{mod } 7) = 1, 2, 4, 1, 2, 4, 1\ldots$$

Restricting the series to mod 7 reveals that it repeats in an infinite cycle. Here, we see that 1, 2 and 4 are not only the first elements in the series. They are the only elements. The placement of God's words, "Let there be" on the first, second and fourth days of creation is now revealed in a new light. The total number of letters in these expressions is 137 (see Chapter 1).▸

The Counter-intuition of Modern Science

Binary logic is common-sense. It allows us to compare or complement two elements. As the basic method of human logic, it formed the working foundation for Newtonian mechanics for centuries. Classical physics holds that velocity is measured with reference to static objects, and space and time are perceived as two separate entities. This means that one object can only be in one position at any one moment in time. This definition is accurate enough for the velocities we use in our everyday world. For example, using it we can determine the time that the next train will arrive at our station, based on knowledge of the speed of the train, the time that it left the previous location and the distance between the two points. However, the velocities of subatomic particles approach the speed of light. At these speeds Newtonian mechanics can no longer accurately determine the position and momentum of a particle at one and the same time. The time has come to discover a new perspective on reality. Enter: counter-intuition.

There are four theories in modern physics that go above and beyond the scope of binary logic. The first two, special relativity and general relativity, were both developed by Einstein. Although their names seem to describe two stages of the same theory, scientists consider them as two distinct hypotheses. The other two theories are quantum mechanics and superstring theory.

The first three theories are accepted almost unanimously by the scientific community. String theory is still undergoing research and it is not accepted by all. At the core of all three universally accepted physical theories lies the unique essence of counter-intuition.

▸ The first, second and fourth days of the week of creation correspond to chesed (the sefirah of loving-kindness), gevurah (the sefirah of might) and netzach (the sefirah of victory). The sum of 1 times the numerical value of chesed (חֶסֶד), 2 times the numerical value of gevurah (גְּבוּרָה) and 4 times the numerical value of netzach (נֶצַח) is: 72 ⊥ 2(216) ⊥ 4(148) is 1,096, which equals 8 (2^3) times 137.

Let's demonstrate the counter-intuition inherent in each of these theories.

Special Relativity

The logic applied by Newtonian mechanics holds that time and space are constant. Special relativity takes us one step out of the limitations of the "real world" described by Newtonian mechanics. It proposes a total reorientation of the universe. It was the first counter-intuitive deviation from the intuition of binary logic.

The motion of an object is measured relative to its distance from a point in space. But, that point in space can sometimes be another moving object. This means that there is no absolute definition of motion. For example, consider a train moving at a constant speed of 120 kmph. To an observer travelling inside the train, the seats and tables inside the carriage appear to be stationary. Trees and buildings outside will appear to be moving backwards at 120 kmph. If the train is travelling parallel to a road on which vehicles are moving at a constant velocity of 70 kmph, the same train passenger will "overtake" cars travelling in the direction of the train at 50 kmph. Vehicles travelling in the opposite direction will speed past at 190 kmph. Newton was aware of the effects of relative motion. He was not happy with the conclusion that there is no standard of rest. Space and time were the context in which everything takes place. They could not change, nor could they affect the events that take place in them.

The counter-intuitive innovation of special relativity is that space and time are an integral part of, and are affected by events. When objects move at velocities approaching the speed of light, time dilates and space contracts (the Lorentz contraction). Time and space are thus variables, not constants! So, time is relative. If you are moving very fast, time goes slower. If you are traveling at the speed of light then time comes to a halt.

Einstein followed the chain of relative motion to its conclusion. He inferred that the one thing that is constantly moving with respect to everything else is light. This means that relative to the speed of light everything is stationary. If light always moves away or towards us at exactly the same speed, no matter what our velocity may be, then everything else, including space-time, is relative. Space and time are purely subjective. When we measure our progression relative to the speed of light, we are not moving at all.▶

General Relativity

Once the theory of special relativity was accepted, Einstein went on to develop the theory of general relativity. General relativity describes gravitation, the most elusive of all four forces of nature.

Einstein realized that from an observer's perspective, the pull of gravity and the experience of acceleration are equivalent. The simplest example of this is traveling up in an elevator. When the elevator accelerates, the passenger experiences a sense of being pulled into the floor. This experience is identical to the effect of a gravitational field. This insight has far-reaching consequences. It is known as the equivalence principle.

Einstein hypothesized that space is curved by the masses that lie in it. The massive objects in space move in straight lines whilst simultaneously following the curvature of space. Einstein came to the realization that gravity is the force that causes space to curve. General relativity thus requires us to re-conceive space as either convex or concave.

General relativity successfully explains the effects of gravitation that Newton described in his equations. But, the idea that space-time is curved is obscure even by the innovative standards of special relativity. Nonetheless, Eddington's observations during the 1919 eclipse proved this second counter-intuitive proposal to be correct. These results were verified in 2005 by findings from the Gravity Probe-B satellite.

▶ Rebbe Menachem Mendel Schneerson, the Lubavitcher Rebbe, taught that we cannot assume that the constants measured in our era have been the same since the beginning of time.[4] In their attempts to unify the four forces of nature, some scientists conjecture that even the speed of light has changed with time. Perhaps at the beginning of creation, the speed of light was much greater than we know it. This idea is gaining popularity; however, it still cannot be proven. Light with a varying velocity would affect the value of the fine-structure constant, which is a derivative of the speed of light.

Quantum Mechanics

Science relates to subatomic particles through quantum mechanics, the third counter-intuitive theory. As sizes drop to the subatomic scale, mechanics based on binary logic are no longer viable. In terms of space-time, a subatomic particle can no longer be understood to be "here" or "there." Indeed, it could be "here" and "there" at one and the same moment.

This strange quality of quantum mechanics is defined by the uncertainty principle, discussed above (Chapter 2), which states that it is impossible to determine the momentum and the position of an electron simultaneously. Because of the diversion in logic necessary to understand the uncertainty principle by Newtonian standards, it was at first thought to be due to the subjective limitations of the observer. However, it was later realized that, absurdly, the uncertainty principle is a purely objective experience—a completely counter-intuitive conclusion that contradicts all of our subjective experience.

M Theory

In the 1980s, a new mathematical model of theoretical physics emerged. It showed how all particles, and all energy in the universe—including the graviton and the gravitational force—could hypothetically be constructed from one-dimensional "strings." These strings are infinitely small building blocks that have length, but neither height nor width. The theory suggests that these strings vibrate in and out of multiple dimensions that transcend three-dimensional space as we know it. Depending on their vibrations, strings are perceived in our world as the various particles defined by the standard model.

Five different versions of string theory were originally proposed. They were unified by Edward Witten in 1995 in what he called M-Theory. The unified theory hypothesizes that strings are 1-dimensional slices of a 2-dimensional membrane that vibrates in 11-dimensional space. This theory is mathematically elegant and relatively simple (for cosmologists, at least), and many believe that it could be the basis for the ultimate "Theory of Everything."

M-theory is counter-intuitive because it employs undetectable extra dimensions. In Chapter 2, we discussed how complementarity supposes that multiple properties of apparently contradictory phenomena all derive from a common origin. Similarly, M-theory proposes that one object in a higher dimensional reality may have multiple manifestations in lower dimensions. Scientists can detect these variations, but their unified, or "super-symmetrical," form in higher dimensional reality is intangible to even the most sophisticated equipment currently available. This makes it impossible to verify or refute the hypothesis. Nonetheless, the theory is mathematically valid and provides an elegant solution that takes us one step beyond the standard model of quantum mechanics.

The Four Worlds

As mentioned above, one of the models used in Kabbalistic correspondences is the model of four Worlds. In the previous chapter, we explained how these Worlds relate to the four forces of nature. The gradual development of counter-intuitive theories beyond the common-sense logic proposed by Aristotle also corresponds to this scheme. In the psyche, these four worlds correspond to four different states of spiritual consciousness. Whichever World my consciousness occupies, I will experience the insights of the next highest level of consciousness as totally counter-intuitive.[5]

General human consciousness is in *Asiyah* (the World of Action). Above that is *Yetzirah* (the World of Formation) and the next highest is *Beriah* (the World of Creation). The common denominator in these three lower states is that human self-consciousness causes us to feel separate from the Almighty. The fourth and highest spiritual World is *Atzilut* (the World of Emanation). The consciousness of *Atzilut* is total oneness with God.

Every World has its own level of intuition or "common-sense."

Sometimes an insight in one of the three lower worlds appears counter-intuitive relative to its innate level of common-sense. This is the effect of light from *Atzilut* illuminating that lower world. The Maggid of Mezeritch, the heir to the Ba'al Shem Tov's spiritual legacy, taught that "*Atzilut* is also here," i.e., the consciousness of *Atzilut* descends into and is present in each of the three lower worlds.

Special Relativity is in *Asiyah*

The first insight that Chassidut offers is that everything is relative—even common-sense! Common-sense itself is relative to the particular mindset that we occupy.

> *Reb Isaac of Homil lived a century before Einstein. He was one of the most intellectually-minded teachers of Chabad Chassidut. Despite his brilliant mind, Reb Isaac said of the renowned spiritual mentor (mashpia), Rabbi Zalman of Zezmir, "Don't say that the difference between his intelligence and my intelligence is like the difference between my intelligence and the intelligence of a cat. Rather, say that relative to the intelligence of Rabbi Zalman of Zezmir, my intelligence and the intelligence of a cat are the same!"*

The most counter-intuitive aspect of special relativity relates to progress. Since special relativity counters our intuition of motion, it affects us physically in *Asiyah* (the World of Action). In our mundane reality, everybody and everything is on the move. Our experience in *Asiyah* is judged with reference to space and time, therefore our most basic intuition is that we are constantly in motion. But, special relativity claims that relative to the speed of light, we are not moving at all! The most challenging notion to our common-sense is that we are stationary.

By the standards of special relativity, space-time is subjective, and the only thing that is objective is light. Anyone who wishes to truly remain in motion must travel on a beam of light. From a spiritual perspective, this means that we must ride upon the beam of Torah, as the verse in Proverbs states, "Torah is light"[6] and the Talmud responds, "Light is Torah."[7]

Breaking the sound barrier was a great human feat. But, although we can move faster than the speed of sound, we can never, physically, hope to move at the speed of light. Yet, while our physical bodies and material objects are limited, our souls are not. When we are asleep for example, our souls are released from the confines of our bodies and travel faster than the speed of light to all ends of the world … and back again. Although reality continues to progress around us, our consciousness rises to a different dimension in which our experiences are not affected by the reality that surrounds our sleeping bodies. The level of consciousness that we achieve in our dreams is counter-intuitive to our daily lives.

This phenomenon echoes the relative state of consciousness achieved by riding the Torah's beam of light. Those Torah scholars who are in tune with this beam and those individuals whose lives are illuminated by the Torah can reach a level of consciousness that is counter-intuitive to the regular fluctuations of space-time. Furthermore, righteous individuals can even influence space-time through their connection with the Divine in prayer and in their Torah study. The miracles worked by the Prophet Elisha, mentioned in Chapter 2, are but one example of many. In the spiritual Worlds above *asiyah* (the World of Action) the light of the Torah is the only constant that affects reality.

The following story, told by Rebbe Nachman of Breslov, illustrates this aspect of special relativity in the spiritual realm from another perspective:[8]

> There was once a righteous Jew who raised himself to great spiritual heights by merely rejoicing in being Jewish. During this state of elation he experienced himself moving thousands and thousands of miles in the spiritual worlds, far away from where he had started. Nonetheless, when he landed, he was greatly surprised to discover that he was still in exactly the same place at which he had begun. Perhaps he had moved a hairsbreadth.

The realization that everything is static relative to the speed of light is the illumination of *Atzilut* (the World of Emanation) within *Asiyah* (the World of Action). One Biblical phrase that illustrates this sense of relativity is God's promise to those who follow His ways and observe His laws, "I will allow you to be walkers amongst these who stand."[9] Rashi explains that, "these who stand" refers to angels, who never sit. By contrast—and counter-intuitive to all common-sense—the souls of the righteous can move around freely in the upper worlds. They are unhindered by the normal limitations of their physical bodies.

General Relativity is in *Yetzirah*

When considered from a Torah perspective, the curvature of space is not surprising, because there are no straight lines in nature. The Hebrew word for "nature" (טֶבַע) shares its root with the word for "ring" (טַבַּעַת), and Kabbalah teaches us that the natural world is one of circles or spirals. Indeed, one Talmudic dictum states that "There has never been a [natural entity that is] square since the six days of creation."[10]

In the Kabbalistic description of the initial stages of creation, straight lines were preceded by the great circle. The straight line of Infinite Light emanated from the great circle and more circles exuded as the ray of light penetrated through the vacuum. This ray of light is the ruler that measures the universe, i.e., everything is measured relative to light.

General relativity proposes that in contrast to our initial intuition that space is a vacuum, in effect, it is a malleable material, "Like clay in the hands of a potter."[11] The art of the potter is to take raw clay and form it into a concave receptacle. The word for "potter" (יוֹצֵר) used in this expression literally means "former" (i.e., one who forms). This idea offers us a profound insight into the meaning of the word "formation."

In the consciousness of *Yetzirah* (the World of Formation), curvature is a statement about human emotions. The heart needs to be bent, shaped and formed (sometimes, it even needs to be broken) in order to become a receptacle.

There is a phrase at the beginning of the Song of Songs that relates to attraction and acceleration: "Attract me [gravitation], and we will run after you [acceleration]…"[12] There is an irregular grammatical feature in this phrase. The first half is in the singular ("Attract me"), but the second half is in the plural ("And we will run after you…"). Chassidic writings[13] explain that the spiritual force of gravity is the revelation of God's love from above. It attracts the spiritual aspect of the person's psyche (the Divine soul). When the Divine revelation is subsequently concealed, its gravitational pull is felt by both the spiritual aspect of the person's psyche (the Divine soul) and their physical aspect (the animal soul). Even though each of the two souls is a different entity from a spiritual perspective, this revelation effects equal acceleration on both aspects as one. (In the context of general relativity, this encapsulates the equivalence principle, which states that gravity accelerates all objects equally, regardless of their masses or the materials from which they are made.) This idea is illustrated by the following parable:

> A father once saw his young son playing with a group of children of his age. When the child became aware of his father's presence, he was instinctively drawn towards his father and he began to run towards him. "Papa!" He called out spontaneously, but he was not yet mentally prepared to detach himself from his childish games. The boy's father, having gained his son's attention, turned away and continued on his way. It was then that the child began to run after him, calling out, "Papa! Papa!" With every step that the child took away from the game, he accelerated towards his father.[14]

The child's initial physical pull toward his father was purely in-stinctive. This corresponds to the non-conscious effects of physical gravitation. However, when the source of attraction is removed (the child's father moves out of sight), and the boy makes an active move (as opposed to instinctive) towards his father, there is a new sense of acceleration relative to the games that he is leaving behind him. The force of the child's acceleration towards his father draws his friends to run after him too, leaving childplay behind them and rising to a new level of maturity. This is expressed in another verse, "In embroidered garments, she will be brought to the King; and maidens in her train, her companions, will be brought to You."[15] When the queen rises to be brought to the King, everyone follows.

Quantum Mechanics is in *Beriah*

Uncertainty is a statement about a state of mind. This corresponds to *Beriah* (the World of Creation). The paradox here is that what we observe to be physical matter is intangible. Kabbalah teaches us that in the consciousness of *Beriah* we can refer to raw material, or potential, but there is no physical reality. Physical matter and events can be conceived, but they do not manifest in mundane reality. Counter-intuition at this level revolves around the question of existence.

God created the world ex-nihilo. At the level of *Beriah*, the nothing that preceded creation still exists. At this level, the physical world is being recreated at every moment, something from nothing.

M Theory is in *Atzilut*

As mentioned previously, the consciousness of *Atzilut* (the World of Emanation) is one of total unity with God. This state corresponds to the search for the origin of all forces and all particles in a single unit. This is the ultimate goal of string theory and its grand successor, M theory.

A Woman's Prerogative

The advance in science from common-sense to progressively counter-intuitive models of understanding nature is reflected by the Talmudic idiom[16] "the opposite is true" (אִיפְּכָא מִסְתַּבְּרָא).[17] This idiom commonly appears when one sage responds to the initial understanding of his peers with counter-intuitive logic. The initial letters of this idiom spell "mother" (אֵם; pronounced "*em*" like the "M" in M-theory), which refers to the mother principle. This phrase, "the opposite is true" (אִיפְּכָא מִסְתַּבְּרָא) appears in the Talmud exactly 19 times, which is the numerical value of Eve (חַוָּה), who Adam called "mother of all life."[18] In Kabbalah, the mother principle is represented by *binah* (the *sefirah* of understanding), which is the source of common-sense logic. Such feminine intuition is often countered when new information is accumulated.

By contrast, *chochmah* (the *sefirah* of wisdom), the father principle, represents direct intuition, which is the experience of absolute objective truth. The female mind is tuned into intuition, but her initial intuition is not always correct in other frames of reference.

The ultimate example of a "change of mind" is when a non-believer repents and takes on the principles of his or her faith, becoming a *ba'al teshuvah* (lit. "master of return [to God]).▶

There are two general types of repentance:

The first type of repentance is associated with the Ten Days of Repentance from Rosh Hashanah to Yom Kippur. During these holy days, we re-evaluate our deeds and character traits. We change those things that are not good by reinforcing our good behavior. This either/or, good/evil perspective on life reflects binary logic.

The second type of repentance is associated with the month of Nisan, the month of Passover when the spring season begins. This type of repentance is associated with a general sense of renewal.[20] It reflects the counter-intuitive metamorphosis of spring. Re-evaluating our judgment of what is right and what is wrong allows us to perceive the world from a totally innovative perspective. This is the type of repentance that is required to bring the redemption. As the sages say, "In the month of Nisan we were redeemed, and in the month of Nisan, we will be redeemed in the future."[21]

▶ The numerical value of *ba'al teshuvah* (בַּעַל תְּשׁוּבָה) is 815. This is the same as the numerical value of "the opposite is true" (אִיפְּכָא מִסְתַּבְּרָא). The *ba'al teshuvah* comes to the realization that all that he or she has believed until now is erroneous, and that, in fact, "the opposite is true." Significantly, the numerical value of "silence" (שְׁתִיקָה) is also 815. Silence is one of the most potent ways to reach a new perception on life. It is a valuable tool for anyone who wishes to penetrate the thoughts of God (see Chapter 9).[19]

In summary:

World	Corresponding mindset	Respective spiritual manifestation of counter-intuition	
Atzilut (**the World of Emanation**)	General unified theory	The unified consciousness of the righteous	The light of *Atzilut*
Beriah (**the World of Creation**)	Quantum mechanics	Constant re-creation	The light of *Atzilut* reflected in *Beriah*
Yetzirah (**the World of Formation**)	General relativity	Bending, or forming the heart	The light of *Atzilut* reflected in *Yetzirah*
Asiyah (**the World of Action**)	Special relativity	The Torah relative to "normal" life	The light of *Atzilut* reflected in *Asiyah*

Intellectual Crochet

> *Trying to understand nature is like watching a game, like chess, for example, where you don't know what the rules of the game are, but perhaps you are allowed to look at a little corner of the board, at least from time to time. And from these observations you have to try and figure out the rules ... The part that doesn't go according to what you expected is the most interesting ... This is an analogy for the laws of physics. They keep on working until a little gimmick shows that they're wrong and then we have to investigate the conditions and gradually, learn the new rules that explain it more deeply.*[22]

Every time a new theory is developed, scientists believe that they are on the verge of understanding all of creation. If only that one little point of concern that cannot be explained would be explained. Humble admission of our ignorance enables us to find the loophole.

This is the point of unexplainable nothingness that constitutes the core of every theory.

Like the thought-experiment of a physicist, every opinion proposed by a Torah authority is a reflection of a particular spiritual realm, or level of consciousness. The Ba'al Shem Tov told his rabbinical peers that he was able to counterintuit any opinion that they might propose. He explained that at the mid-point of each World lies an essential point of vacuum that links all levels of consciousness together. The Ba'al Shem Tov was in tune with that zero-point. He had the ability to ascend and descend through all the Worlds via the channel that is created by these loopholes. The Ba'al Shem Tov's phenomenal humility led him to find that loophole in everything.

This profound idea is a realization that some of the more honest scientists, like Feynman, quoted above, have admitted is true. In the last few centuries, time and again, new theories have emerged that completely override previous scientific notions. In every theory, there has always been a "loophole" that could not be ignored. This point of non-knowledge led to the creation of a new, more encompassing theory that superseded the previous theory. The process continues as long as a loophole in the new theory remains unexplained. This is a kind of intellelctual "crochet"—a process of creating a garment or work of art by interlocking loops.

Applying this "crochet" principle to our personal lives, means that regardless of how we live, there is always a loophole that we can access.▶ Through that opening, we can extricate ourselves from whatever problems we encounter. The loophole becomes a conduit through which we can ascend to a higher level of consciousness that will breed new solutions. The loophole to repentance can be accessed through the silence of the humble admission that we do not know everything.

▶ "Holes" (חורים), also means "free." This word appears twice in the Bible,[23] both times with reference to the free social elite. Another way of writing this word substitutes the final letter *mem* with a final *nun* (חורין) [as in the expression "free men" (בְּנֵי חוֹרִין)]. The numerical value of this second spelling is 274, which equals 2 times 137.

Another reference to a "hole" is "My beloved sent his hand through the portal/hole" (דּוֹדִי שָׁלַח יָדוֹ מִן הַחֹר).[24] The numerical value of this phrase is 685, which is 5 times 137. When rearranged, the final letters of the words spell "holes" (חורין), or "free." The sum of the numerical values of the remaining letters in this phrase is thus 411, the numerical value of "chaos" (תהו), or 3 times 137. Since the phrase contains five words, the average numerical value of each word is 137.

149

REFERENCE NOTES FOR CHAPTER 5

1. As reflected by the Aramaic translation of *Bereishit* (בְּרֵאשִׁית) as "with wisdom" (בְּחוּכְמָה); see Chapter 1.

2. Genesis 2:4.

3. Deuteronomy 6:4.

4. See http://www.chabad.org/library/article_cdo/aid/112083/jewish/Theories-of-Evolution.htm.

5. This idea is expanded on with reference to evolution and the time factor in our book, *The Breath of Life*.

6. Proverbs 6:23.

7. *Megillah* 16b.

8. See our book, *A Sense of the Supernatural*, Letter 1, Endnote 18.

9. Zachariah 3:7.

10. *Yerushalmi Ma'asrot* 5:3; *Y. Nedarim* 3; *Y. Shvuot* 3.

11. Jeremiah 18:6.

12. Song of Songs 1:4.

13. *Likutei Torah, Vayikra* 2:4.

14. See *Or Torah, Shir Hashirim* 188.

15. Psalms 45:15.

16. *Berachot* 45b; *Shabbat* 3b; 20a; 69b; *Pesachim* 28a, etc.

17. Alternative spellings in the Talmud are איפכא מסתברא or אפכא מסתברא מיסתברא.

18. Genesis 3:20.

19. See *Lectures on Torah and Modern Physics*, Lecture 1. See also our article "Kabbalah and Telepathy" http://www.torahscience.org/communicat/kabbalah_and_telepathy.htm.

20. See *Torat Menachem, Parashat Tazria* 5717 (a) 6.

21. *Rosh Hashanah* 11a.

22. Richard Feynman speaking in BBC interview (paraphrase), 1981.

23. Nechemiah 2:16 (וְלַחֹרִים); Ecclesiastes 10:17 (חוֹרִים).

24. Song of Songs 5:4.

Physics and the Psyche

*"The conception of objective reality ... has thus evaporated
... into the transparent clarity of mathematics that represents
no longer the behavior of particles but rather our knowledge
of this behavior."[1] (Werner Heisenberg)*

The Reality Test

Until the beginning of the 20th century, it was clear to everyone that all objects have tangible reality (except certain Chassidim, who saw their own existence as questionable). Then came quantum mechanics, which brought with it a multitude of problems concerning the essence of reality.

As mentioned in the Introduction, scientific analysis incorporates three stages of research. First, experiments are carried out and observations are made. Once enough facts have accumulated, scientists propose a theory that seeks to explain all the data. They formulate a hypothesis in the form of a mathematical model to see whether or not it forecasts new phenomena. In the final stage, new experiments are invented to test the results of these forecasts. As long as the theory satisfactorily explains the facts and correctly forecasts new phenomena, it is valid. If either of these criteria is not upheld, the theory—however elegant it may seem—must be discarded.

Regarding quantum mechanics, this three-stage process has resulted in three different concepts, each associated with one stage of the process: 1) Quantum facts, 2) Quantum theory, and 3) Quantum reality.

Quantum facts are clear-cut. The data that has accumulated over the decades is extensive. Similarly, quantum theory is unequivocal. Scientists believe that there has never been a more successful theory that describes the facts and also forecasts the results of new experiments so precisely. The main riddle concerning modern physics is quantum reality. Once something has been observed, it becomes real. The question is, does something that has not been measured or observed really exist?

Quantum Reality

When physicists began to research the micro-cosmos, they observed a variety of counter-intuitive phenomena that apparently jeopardize the very core of physical existence. These paradoxes could by no means be described by the previous stage of physics. It took decades before a scientific model was developed that could explain them.

Heisenberg began by introducing matrix mathematics. Schrodinger followed with wave mechanics. This model explained quantum mechanics by proxy waves. A proxy wave is not a physical wave, but a wave of possibilities, a concept related to probability theory. Dirac developed a third, more sophisticated model. His model involves arrows and vectors that describe the forces acting on a quantum object.◄

A few decades later, Feynman devised his "sum over histories" approach. The moment a quantum object has been observed in one location, it cannot be anywhere else. Nonetheless, this method supposes that it takes all possible routes to reach its destination. Common-sense holds that if something is at one location, it cannot be in another location at the same time. Yet, according to Feynman, before it is observed or measured, it travels in infinite directions. Feynman's theory is viable because all the routes except one cancel each other out.

Feynman's approach defies the basic logic that proves whether a suspect in a crime is guilty or innocent. If the accused has a valid alibi, he could not have been at the scene of the crime. He must therefore be innocent.[3] But, if he lives in the quantum world, one can never know…

These four different mathematical models (Heisenberg's,

► Dirac's notation illustrates quantum reality using revolving vectors. This is reminiscent of the phrase in Genesis, "the blade of the revolving sword" (לַהַט הַחֶרֶב הַמִּתְהַפֶּכֶת), which guards the path to the Tree of Life.[2] The numerical value of the final letters of this phrase (ט־ב־ת) is 411, the numerical value of "chaos" (תֹהוּ), or 3 times 137. The Garden of Eden is the epitome of unified reality, where God is all and all is God. Our intellectual limitations prevent human access to that reality.

Schrödinger's, Dirac's and Feynman's) are parallel theories, all of which satisfactorily explain the quantum facts. They have been tested through experimentation and proven true at the most precise levels, such that quantum theory is almost unequivocal. No theory has ever forecast new phenomena so precisely. Sophisticated technologies that have significantly advanced mankind have been developed using this knowledge.

Yet, to the extent of its huge predictive success, quantum theory remains an enigma regarding what it describes. What reality lies beyond the mathematical models? Discovering the answer to this question is the final stage necessary to complete the scientific method.

There are six principal theories of reality taught today:

♦ The interpretation of quantum mechanics most prevalent among physicists is **Bohr's** Copenhagen interpretation, which concludes from the mathematical data that there is no reality at all until the act of measurement. The moment of observation (by a measuring instrument of any kind) collapses the wave function and fixes the particle's reality at a particular location.

An alternative to Bohr's Copenhagen interpretation is **Von Neumann's** interpretation. This opinion claims that "consciousness causes collapse." It is not enough for an instrument to measure the quantum phenomena. In order for the wave function to collapse, it is essential for the measurement to be integrated by the mind of a conscious observer. Reality is created in the conscious mind.

Another interpretation is a compromise between these two opinions within the Copenhagen interpretation. According to this interpretation, the measuring instrument, or observation by an animal without intelligent consciousness, creates the property, but not the details of that property. The measuring apparatus creates the space-time in which the particle acts. It is enough for the wave to pass through the instrument to define one particular property. However, the specific result of the measurement, such as where in space it exists, or at what velocity it is travelling, can only be established by human consciousness.

- **Werner Heisenberg** perceived quantum reality as a function of probabilities. He stated: "The quantum world is not a world of actual events like our own but a world full of numerous unrealistic tendencies for action. The foundation for our everyday world … is no more substance than a promise." This ghostlike reality exists as an urge, or tendency towards a certain type of physical reality. Measurement, or observation of the phenomenon, completes the materialization process. Heisenberg surmised that, before observation, there are no particles and waves as we know them, but something dreamlike.

The difference between Heisenberg's interpretation and Bohr's is that for Heisenberg the wave function represents a probability, but does not describe objective reality. For Bohr, the concept of a physical state is not well-defined beyond the conditions of its experimental observation.

- Although he passed away in 1955, **Einstein** remains the most renowned scientist of modern science. His views were conservative, and he searched for a solution to the paradoxes of the quantum world within the existing framework. In order to do so, he performed "thought experiments" and concluded that the Copenhagen interpretation was absurd. He stated: "The world is made up of ordinary objects. An ordinary object is an entity which possesses attributes of its own, whether observed or not." Einstein assumed that the absence of a satisfactory description of reality according to quantum mechanics was not because there is no reality, but because the theory is incomplete. He believed that all paradoxes would be solved once the rest of the theory was formulated. This view led to the neo-realistic approach.

In the last few decades it has become clear to science that, if material reality exists, its properties are even more astonishing than Bohr's counter-intuitive wave-particle. The neo-realistic approach requires that quantum objects move backwards and forwards in time. According to Bell's theorem, there must also be the possibility of non-local communication between every particle in the universe. This contradicts Einstein's own finding that nothing can travel faster than the speed of light.

- The fourth method claims that our problem with reality is that we try to relate to it with Boolean logic. **David Finkelstein** states, "To cope with the quantum facts we must scrap our mode of reasoning for new quantum logic." Just as classical physics is no longer viable in the quantum world, so too a new type of logic is needed to explain quantum reality. Adopting this new logic will enable us to understand the apparent paradoxes.

- Proponents of the many-worlds theory, initially proposed by **Hugh Everett**, hypothesize that many parallel worlds exist. This fifth interpretation of quantum reality is similar to Feynman's "sum over histories" approach. However, instead of the probabilities cancelling each other out, it suggests that each possible outcome occurs in a different world. Rather than the wave function collapsing, it continues to obey the same deterministic, reversible laws at all times. The phenomena associated with measurement are explained by decoherence; i.e., loss of energy through interaction with the environment. In this interpretation, the wave function has objective reality at all times. However, our consciousness only perceives one of the many worlds that it occupies. Since the beginning of creation to the present, the number of worlds is constantly increasing and has reached figures that defy human reasoning. The number of "universes" that exist according to this theory may be 10^{100} (10 with one hundred zeros following), or even as many as 10^{500}.

- A sixth interpretation sees quantum reality as undivided wholeness. It has been advocated by **David Bohm** and Walter Heitler. Their claim is that the world cannot be divided between objective reality and a subjective, conscious onlooker. Before measurement, the measurer and what is observed are one and the same. The quantum facts can be neatly explained if the observer and the observed object are one unified essence; this is also true of the entire universe.

From Darkness to Light

Quantum theory relates to three moments in time: before measurement, during measurement and after measurement. These three states of reality correspond to the three lower spiritual worlds, *Beriah*, *Yetzirah* and *Asiyah*. In *Beriah* (the World of Creation), before measurement, reality is in total darkness, as the verse in Isaiah states, "He [God] creates darkness."[4] In *Yetzirah* (the World of Formation) light begins to shine through as measurement illuminates the phenomenon, but it is still limited by the uncertainty principle. At this stage, the general properties of the object are revealed, in keeping with the third version of the Copenhagen interpretation, as mentioned above. In *Asiyah* (the World of Action) light illuminates the specific state of reality.

Above the three lower Worlds is *Atzilut* (the World of Emanation). This is a world of pure Divinity, where everything exists in a unified abstract state of being.

System of Worlds	The unfolding of quantum reality
Atzilut (Emanation)	Metaphysical interpretation
Beriah (Creation)	World of possibilities (pre-measurement)
Yetzirah (Formation)	Wave collapse (measurement)
Asiyah (Action)	One actuality (post-wave collapse)

Atzilut comprises six complete *partzufim*, which are the root of the three lower worlds and their *sefirot*. The six different interpretations of quantum reality mentioned above correspond to the six *partzufim* in *Atzilut* from the lowest to the highest level. These six *partzufim* are called *Nukva, Zeir Anpin, Ima, Abba, Arich Anpin,* and *Atik Yomin.*

Nukva: **Nothing Exists without Consciousness**

The first *partzuf* is *nukba* (נוּקְבָּא), the feminine persona. It is likened to the moon, which reflects sunlight, but is not a source of light in its own right. This *partzuf* "has nothing of its own."[5]

Another name for this *partzuf* is *malchut* (kingdom), because it governs all of reality. Like an empty receptacle that reflects light that is shone into it, *malchut* shapes reality, gives it color and affects the way it is perceived. *Malchut* is also associated with the mouth in general and with speech in particular. A Talmudic teaching states that successful dream interpretation depends on the mouth that pronounces it.[6] This is reminiscent of the Copenhagen interpretation, which claims that following observation or measurement, the physicist's statement about reality establishes it and anchors its material existence. Thus, the quantum world is really nothing until it is defined by the measuring apparatus (according to Bohr), or deposited in the observer's consciousness (according to Von Neumann).

Measuring something means defining it, classifying it, or giving it a name. This process is a function unique to human consciousness. The ability to give something a name is the power of governance/ kingdom. Thus, although this level appears to be passive, in fact it defines and governs reality. ▸

Zeir Anpin: **Potential that May become Manifest**

Above the *partzuf* of *Nukva* in the World of *Atzilut* is the *partzuf* of *Zeir Anpin* (the "Small Countenance"). This *partzuf* relates to the transient world of emotions. It is like a child's face that in a fleeting moment can transform from anger to laughter. *Zeir Anpin* is associated with the emotions before they manifest in action. These emotive tendencies direct themselves along different paths, but like streams of water that may converge or dry up before they reach the sea, they may never come to fruition. Emotions motivate an individual to act in a particular way, but not all of them will be realized. Heisenberg's interpretation

▸ From a psychological perspective, individuals create their own reality either by deciding by which standard to measure themselves, or by creating their own reality in their consciousness. In this context, the Tzemach Tzedek, the third Rebbe of the Chabad Movement, upheld the power of positive thinking, stating: "Think good and it will be good." The concept that consciousness changes reality is prevalent in Kabbalah. *Sefer Yetzirah*, the earliest known Kabbalistic text, states that there are three defining elements to reality: space, time and consciousness.[7] Space and time are sculpted and defined by the consciousness of the observer. This accords with the Von Neumann interpretation.

suggests that before it is measured, reality exists as an innate tendency, or a "promise," held in potential as a provisional gesture. Just like we overcome negative emotions without spontaneously articulating them or acting on them, so too the promise may never be realized.

Of the six emotive characteristics in the soul, *yesod* (the *sefirah* of foundation) is the channel that conveys the other emotions into reality, transforming them from subjective sensations into active expression. *Yesod* is closely identified with the covenant of circumcision (*brit milah*). Like a promise, a covenant is a provisionary gesture that binds two people together. (This is especially true of the marriage covenant.) A covenant serves as a temporal bridge that connects the moment of making the promise (in the past) to its fulfillment (in the future). According to this interpretation, quantum reality is a predetermined, ghost-like reality that is not yet fully completed. The observer participates in fulfilling the promise of that reality by choosing which way to direct it. This is similar to the Copenhagen interpretation, which states that what tips the scales is the measurement, or how one articulates it. In this case, reality exists in spirit until we consummate the covenant between subjective and objective reality, between observer and space-time.

Ima: Something Must Exist

Above *Zeir Anpin* is *Ima*, the mother principle. From the moment of conception, a developmental process begins in the mother's womb. Although initially it appears that a relative state of nothingness exists, she adamantly sets out to prove that indeed something does exist. She is not satisfied until the fetus is fully developed. She constructs it, stage by stage, into complete human form, as mothers have done since time immemorial.◄

Neo-realism takes the approach that the new quantum data must in some way be founded on classical knowledge. Just as other scientific paradigms were not understood until new apparatus were devised to probe the phenomena, so too, quantum mechanics is inconclusive.

▶ *Ima* corresponds to *binah* (the *sefirah* of understanding, בִּינָה) which shares its three-letter Hebrew root with *bein* (בֵּין) meaning "between."[8] This relates to the deductive/comparative reasoning involved in analysis, i.e., deducing something from an existing phenomenon, or comparing and contrasting between different entities. However, the two-letter root of *binah* (בִּין) is found in the infinitive form of the verb "to build" (בְּנִיָּה), which relates to a process of synthesis. Thus, the faculty of *binah* relates to both deductive and inductive reasoning.

One biological example is the cell. Initially, when human cells were viewed under a microscope, no deep structures were discerned inside the cell. Therefore, cells were believed to be amorphous. But, when stronger microscopes became available, the subcellular structures became apparent. Following this logic, Einstein assumed that quantum theory was not yet complete. He was not prepared to compromise until a whole model was developed. He thought that, although atoms appear to be an energy cloud, if more knowledge and stronger instruments became available, we could determine the inner structure in a more precise fashion that would redefine it in classical terms. Thus, Einstein proposed further analysis of the quantum data to discover its basis in classical reality.

However, as data accumulated, it became increasingly clear that quantum mechanics precludes the possibility of ever being defined by classical terminology. As a result of the exclusion principle, the more precisely one quality is measured, the less we can know about another. Dealing with both is impossible; therefore, no classical definition can ever suffice for both together. Nonetheless, the neo-realists refuse to abandon classical logic and their claim that reality exists. Therefore, they suggest that our definition of reality requires modification. ▶ The stubbornness of wanting reality, even if it is an inconceivable, miraculous type of reality, corresponds in Kabbalah to *Ima/binah*.

Abba: **Nothing is Really Something**

In contrast to *Ima*, which relies on there being something to analyze and synthesize, *Abba*, the father figure, is associated with *chochmah* (the *sefirah* of wisdom), which initiates new insights from nothing. When classical logic is no longer viable, we need to invent a completely new system of quantum logic. Finkelstein contended that we don't need to redefine reality. Instead, we need to devise a new way to think about it. This type of revolution has already occurred in mathematics. Euclidian (classical) geometry implied that space had a linear, flat character that is defined by points, lines and areas. However, the

▶ This type of logic is apparent regarding the speed of light. Einstein "proved" that the speed of light must be constant. However, experimental data showed that the speed of light could not be precisely defined. Instead of revising the mathematics behind Einstein's proof, scientists have preferred to solve this problem by redefining the length of the meter according to the speed of light such that the speed of light in a vacuum is always 299 792 458 meters per second.

Euclidean rules no longer apply under general relativity's assumption that space is curved. In modern science, the logic for thinking about space has thus been transformed. Another example is the changing mode of logic in mathematics. In arithmetic logic, the whole is the sum of its parts. But, there is a different type of logical principle that states that the whole is greater than its parts. There are also instances in modern mathematics in which the result of multiplying two negatives is a greater negative, not a positive. This is the father principle, by which a new premise or hypothesis is introduced that is not grounded in current understanding and cannot be deduced from our previous level of comprehension.

Thus, in *Atzilut*, neo-realism and new quantum logic represent mother logic and father logic.◀

Arich Anpin: All Realities Exist

Arich Anpin (the Long Countenance) is in the realm of *keter* (the *sefirah* of crown) in *Atzilut* (the World of Emanation). *Keter* is divided into an inner manifestation (*Atik Yomin*) and an outer manifestation (*Arich Anpin*), which is the lower aspect. The crown represents the super-conscious mind of *Atzilut* itself, meaning that it cannot be directly experienced. The length of *Arich Anpin* (the Long Countenance) refers to spatial length i.e., it covers a lot of territory. It also relates to tempo-ral length, and to infinite patience. Kabbalah and Chassidut teach us that a multitude of worlds exist within *Arich Anpin*. This idea matches the many-worlds interpretation of quantum reality. In one context, we are taught that there are eighteen thousand worlds that populate *Arich Anpin*;[9] in another, there are tens of thousands of worlds; in a third, there are infinite worlds. The many worlds that derive from *Arich Anpin* are often associated with the hairs of the head in which each hair connects to a different universe and nourishes it.

This level of *keter* is motivated by *ratzon* (will). When we desire something, we are not satisfied with achieving our desire in one way; we are motivated to try out every option. Like a musical composition,

▶ Modifying the definitions of classical reasoning and new quantum logic can be compared to the difference between two primary types of head *tefillin*. Rashi and Rabbeinu Tam (Rashi's grandson) differed in their opinions concerning how the four scrolls in the head *tefillin* should be ordered. The Lubavitcher Rebbe, Rebbe Menachem Mendel Schneerson strongly advocated that men don Rabbeinu Tam *tefillin* in addition to the classical Rashi *tefillin*. The two types of order reflect different logic gates (to coin a phrase from computer science) through which one may comprehend reality. Rashi *tefillin* are worn first. They reflect the mother mentality, relating to classical logic, as described above. Rabbeinu Tam *tefillin* relate to the father principle and to counter-intuitive quantum logic. This suggests that Rashi *tefillin* suffices for classical mechanics, but one who wishes to understand quantum mechanics should also wear Rabbeinu Tam *tefillin*.

the same piece can be played with infinite variations. The will wants to catch hold of every facet of the infinite abundance of options simultaneously so that each one is anchored in reality. However, despite the desire for a multitude of worlds, as one of the options manifests in reality, none of the other worlds are ever realized.

Atik Yomin: **All Realities Are One**

Atik Yomin is the inner aspect of *keter*, the super-conscious crown. It represents ultimate unity. One interpretation of *Atik Yomin* (עַתִּיק יוֹמִין; lit. Ancient of Days) is that it relates to a time period that is separate and removed from current history (נֶעְתָּק מִן הַיוֹמִין).

In Kabbalah, all dichotomies—such as self and other, male and female, etc.—are exemplified by the tension between left and right. But, at the level of *Atik Yomin*, the left is included in the right and is thus indistinguishable from it. One example of this is the eventual rise of the Levites above the level of the *kohanim*. The *kohanim* are renowned for their loving-kindness, their ability to bring people together, and their directive to bless the people with love. Therefore, they are associated with *chesed* (loving-kindness) on the right. By contrast, the Levites, the Temple guardians, are associated with the side of *gevurah* (might and restraint) on the left. Yet, Kabbalah teaches us that, in the future, the Levites will ascend to a level at which they are included in the right.

Similarly, the current distinction between body and soul will eventually be annulled.[10] This can be compared to the notion in modern science that all physical substance is concentrated energy. Just as concentrated energy becomes physical reality, so too, intensifying the power of the soul generates the creation of a physical body that is indistinguishable from its light source. This idea is expressed in the Kabbalistic expression, "the condensation of the lights (spiritual energy) produces the vessels."[11] At the level of *Atik Yomin*, the vessels are in total harmony with the lights.

Arich Anpin corresponds to the level of *ratzon* (will) in the soul, as mentioned. Similarly, *Atik Yomin* is motivated by *ta'anug* (pleasure). Ultimate pleasure is a sense of wholeness and completion. This highest and most primal level is the innermost aspect of the crown. The insight that matter and energy are interchangeable eliminates all dualities. Thus, at this level of total unity, there is no distinction between God and the different creatures that populate physical reality. Reality can no longer be broken down into distinct constitutive parts. Therefore, the interpretaton that everything is undivided wholeness, and that the observer and the observed are one, corresponds to *Atik Yomin*.

Theory of quantum reality	Corresponding *partzuf* in *Atzilut* (the World of Emanation)
Undivided wholeness	*Atik Yomin*
Many-worlds	*Arich Anpin*
New quantum logic	*Abba*
Neo-realism	*Ima*
Ghost Reality	*Zeir Anpin*
Wave collapse	*Nukva*

To conclude, *Atzilut* (the World of Emanation) is the highest of the four spiritual Worlds. The six *partzufim* of *Atzilut* represent the metaphysical expression of what will eventually materialize in the three lower worlds. The three lower *partzufim* of *Atzilut* manifest in the three lower Worlds, *Beriah*, *Yetzirah* and *Asiyah*. However, their order is reversed so that the highest, most abstract representation drops to the lowest level. By this standard, the level of *Ima* (corresponding to neo-realism) manifests in *Asiyah* (the World of Action) as the theory that demands the most particular, physical reality. The next *partzuf* of *Zeir Anpin* (corresponding to ghost reality)—where the reality that precedes measurement is comprised of abstract urges—manifests in *Yetzirah* (the World of Formation). The lowest *partzuf* (*Nukva*) of *Atzilut*, which by definition has nothing of its own, is reflected in *Beriah* (the World of Creation). There, reality becomes a set of speculative probabilities, which is the most abstract of these three theories.

The correspondence between the different interpretations of reality according to quantum theory and the six *partzufim* of *Atzilut*, shows us how each interpretation is valid from its own unique perspective. As mentioned in the Introduction, building models such as this is the first level of the second stage of uniting Torah and science.

Mind and Body

Until the advent of quantum mechanics, physicists believed that the physical world was pre-determined. They surmised that, if all the conditions of a system at a particular moment are known, the outcome at any point in the future is completely predictable. Quantum mechanics proved determinism to be a fallacy. Whether the effects of an experiment will determine if light is a wave or a particle depends entirely upon the conscious choice of the scientists. If they choose to perform an experiment that will measure the wave property of the photon, they will perceive it as a wave. If they choose a similar experiment that measures the particle property, they will perceive it as a particle. Thus, the state of a system is determined by conscious human observation and measurement.[12] Before the system has been measured, its state cannot be determined and all options remain equally viable.

The Copenhagen interpretation of quantum mechanics holds that so long as the onlooker does not observe (or interfere) with a given system, the system exists in all possible states at one and the same moment. Scientists refer to the system's pre-observed state as superposition. In this state, light (for example) is both a wave and a particle at one and the same time. The baffling phenomenon of superposition diverts the focus of theoretical physics from the physical world as a deterministic set of rules and brings human consciousness into focus, with an emphasis on free choice. As one contemporary quantum physicist described it: "What we perceive as reality now, depends on what earlier we decided to measure."[13]

Most scientists are skeptics when it comes to consciousness, especially when it seeps into hard science like physics. In the minds of these skeptics, matter cannot have consciousness. Yet, all physical substance—including stones, trees, cats and humans—is composed of matter. The average scientist would prefer that human consciousness be mechanistic, driven by DNA, hormones, instincts and the like. The inclusion of human consciousness as a variable in physical measurement is mind-blowing. It alludes to the involvement of consciousness in the creation of the world—a fact that most scientists have great difficulty in accepting, as it clearly points to the existence of a conscious Creator.

The introduction of human consciousness as a deciding factor in scientific calculation is a counter-intuitive development for which scientists were unprepared. But, many great physicists, particularly those of the previous century, believe that the cutting edge of science is to discover the connection between physics and psychology. Niels Bohr◄ was one such scientist. He said that science will not reach its ultimate purpose until it bridges the subatomic world of elementary particles with the human mind.

Human consciousness dwells in the mind. It cannot escape scientific notice that the mind controls the body via electromagnetic impulses. One of the great challenges of our generation is to bring the science of physical light together with the science of the psyche.

"Let Us Make Man"

We have seen that the creation of light is an important link in understanding the place of *alpha* in physics. The numerical value of the phrase "'Let there be light,' and there was light" (יְהִי אוֹר וַיְהִי אוֹר) is 470. This is the same as the numerical value of the phrase that introduces the creation of man, "Let us make man" (נַעֲשֶׂה אָדָם). This equality is the key to the connection between physics and the psyche.◄◄

Man was created in the image of God.[14] His physical attributes correspond to God's actions, as the verse in Job states, "From my flesh I envision God."[15] The human psyche also reflects His ways. This is why the sages teach us to emulate Him, "Just as He is compassionate,

▶ Bohr's mother was from a Sephardic Jewish family. He was therefore a Jew according to Jewish law, but he was unaware of his Judaism.

▶▶ 470 is the fifth number in the linear progression that begins with 26 and is followed by 137 etc. where the base of the series is 111, the numerical value of *alef* (אָלֶף) and "wonder" (פֶּלֶא). See Chapter 4.

you too should be compassionate, etc."[16] Thus, both man's body and his mind reflect Divinity.▼

It is well accepted today that the body influences the mind and the mind influences the body. In Hebrew, the sum of the numerical values of the two words, "mind" (48 ;מֹחַ) and "body" (89 ;גּוּף) is 137, the numerical value of Kabbalah (קַבָּלָה), the facet of the Torah that unites energy and matter, mind and body, physics and psyche.▶▶

Measurement and Blessing

In quantum phenomena, conscious choice and observation cannot predict the outcome of a system in superposition. They do objectively establish the outcome once the system has been observed in the present. We can go further to say that, by Torah standards, the intention of the observer does predict the quality of the outcome. The statement, "When a righteous individual makes a decree, the Almighty carries it out,"[19] relates to the ability of the righteous observer to influence reality.[20] This is also apparent in the Torah approach to dream interpretation, which holds that the verbal interpretation of the dream establishes its meaning, as mentioned.[21]▶▶▶ Similarly, our intentions during prayer have the power to affect reality etc. So, quantum weirdness is a familiar concept to any believing Jew.

Quite simply, the Torah teaches us that measuring things is the opposite of blessing. Blessing is abundance derived from a spiritual

▶▶ The perfect ratio between the head/mind and the torso in humans is the golden section, which appears in many places throughout creation. The golden section is closely related to the Fibonacci series of numbers, which is called the series of love numbers. [18]

▶▶▶ Once, two sages dreamed exactly the same dream and went to the same dream interpreter. One of the sages gave the interpreter a coin, while the other did not. The interpreter positively interpreted the dream of the sage who paid him, and his interpretation proved true. However, to the sage who did not pay he offered a bad interpretation. Unfortunately, his bad interpretation also came true.[22]

▶ In the primordial numbering system, the numerical values of "mind" (מֹחַ) is 181 and "body" (גּוּף) is 432. The sum of these two numbers is 613, the number of commandments in the Torah. This reflects the Torah as the unifying factor between mind and body that manifests the image of God in this world.

Thought, the power of the mind, is represented by Leah, the older sister whom Jacob married first. Speech, the physical power of expression, is represented by Rachel, the younger sister, whom Jacob married a week later.[17] In Kabbalah, we are taught that ultimately, Rachel's power of speech will mature to such an extent that it will include the hidden thoughts of her older sister. Then she will be able to express that which has until now been concealed. This idea is

indicated in the letters that precede the letters of "mind-body" (מֹחַ גּוּף), which are לד־בהע, whose primordial value is 476 (613 minus 137). This number is twice the numerical value of Rachel (238 ,רָחֵל), referring to her mature state. In this way Rachel is the average value of "mind-body." If we construct a linear series beginning with 137, adding 476 each time, the next number will be 613, which is the primordial value of the words, as mentioned. The next number in the series will be 1,089, which is 33^2. The 14th number in this series will be 6,325 which is 25 (5^2) times 253 (the triangle of 22), the numerical value of "the origin of God's mouth" (מוֹצָא פִי הוי) as explained in Chapter 4. This further relates to the power of speech, the power of Rachel.

source, as the Prophet Malachi states: "I will pour out blessing to you until there is no room to suffice for it."[23] By contrast, the moment something is measured, it becomes finite and limited and must follow the laws of nature that relate to it.

Another blessing in Deuteronomy states, "and God will order the blessing to be with you in your granaries, and in every one of your endeavors."[24] The sages interpret the word "granaries" (אֲסָמֶיךָ), as a conjugate of "hidden" (סָמוּי). They learn from this that "a blessing is present only in something that is hidden from view."[25]

The Talmud expounds on this issue: "One who enters to measure [the grain in] his granary says, 'May it be Your will that You send blessing in our endeavors.'" Before the farmer measures the amount, it is still concealed, therefore he prays that God bless him with abundance. "Once he has begun to measure [but has not yet completed the measurement] he says, 'Blessed is He who sends blessing in this heap of produce.'" At this stage he does not pray, but blesses God for His blessing. "Once he has completed the measurement, if he makes a blessing, it is a futile blessing." Indeed, making a blessing at this stage is considered a transgression. The Talmud explains that this is, "Because blessing is not present in something that is measured, nor in something that is counted, nor in something that is weighed, but only in something that is concealed from view."

Once the grain has been measured, it comes under the deterministic auspices of Newtonian mechanics. From then on, it can only reduce in size (through decomposition etc.) and the supernatural blessing departs from it. There is no longer room for surprises—so none will be found. Before it has been measured, the grain is still adaptable and open to change. It remains in the unlimited spiritual realm of superposition—in the realm of probability, where surprising results can happen. From this perspective, measurement is the "evil eye" that brings what was a previously unlimited blessing (in potential) into the finite, limited realm of reality.

When we refrain from limiting our every act with the narrow view of the "evil eye," something remains concealed from view. This positive mentality leaves more room for blessing. When we act generously, God reciprocates by pouring down blessing from above.

The numerical value of the Hebrew word for "granary" (אָסָם) is 101, which is equal to the numerical value of the word, "from nothing" (מֵאַיִן). This offers us a clue to the source of blessing, i.e. the Divine nothingness, which is always hidden from view. This corresponds to the super-conscious in the psyche (as discussed in Chapter 1). Indeed, Rebbe Levi Yitzchak of Berditchev wrote that in order to connect to the Source of all blessing, one must cleave to the "nothing" (אַיִן) that is above the observing "eye" (עַיִן). [Note that the difference between these two words is the first letter—alef in the former word and ayin in the latter. Phonetically the two letters are related to one another.] The numerical value of "your granaries" (131, אֲסָמֶיךָ) is equal to the numerical value of "humility" (עֲנָוָה), which is the key to selflessness. Humility breeds a connection with Divine "nothingness."

Another quantum-like phenomenon occurred when the Children of Israel collected the manna in the desert. They were told that the measurement was "an *omer* [a fixed quantity] per head."[26]▶ There were some who gathered more and some who gathered less. Nonetheless, when they measured their harvests, all of them had exactly the same amount—one *omer* each.[27]

It makes no difference how much we imagine we have earned. It is the Almighty who defines the measure. Mundane reality conceals God's presence at the supernatural level.▶▶ But, true blessing is defined by the extent to which His presence is revealed, regardless of our efforts in the physical world.

This is also true with regard to our spiritual endeavors, concerning which the sages stated, "Whether a person does much or little is the same, as long as he directs his heart heavenwards."[28]

Moses' Blessing

The general Torah rule is that it is forbidden to count the individuals

▶ The numerical value of "and they measured by the omer" (וַיָּמֹדּוּ בָעֹמֶר) is 378, which is the numerical value of "your endeavors" (מִשְׁלָח). The letters of this word are a permutation of the letters of the mysterious word *chashmal* (חַשְׁמַל) discussed in Chapter 1, etc. Since these words share the same letters they are numerically equivalent.

The numerical value of *omer* (עֹמֶר) is 310, which is the numerical value of "something," or "there is" (יֵשׁ). The numerical value of "per head" (לַגֻּלְגֹּלֶת) is 496, the numerical value of "kingdom" (מַלְכוּת).

▶▶ "World" (עוֹלָם) is conjugate to "concealment" (הֶעְלֵם).

in a group of people.[29] When God commanded Moses to take a census of the Jewish People in the wilderness, He told him to collect silver coins from each individual involved in the census. The coins were then counted—not the people—to determine the number of people in the group.

Counting people is relating to them as inanimate objects that have no life-force. Conversely, a blessing expresses the flow of life, which constantly increases (see Chapter 3). ◄ Moses' census was thus a blessing. The Zohar[30] offers the following explanation:

> *Rabbi Yitzchak asked Rabbi Shimon: "We have learned that blessing is not present in something that can be measured and calculated, but, here in the Mishkan, why was everything accounted for?" … [Rabbi Shimon replied] "If the account comes from the side of holiness then blessing is constantly present in it and never ends from it … but worldly things that do not come from the side of holiness have no blessing present in them when they are accounted for, because the 'other side,' which is the evil eye, can overcome it…"*

In contrast to Moses, King David was seduced by Satan to take a census of the people.[31] The result of his census was that a plague struck, killing seven thousand individuals. Nonetheless, David's census brought hidden blessing to the entire nation. As a result of it, David purchased the granary of Aravnah[32] in the ancient city of Yevus, which would later become the location of the Holy Temple in the city of Jerusalem.

Divine Providence

Wolfgang Pauli ◄◄ was one of Bohr's contemporaries and a renowned physicist in his own right. Pauli went through a traumatic period during which he sought psychological help from the famous psychoanalyst, Carl Jung.

Physics and science were in Pauli's field, while psychology and mysticism were Jung's territory. This became the initial meeting-ground

---

► This idea is indicated by the letters of the word "bless" (בְּרֵךְ). The values of each of the three letters in the word allude to plurality. They are all multiples of 2 times 10^n. The numerical value of the first letter *bet* (ב) is 2 (2 times 10^0), the numerical value of the next letter, *reish* (ר) is 200 (2 times 10^2), and the numerical value of the final *kaf* (ך) is 20 (2 times 10^1).

►► Pauli was of Jewish ancestry. However, since his maternal grandmother was not Jewish, he was not Jewish by Jewish law. Nonetheless, he was deemed three-quarters Jewish by German law. Because of his Jewish roots he had to flee from Nazi Germany.

of these two great minds. Pauli was intrigued by the mystery implied by relativity and quantum mechanics. He was similarly puzzled by the enigmatic number 137 that governs the quantum riddle.

Jung had a penchant for researching coincidences that had profound psychic significance for the individual involved. He called this concept synchronicity. One example of such synchronicity was a patient who had dreamed of a golden scarab. As the patient was relating her dream to Jung, the psychoanalyst heard a tapping at the window behind him. Upon opening the window, in flew a beetle that was the local species most similar to the type of scarab that his patient was describing. Following certain mystical systems that he had studied, Jung interpreted this to have particular significance with regard to the patient's psyche.

Pauli found Jung's theory of unexpected synchronicity compatible with the new scientific revelations that indicate that human observation may possibly be a variable in the equations.

Pauli and Jung became close friends. They eventually published a book together on the connection between physics and psychology.[33] Their quest for the bond between the two sciences was riddled with the number 137. Pauli and Jung believed that in some mysterious way, *alpha* serves as a missing link between elementary particles and the psyche. Pauli himself experienced the synchronicity that Jung spoke of when he was hospitalized in room number 137. When he saw the number on the door he groaned, "I'll never leave this place alive!" Indeed, he died in that room a few days later.

Synchronicity is a common concept in Jewish teachings. The sense that God is synchronizing our every step is one that accompanies all believing Jews throughout their lives. According to Kabbalah and Chassidut, every event in the world occurs in many realms. ▶ Synchronicity is experienced when we tap into more than one of these planes simultaneously. Such Divine Providence inherently attests that the quality of nature is determined by man's conscious interpretation. Events are "meaningful coincidences" if they occur with no apparent

▶ Divine Providence is experienced at five different levels, depending on one's level of consciousness. These five levels correspond to the five levels of the soul: *nefesh*, *ruach*, *neshamah*, *chayah* and *yechidah*. At the level of *neshamah*, for example, Divine Providence is experienced as a constant stream of events that directly relate to one's Torah study. The Ba'al Shem Tov taught that if a person studies Torah regularly for its own sake, he will experience phenomena that relate to what he has just learned.

causal relationship, yet seem to be related. The relationship between the events is significant only in the presence of a conscious mind. Living in such a state of synchronicity is described in Chassidic literature as "Jewish nature," or "natural consciousness."[34]

A person who is in tune with God and chooses His Torah, will constantly experience Divine Providence in mysterious ways throughout life. The most obvious type of synchronicity appears to be absolutely miraculous, when seemingly unrelated events occur simultaneously. The ultimate form of Divine Providence is when the person experiences the re-creation of reality as "something from nothing" at every moment. ◄

▸ As mentioned above, the numerical value of "something from nothing" (יֵשׁ מֵאַיִן) is 411, the numerical value of "chaos" (תֹּהוּ), which equals 3 times 137.

All or Nothing

Causal determinism is ruled out by both synchronicity and quantum mechanics. Herein lies the similarity between the two. From the staunchly atheistic perspective of science, this puts science in trouble. Without causality, the only two fundamental laws of science that remain are probability and energy conservation.

In the type of deterministic world described by Newtonian mechanics, science should be able to predict the outcome of any macroscopic or microscopic interaction. However, it is impossible to know all the minute details of every force that affects each interaction. Therefore, science requires probability to allow for equilibrium in the causal assessment of nature. In the words of Richard Feynman:[35]

> It is true classically that if we knew the position and the velocity of every particle in the world, or in a box of gas, we could predict exactly what would happen. And, therefore, the classical world is deterministic ... if we start with only a tiny error it rapidly magnifies to a very great uncertainty. To give an example: if water falls over a dam, it splashes. If we stand nearby, every now and then a drop will land on our nose. This appears to be completely random, yet such a behavior would be predicted by purely classical laws. The exact position of all the drops depends upon the precise

wigglings of the water before it goes over the dam. How? The tiniest irregularities are magnified in falling, so that we get complete randomness. Obviously, we cannot really predict the position of the drops unless we know the motion of the water absolutely exactly.

Speaking more precisely, given an arbitrary accuracy, no matter how precise, one can find a time long enough that we cannot make predictions valid for that long a time ... in only a very, very tiny time we lose all our information ... already in classical mechanics there was indeterminability from a practical point of view.

By contrast to the deterministic view, synchronicity specifies that every drop of water, indeed, every particle, is projected towards a precise objective. Whereas probability means that there is a very slim, or even negligible chance that you will win the lottery, synchronicity means that if you are not intended to win the lottery, your chances of doing so are zero. Conversely, if you are meant to win, your chance of winning is one-hundred percent. The only way to know which possibility is true is through prophecy, or after the lottery has taken place. Before the lottery, you win and lose simultaneously.

Creating New Energy

Any proof that the total energy in the universe changes is tantamount to blasphemy in the scientific community. Such a discovery might inadvertently prove the presence of a Creator, who creates and inputs new energy into the world.

Yet, according to special relativity, the total energy of a system remains constant only as long as the observer's reference frame does not change. Different observers will disagree regarding the system's energy value. Like the dual wave-particle quality of energy, conservation of energy thus ultimately depends on the consciousness of the observer. Even general relativity has failed to establish whether or not the energy of the entire universe remains constant. Therefore, relativity suggests that new energy can be formed.

However, in order to guard conservation of energy at all costs, science has developed a new type of vector mathematics called tensor analysis. Following this line of logic, the law of energy conservation is duly conserved.

According to Kabbalah and Chassidut, unlimited new energy is constantly being introduced into the universe as a result of our good deeds.[36] The current state is that our good deeds and sins are balanced,[37] creating an even state that appears to be probabilistic. Yet, every good deed has the power to tip the scales and advance the world to a better state of affairs. For this reason, discovery of any physical phenomenon that defies the law of energy conservation should not come as a surprise.

The Creation of Consciousness

The human psyche is a miniature universe. In order to understand how consciousness works, we need to delve into the secrets of how God created the universe. Kabbalah teaches us that before God created the world, His infinite light was omnipresent, so there was no room for anything to be created. God contracted His infinite light to make way for the world. He withdrew His brilliance by creating a vacuum that would allow creation to begin. The Talmud states that God stored away the secret of the original light for the righteous of the future.[38]

The initial contraction is one of the most profound innovations of the Arizal. He taught that the apparent vacuum is not truly empty. Think of a goblet of wine, for example. If the wine is poured out of the goblet, the goblet appears empty. But, an impression—for example the aroma of the wine—still fills the cup. According to this understanding of the contraction, the vacuum is dark. Nonetheless, an impression of God's infinite light remains that still fills the entire universe. It was from this impression of light that the physical universe was created. The impression of God's light is the creative energy that continues to manifest in reality.

The Ba'al Shem Tov taught that the contraction should not be taken literally. He taught that the vacuum only came into existence from the relative perspective of human consciousness. From God's perspective, so to speak, nothing changed at all. Returning to our example of the cup of wine, this would mean that the wine was not spilled out of the cup in any way. Rather, through the contraction of God's infinite light the "wine in the cup" became invisible. From this perspective, the Divine light is present in all its glory, filling every point of existence. It remains available for those who merit to see it.

Despite its enormous expanse, the apparently invisible impression is referred to in Kabbalah as a "point." Like a point, it has no measurable dimensions and is thus essentially chaotic. The lack of dimensions reflects the multiple possibilities that are available simultaneously in the spiritual realm. Through this point, God projects a line of light— His own Divine consciousness—using it as His measuring apparatus, as it were. We have seen that introducing human consciousness into a system defines it by measuring its quantum state. So too, the introduction of Divine consciousness creates reality in the residue of God's infinite light.▸

Unlike a zero-dimensional point, a line is differentiable and contains a two-way consciousness of "above" and "below." In this case, the line represents the transition from the spiritual to the physical. The "top" of the line represents spiritual reality, while the "bottom" of the line represents physical reality. The line is projected through the point of raw, formless matter of the impression. This projection sculpts a system of phases that become progressively more physical. The gradual transition from spirituality to physicality is defined in Kabbalah as four "Worlds." As seen above, these four Worlds are *Atzilut* (the World of Emanation), *Beriah* (the World of Creation), *Yetzirah* (the World of Formation), and *Asiyah* (the World of Action).▸▸

Before the contraction of God's infinite light and before the line descends, there is only Divine consciousness. This "prehistoric" stage that precedes the manifestation of the four Worlds is referred to in Kabbalah as *Adam Kadmon* (Primordial Man). This stage corresponds to *keter* (the *sefirah* of the super-conscious crown).

▸ From God's perspective, as it were, the line entered the point, but from our perspective, it appears that the line emanates out of the point. In actual fact, the line is above the point and passes through the point in order to measure it. The line thus sculpts the undifferentiated point into its own form, with a consciousness of above and below. In infants, cognition of above and below precedes the development of lateral consciousness. This reflects the line as it manifests in psychology

▸▸ *Asiyah* (the World of Action) contains the physical element that we experience through our senses, but also includes a relatively spiritual element. As mentioned in the previous chapter, the four spiritual Worlds correspond to the four letters of God's essential Name, *Havayah*. The physical element of the World of Action corresponds to the Name *Elokim*.

In the human psyche, the four Worlds correspond to the *sefirot*, which manifest as different states of consciousness. *Atzilut* corresponds to *chochmah* (the *sefirah* of wisdom); *Beriah* corresponds to *binah* (the *sefirah* of understanding); *Yetzirah* corresponds to the six emotive *sefirot* (*chesed, gevurah, tiferet, netzach, hod* and *yesod*); and *Asiyah* corresponds to *malchut* (the *sefirah* of kingdom).

In the as yet unformed, or unmeasured, impression of God's infinite light, our consciousness is merged with Divine Providence. The more we refine our consciousness, the higher it rises and the more influence we have on forming reality together with God, as it were. Primordial Man, *keter*, is the super-conscious source of our thoughts and actions. In general, we cannot access this level of our psyche.

In Kabbalah, *keter* is comprised of three heads. The highest head, wherein resides "the essence of the soul," corresponds to faith. The second head corresponds to the spiritual pleasure principle. The third head of *keter* corresponds to pure will.

Usually, the relationship between faith, pleasure and will is described metaphorically as a soul relates to a body. Each level is "enclothed" within the level below it. The higher, inner level is "soul" and the outer level is the "body."

Faith

Pleasure

Will

Alternatively, the three "heads" of *keter* (the crown) can be depicted as a triangle with faith at the top, pleasure to the right and will to the left. These define the three lines of the "Tree of Life" — right, left and center.

Faith

Will   Pleasure

Faith

To have faith means believing in the unknown; this is referred to as "dark," not yet illuminated by intellectual perception. As Rebbe Nachman of Breslov wrote, "…as long as the rational mind understands something, faith is not necessary; and the essence of faith is only where the rational mind ends and the matter cannot be rationally understood."[39] Richard Feynman, although Jewish by birth, defined himself as atheist. Yet, he instinctively echoed this inherently Jewish tenet when he said, "God is always associated with those things that you do not understand." Once something has been illuminated by being proven intellectually, we no longer need to have faith in it.

Quantum physics necessitates the existence of uncertainty. When this principle was originally discovered, scientists thought that there was a problem—the result of the human factor that is unable to observe two properties simultaneously. A decade or so later, it became clear that the problem is not related to the way humans observe reality. The uncertainty principle is objectively inherent in reality.

One would imagine that a person who has faith has certainty. Paradoxically, in his writings, the Arizal enumerates five existential uncertainties with regard to faith.▼ At the very level at which we believe in God, we are aware that we can have no certain knowledge of Him. This is the root of the uncertainty principle innate in creation.

▶ The five uncertainties refer to the different possible hypotheses, or "doubts," regarding the definition of faith in the soul. Measuring the point of the impression separates the conscious from the unconscious in the psyche. This renders each of them imperceptible while the other is being contemplated. Thus, for example, someone who is hypnotized to reveal what is going on in his unconscious mind, will not be aware of his "conscious" experience at the time. This is similar to what we are taught regarding the location of Moses' grave[40]: "The evil empire [the Roman emperor, Adrianus] sent [emissaries] to the military chief of Bet Peor [saying] 'show us where Moses is buried.' When they stood above, it appeared that it was below, [and when they stood] below it seemed that it was above. They split into two groups. To those who stood above, it appeared that it was below, [to those

who stood] below it appeared that it was above… Even Moses did not know where he was buried." The phrase relating to Moses' burial by God reads, "and He buried him in the valley" (וַיִּקְבֹּר אֹתוֹ בַגַּי). "Valley" is usually spelled with an additional *alef* (גיא). The unusual spelling of the word in this phrase indicates that Moses was buried in a place that relates to the three (ג) "heads of heads"—including the unknowable head of *Radla*—that are above the ten (י) *sefirot*. The numerical value of "and He buried him in the valley" (וַיִּקְבֹּר אֹתוֹ בַגַּי) is 740, which equals 20 times 37, the companion number of 137. This indicates an aspect of 37 that represents a higher phenomenon than 137. 137 is the number of interincluded *sefirot* (see Chapter 1), while 37 in this case relates to three superconscious heads above the ten *sefirot*. This higher aspect of 37 is the source of the intellectual and emotive powers of the soul.

The uncertainty principle manifests in the two powers below faith. Just like uncertainty in physics concerns position or momentum, so in Kabbalah uncertainty concerns right or left, pleasure (position) or will (momentum).

Pleasure

Early founders of modern psychology identified the pleasure principle as the highest motivation of the psyche. In the unconscious mind , pleasure derives from the higher level of faith. ◀ In the terminology of Kabbalah and Chassidut, pleasure is the light that shines from the essence. Towards the end of his life, Freud became aware of something beyond the pleasure principle, which he called the "death drive." We can uncover the grain of truth that lies dormant in the ideas that Freud and others like him have proposed by delving into the inner mysteries of the Torah as they are taught in Kabbalah and Chassidut.

The greatest spiritual pleasure is basking in God's light. This is why the origin of physical pleasure is the sense of sight. Indeed, seeing the forbidden fruit was the first stage of the primordial sin, "And the woman saw that the tree was good to eat and desirable to the eyes…"[41] Any pleasure that we experience via other senses is a facet of the ultimate pleasure of perceiving Divinity. Sight, and therefore pleasure, correspond to the ability to identify ("see") and measure the position of a particle.

There are two types of spiritual pleasure, simple pleasure and compound pleasure. Both of these are super-conscious experiences.

Simple pleasure is the pleasure of living. Many people testify that the most uplifting moment of their life was being released from an almost fatal experience. In extreme cases, there are individuals who artificially recreate the thrill of living by engaging in life-threatening sports. Nonetheless, in general, the sense of pleasure in being alive is beyond our conscious experience.

Compound pleasure is the result of actualizing our will. This type

▶ If I believe in God (the belief system of the Divine soul) then the ultimate pleasure is to bask in God's presence and to achieve union with God. In the intellectual soul, the ultimate pleasure is to know everything that God teaches me. If my belief system is that of the animal soul, then the ultimate pleasure is to unite physical pleasure with spiritual pleasure. True rectification is achieved by inspiring the intellectual and animal souls with the Divine soul. When the ego, which resides in the intellectual soul, is nullified and true selflessness is achieved, the intellect serves as the unifying link between the two other souls. This union results in the synchronicity of "natural consciousness" by which the Divine soul and the animal soul have the same desires and purpose, i.e., the animal soul will desire to perform God's commandments of its own will.

	Faith	Pleasure	Will
Divine Soul	Belief in God's omnipresence	Basking in God's light	To unite with God
Intellectual Soul	Belief in God's omniscience	Knowledge of God	To study God's teachings in the Torah
Animal Soul	Belief in God's omnipotence	Realizing God's will	To work with God by performing His commandments

of pleasure remains in the realm of the super-conscious, but at a lower level than simple pleasure. Therefore, it more commonly manifests in our consciousness. It is experienced as an initial moment of insight or inspiration that enables us to realize an innate talent or desire. Some examples of compound pleasure are music ("the pleasure of the soul"), or the pleasure of understanding science or mathematics. The most profound example of compound pleasure is experienced through revealing a new Torah insight.

Will

Will is the raw energy that motivates the soul. It is based on the pleasure principle and fired by faith. Will motivates us to achieve the type of pleasure that we appreciate according to our belief system.

Will is the super-conscious force of the soul's essence, the motor that drives all the conscious powers of the soul. "Will" (רצון) is related to "running" (רץ). One who truly desires to achieve a particular objective is naturally motivated to pursue it. Like a soul without a body, or a body without legs, the super-conscious can achieve nothing in reality without will.

When a particle is running, or moving so fast that its position cannot be determined, its force can still be felt. Therefore, will corresponds to the uncertainty of a particle's momentum.

Like pleasure, will also has two levels—the simple will to live and the compound will to realize one's talents. These two correspond to the two levels of pleasure.

In general, our will is motivated by whatever gives us pleasure. Pleasure permeates will just as the soul permeates the body. Nonetheless, in Kabbalah we are taught that there is an exception to this rule. Just as in physics, position and momentum are two different properties that cannot be known simultaneously, the uncertainty principle of the soul manifests when will defines the pleasure principle.

The classic, most extreme case is an individual who wants to suffer, or even wishes to die. This can result from the most severe psychological illness, e.g., when a person has suicidal tendencies (similar to Freud's "death drive" hypothesis). However, it also manifests as the noblest trait of humanity, when an individual is prepared to relinquish his life in order to uphold higher ideals.

As a rule, the Torah teaches us to live by its precepts and not die for them. The exception to this rule is the obligation to die rather than transgress any of the three most severe prohibitions: idolatry, sexual immorality, or murder. In such a dire situation, even those who are not at a high spiritual level can consciously influence their will to relinquish their pleasure in living. Under such circumstances, they will choose to sacrifice their life rather than transgress God's commandments.[42] In a similar situation, the will of a *tzadik*—a truly righteous soul—to die under those circumstances becomes so dominant that it overrides the pleasure to live. Death then becomes the greatest pleasure.

The classic example of a *tzadik* who reached the level at which death was more pleasurable to him than life without Torah is Rabbi Akiva. He felt like a fish out of water without it.▶ The Talmud[43] relates that at the moment before his soul left him, as he was being tortured to death by the Romans, he recited the ultimate proclamation of faith, "*Shema*… Hear O' Israel, *Havayah* is our God, *Havayah* is one." His students were so startled by his joyful proclamation even as he was suffering so greatly, they asked him, "Even now?!" Rabbi Akiva's astounding reply echoes throughout all of Jewish history, "All my life," he said to them, "I waited for the opportunity to show how much I love God, and now that I have the opportunity should I forgo it?" Rabbi Akiva's soul left him climactically as he reached the end of the verse, "*Havayah* is one."

Another example of this phenomenon is that most people will avoid experiencing pain, the opposite of pleasure (unless they are psychologically unsound). However, certain *tzadikim* may have an unconscious desire to feel suffering, whether to atone for their own

▶ Without the Torah we are all like fish out of water (*Avodah Zarah* 3b). The Torah verse quoted in the Talmud that describes this is, "And You have made man like the fish of the sea." (Chabakuk 1:14) The numerical value of the phrase, "Man like the fish of the sea" (אָדָם כִּדְגֵי הַיָּם) is 137. The numerical value of "like the fish" (כִּדְגֵי) is 37.

sins in this world (perhaps in previous incarnations), or for the sins of their generation, and they accept their suffering with joy.

> *A man once came to the Maggid of Mezeritch and complained about his terrible poverty and suffering. He asked the Maggid how to cope with it, but the Maggid told him he couldn't help him. Instead, he sent him to Rebbe Zusha of Anipoli. The man visited Rebbe Zusha and saw that he too, lived in dire poverty. When he asked Rebbe Zusha how he coped with such suffering, Rebbe Zusha replied that he didn't know what the man was talking about. "Zusha doesn't know what it means to suffer," the tzadik concluded, referring to himself.*

From these examples, we can see that a mysterious dynamic oscilates between pleasure and will. In general, our expectation of pleasure defines our will, but there are exceptions to that rule in which will overrides, or even defines pleasure.

In our current state of consciousness, the body is sustained by the soul, i.e., will is sustained by the pleasure principle. However, in the consciousness of the World to Come, the soul will be sustained by the body. This represents a higher level of consciousness, wherein the pleasure principle is defined by the will. This can happen when will is defined directly through our faith.

In terms of classical logic, this paradox is similar to the rule that there is no rule that does not have an exception. Since that in itself is a rule to which there can be no exception, an inherent paradox is present in the rule itself! Is the exception defined by the rule or is the rule defined by the exception? Such paradoxical logic, or super-logic, is an essential characteristic of the super-conscious crown.

As mentioned above, the higher powers enclothe themselves in the power below them and thus become revealed. In the psyche, this begins by refining one's faith until one creates a clear picture of the future as God would like it to be. This arouses the will to want to realize the vision in its entirety, which in turn arouses the mind to conceive more and more ways to bring the vision into reality.[44]

Those who act in accordance with the above, will experience synchronicity in every move they make. Divine Providence will constantly guide them to realize their will, because it is God's will. Sometimes Divine Providence will appear via natural synchronicity; at other times, it may materialize via a miracle. The various levels of Divine Providence are enclothed in nature relative to the ability of the natural world to experience the vision. The more congruent the material world is with God's will, the more natural the revelation will appear. In this way, a miracle that is hidden within nature is a higher form of miracle than a sensational miracle such as the Splitting of the Red Sea.

Miraculous and Natural Miracles[45]

Chassidut explains that faith, pleasure and will are the inner powers of *keter*, the super-conscious crown that hovers above conscious intellect. They are the source of the supernatural power that works miracles in the world.

God works miracles by the power of His will. This is also how He created nature (with all of its Newtonian laws). The miraculous power of His will manifests at three different levels, each of which overrides the laws of nature to varying degrees. This creates the possibility of scientifically inexplicable phenomena appearing in our lives.

The first level includes supernatural miracles that completely defy the laws of nature, such as the Splitting of the Red Sea. Such miracles are extremely rare, because God prefers to work within the laws of nature rather than against them. However, sometimes—for reasons known only to God—He uses His omnipotent prerogative to override those laws.

The second level includes miracles that occur within the boundaries of nature, but against all statistical odds. Such miracles appear to be completely natural. However, on further examination, it becomes clear that the turn of events was completely unfathomable. Purim is an example of one such miracle. The date had been set for the extermination of the Jewish People at the hands of the non-Jews. Yet, salvation had already been set in motion thirteen years earlier, when King Ahasuerus fell in love with Esther (who did not reveal to him her Jewish identity), and he made her his queen. At the same time, Esther's uncle, Mordechai, uncovered a plot to assassinate the king, saving the king's life. When the genocidal decree was proclaimed, Queen Esther made a feast for Ahasuerus and Haman, the minister responsible for the decree against the Jewish nation, and salvation came about – all through natural occurrences, albeit with well-timed synchronicity. Eventually, the tables were turned and not only did the Jewish People survive, they even rose to grandeur. The same is true of the victory of the Jews against Hellenism, commemorated by Chanukah.

There are a number of Chassidic holidays on which we commemorate miraculous events that occurred to various *tzaddikim*. For instance, during the early years of the Soviet regime in Russia, Rabbi Yosef Yitzchak Schneerson was charged with treason and sentenced to death. By natural standards, there was no way by which he could be freed. Yet, through the efforts of his students, who put themselves at great risk, his release was attained and the hand of God at every step of the way became apparent to everyone. This type of miracle is enclothed in nature, without defying it. But, it could never have transpired if nature was merely a collection of deterministic laws.

The third level includes those miracles that happen at every moment of our lives. They are so deeply rooted in nature that most people are totally unaware that God is performing a miracle. For example, a person crosses the road, failing to see an approaching bus. Yet, the bus miraculously passes by, missing him by a hairsbreadth, without anyone noticing the narrow escape that just occurred. The person goes on with his life, unconscious of the workings of Divine Providence. But, God is watching over him, every step of the way.[46]

One example of a perpetual ongoing miracle is the continued existence of the Jewish People. In the renowned and eloquent words of non-Jewish literary writer, Mark Twain,[47]

> *...If statistics are right, the Jews constitute but one percent of the human race. It suggests a nebulous dim puff of stardust lost in the blaze of the Milky Way. Properly, the Jew ought hardly to be heard of, but he is heard of, has always been heard of. He is as prominent on the planet as any other people, and his commercial importance is extravagantly out of proportion to the smallness of his bulk. His contributions to the world's list of great names in literature, science, art, music, finance, medicine, and abstruse learning are also away out of proportion to the weakness of his numbers. He has made a marvelous fight in this world, in all the ages; and has done it with his hands tied behind him. He could be vain of himself, and be excused for it.*
>
> *The Egyptian, the Babylonian, and the Persian rose, filled the planet with sound and splendor, then faded to dream-stuff and passed away; the Greek and the Roman followed and made a vast noise, and they are gone; other people have sprung up and held their torch high for a time, but it burned out, and they sit in twilight now, or have vanished. The Jew saw them all, beat them all, and is now what he always was, exhibiting no decadence, no infirmities of age, no weakening of his parts, no slowing of his energies, no dulling of his alert and aggressive mind. All things are mortal but the Jew; all other forces pass, but he remains. What is the secret of his immortality?*

In the current context, we can say that the answer to Mark Twain's rhetorical question is God's special providence over the Jews, His chosen people.

These three levels of miraculous occurrences, together with the mundane level of natural causation, correspond once again to the four spiritual Worlds.

Spiritual World	Type of miracle/nature	Human experience
Atzilut	Against the laws of nature	Miracle in miracle (Splitting of the Red Sea)
Beriah	Against statistical odds	Nature in miracle
Yetzirah	Imperceptible as a miracle	Miracle in nature (Purim)
Asiyah	Natural causation	Nature in nature (normative life)

To recap from Chapter 4, the verse from Isaiah that alludes to the four Worlds is: "Everything that is called by My Name and for My honor, I have created it, formed it, and even have I made it."[48]

"All that is called by My Name," refers to *Atzilut* (the World of Emanation), in which the unity of the essential Name of God, *Havayah*, is clearly perceived. This World was revealed to the Jewish People for the first time through the miracles that facilitated their redemption from Egyptian slavery.[49] "I have created it," refers to *Beriah* (the World of Creation), in which the Name of God is concealed, as in the Book of Esther, in which God's Name does not appear even once. "I have formed it"—referring to *Yetzirah* (the World of Formation)—employs the same verb as used to describe the creation of man, "And *Havayah*, God, formed man from the dust of the earth, and He breathed into his nostrils the spirit of life."[50] *Asiyah* (the World of Action) is represented by the phrase "I have made it." The references to *Atzilut*, *Beriah* and *Yetzirah* (the three upper Worlds), are separated from *Asiyah* by the Hebrew word "even" (אף, which also means "wrath"). Thus, the three higher levels of miracle are separated from the level of natural causation.

The first time "even" (אף) appears in the Torah is in the first word spoken by the serpent to Eve, "Even has God said, 'You shall not eat from all the trees in the Garden?'" Man's subsequent fall into sin that resulted from heeding the serpent's words corresponds to the fall of consciousness from the three Divine levels of *Atzilut, Beriah, Yetzirah,* into the lowest, separate level of human consciousness in *Asiyah*.

In the psyche, the three levels of miracles correspond to faith, pleasure and will, the inner qualities of *keter* (the *sefirah* of the super-conscious crown). Natural causation corresponds to human intellect.

REFERENCE NOTES FOR CHAPTER 6

1. Werner Heisenberg, "The Representation of Nature in Contemporary Physics," *Daedalus* (1958), p. 87; pp. 95-108, as cited in Karl Popper, *Quantum Theory and the Schism in Physics* (1992), p. 85.

2. Genesis 3:24.

3. See also: Maimonides, *Hilchot Eidut* 18:5-8.

4. Isaiah 45:7.

5. *Zohar Bereishit* 249b; see also *Sefer Halikutim, Yeshayahu* ch. 38.

6. *Berachot* 56a.

7. *Sefer Yetzirah* 6:1.

8. See Rabbi David Kimchi *(Radak)*, *Sefer Hashorashim*.

9. *Zohar Chadash, Parashat Acharei*.

10. *Etz Chayim, Sha'ar Ha'akudim* Ch. 3

11. Ibid.

12. These phenomena are described at length in *Quantum Enigma*, by Bruce Rosenblum and Fred Kuttner. See also in their article, "The Observer in the Quantum Experiment."

13. *From Einstein to Quantum Information: An Interview with Prof. Anton Zeilinger* (University of Vienna), Centre for Mathematical Sciences, Cambridge, 2004.

14. Genesis 1:27; 9:6.

15. Job 19:26.

16. *Yerushalmi Peah* 1.

17. Genesis 29:27-28.

18. See our book *913: The Secret Wisdom of Genesis* p. xix; or, in Hebrew, *Einayich Breichot Becheshbon* pp .36-39. See also *The Twinkle in Your Eye*, Appendix B.

19. See *Moed Katan* 16b.

20. The ability of the conscious observer to influence reality is discussed in depth in our article, *Emunah Vebitachon*, in our book *Lev Lada'at* (Hebrew).

21. For more on dream interpretation, see our book, *A Sense of the Supernatural*.

22. *Berachot* 56a.

23. Malachi 3:10.

24. Deuteronomy 28:8.

25. *Ta'anit* 8b; *Baba Metzia* 42a.

26. Exodus 16:16.

27. Ibid 16:17-18.

28. In various places in the Talmud, e.g., *Berachot* 5b; 17a, etc.

29. Maimonides, *Hilchot Temidin Umusafin* 4:5.

30. *Zohar Shemot* 225a.

31. II Samuel 34; I Chronicles 21.

32. In Chronicles, the name is Arnan.

33. *Interpretation of Nature and the Psyche*. In addition, Jung published his analysis of Pauli's dreams in *Psychology and Alchemy*. Later, the written correspondence between the two friends was published in another book, entitled *Atom and Archetype*.

34. For a partial translation of one of our Hebrew books, *Mudaut Tiveet*, relating to natural consciousness, see:

http://www.torahscience.org/psychology/consciousness1.html

35. *The Feynman Lectures*, Chapter 2.

36. See for example *Derech Mitzvotecha, Shoresh Mitzvat Hatefillah* 13 (p. 121b).

37. See for example *Kidushin* 39b.

38. *Chagigah* 12a.

39. *Likutei Moharan* II, 8:7.

40. *Sotah* 14a.

41. Genesis 3:6.

42. *Tanya* Ch. 18.

43. *Berachot* 61b.

44. See our book in Hebrew, *Lev Lada'at* p. 90-93.

45. See our book in Hebrew, *Sod Hashem Liyreiav* pp. 543 ff.

46. See *Niddah* 31a.

47. Mark Twain, "Concerning The Jews," Harper's Magazine, 1899.

48. Isaiah 43:7.

49. See Exodus 6:3.

50. Genesis 2:7.

137 AND AGE

Rectifying Chaos

137 is related to particle decay. As mentioned in the Introduction, the probability that a charged particle (an electron in particular) will emit a photon is proportionate to the fine-structure constant. This teaches us that there is an essential relationship between 137 and particle life.

Human life is possible as long as our physical body is fused with a soul; this corresponds to the electron/photon relationship. Like a particle charged with a photon of light, we are empowered with life force. Once the bond between body and soul reaches its conclusion, the body "emits" the soul, like a particle that emits a photon of light, and human life comes to an end.

The most common lifespan recorded in the Torah is 137 years.[1] There were three individuals in the Torah who lived for this period of time. No other lifespan appears more than twice.

The first individual to live to the age of 137 was Ishmael.[2] Ishmael was not Jewish, and his modern-day descendants are primarily Muslims. He was born from Abraham's concubine Hagar before Abraham carried out the commandment of circumcision. The other two people who lived for 137 years were Jacob's son Levi,[3] the progenitor of the Levites and the priestly sect of *kohanim*, and Levi's grandson, Amram,[4] who was Moses' father. The Torah describes Ishmael as "a wild man,"[5] while Levi and Amram are described as righteous men and leaders of their generations.

Since there were three individuals with a lifespan of 137 years, the sum of their years is 411, which is the numerical value of "chaos" (תהו). A common phenomenon in the Torah is that the first time a particular concept is mentioned it appears in a chaotic state. Only later does it

▸ The same order appears in the statement of Sarah's lifespan, but this is because she reached a level of the "Tree of Knowledge of [only] Good [and not evil]". Through its clarification, the Tree of Knowledge will ultimately rise to a level higher than the Tree of Life.

▸▸ "Year" (שָׁנָה) appears a total of four times in all three cases. "Years" (שָׁנִים) appears once in the verse that refers to Ishmael. The numerical value of "year" (שָׁנָה) is 355. Four times 355 equals 1,420. The numerical value of "years" (שָׁנִים) is 400. The sum of these two numbers is 1,820, which equals 70 times 26, the numerical value of *Havayah*. There are exactly 1,820 appearances of this Divine Name in the entire Torah.

The total numerical value of the three phrases that refer to the age of the three people who lived to 137 years is 8,400. There are 24 words in all three phrases, thus, the average value of each word is exactly 350, which is the numerical value of Amram (עַמְרָם), the third of the three people who lived to the age of 137 years.

The number 8,400 is also equal to 70 times 120, the lifespan of David times that of Moses, the two most general lifespans in the Torah.

reappear in a state of rectified sanctity. God created chaos before order so that ultimately, the potency of the great lights of chaos manifest their intrinsic, hidden order within the stable rectified framework that later ensues. Today, Ishmael's descendants are an enemy of the Jewish People (and of the non-Muslim world in general). This fact reflects the chaos inherent in Ishmael, who is the first person mentioned in the Torah who lived to the age of 137.

When Sarah told Abraham to expel Ishmael with his mother, it did not please him. The verse in Genesis states, "The matter was very bad in the eyes of Abraham regarding his son."[6] This is the first time that "regarding" (אוֹדֹת) appears in the Torah. The numerical value of this word is 411, the numerical value of "chaos" (תֹּהוּ). In this context, God told Abraham, "Listen to her [Sarah's] voice."[7] "Her voice" (בְּקֹלָהּ) is a permutation of Kabbalah (קַבָּלָה) and has a numerical value of 137.

Ishmael's affinity to chaos is further alluded to in the way his lifespan is presented in the Torah. The verse in Genesis states, "And these were the years of the life of Ishmael, one hundred years and thirty years and seven years."[8] The verse begins by stating the general ("one hundred years") and then goes into greater detail ("thirty years and seven years"). But, concerning Levi and Amram, the Torah states, "and the years of the life of Levi/Amram were seven and thirty and one hundred years,"[9] going from the detail to the more general. ◂ The former reflects the associative deductive reasoning that emanates from the scientific logic of the Tree of Knowledge of Good and Evil.[10] In general, such deductive reasoning leads to the discovery of infinitely more division and consequently, more chaos. The latter reflects the inductive reasoning of the Torah, the Tree of Life, which leads to greater unity and thus, more order. The fact that each segment of Ishmael's lifespan is followed by, "years," further emphasizes this entropic tendency. With reference to Levi and Amram, "years" appears as the summary of all the years. ◂◂

The ultimate rectified state of chaos is when the lights of chaos are revealed in rectified vessels. In this case, the 137-year lifespan represents the paradoxical power to contain chaos while simultaneously making use of the full force of its energy (as explained in Chapter 2).

In this triplet, Ishmael represents the potent chaotic energies; Levi and Amram represent the refined vessels able to contain them.▸

The Lifespan of Man

"All sevens are dear."[16] One example of this is Moses, who was dear because he represented the seventh generation from Abraham.

Let's consider the lifespans of these seven generations from Abraham, the first Jew, to Moses, the great leader of the Jewish People.[17] These seven correspond to the seven attributes of the heart as shown in the table below:

	Isaac		Abraham
	180		175
	gevurah		*chesed*
		Jacob	
		147	
		tiferet	
	Kehat		Levi
	133		137
	hod		*netzach*
		Amram	
		137	
		yesod	
		Moses	
		120	
		malchut	

Moses is usually associated with *netzach* (the *sefirah* of victory). Here, because he was the first king, or leader, of the Nation of Israel, he is associated with *malchut* (the *sefirah* of kingdom).

▸ Ishmael's age at his death appears in the 676th verse from the beginning of the Torah—676 equals 26^2. The numerical value of the essential Name of God, *Havayah*, is 26. A square number represents perfect inter-inclusion of each of the units (thus symbolizing order). This reflects the fact that Ishmael, as Abraham's son, had a lofty—albeit chaotic—soul. Indeed, the sages teach us that he repented for his evil ways before Abraham's death.[11] The sum of Ishmael's age (137) and 676 (the ordinal number of the verse in which it appears) is 813, which is the sum of the numerical values of three important phrases in creation:

And God said, 'let there be light,' and there was light[12]	וַיֹּאמֶר אֱ־לֹהִים יְהִי אוֹר וַיְהִי אוֹר
And God separated between the light and the darkness[13]	וַיַּבְדֵּל אֱ־לֹהִים בֵּין הָאוֹר וּבֵין הַחֹשֶׁךְ
And God said, 'let us make man'[14]	וַיֹּאמֶר אֱ־לֹהִים נַעֲשֶׂה אָדָם

These three verses relate to the three stages of rectification that are enumerated in the teachings of the Ba'al Shem Tov: submission, separation and sweetening.[15] The first verse represents the first stage of total submission to God's commandments (light). The second verse clearly involves the stage of separation. Finally, the stage of sweetening is the creation of man, who is able to contain the chaotic lights in rectified vessels. This is achieved through the power of repentance, as indicated by the numerical value of "repentance" (תְּשׁוּבָה), which equals 713, the sum of "lights" (613, אוֹרוֹת, the number of commandments in the Torah) and "vessels" (כֵּלִים, 100, i.e., the inter-inclusion of all ten *sefirot* within each other).

The sum of these seven lifespans is 1,029. It is interesting that this number divides exactly by 7, the number of elements from which it is derived, yielding an average of 147 years—the length of Jacob's lifespan. He represents beauty and compassion, the central qualities of the heart. But, not only is 1,029 divisible by 7, it equals 3 times 7^3. This surprising equality emphasizes the interactions between these lifespans and their inherent inter-inclusion. The number 137, appearing twice, plays a significant role in arriving at the average value of all seven numbers.

Sarah's Lifespan

In order to better understand the unique meaning of 137 as the ideal lifespan, let's take a look at the lifespan of the Matriarch Sarah. Sarah's is the first lifespan of a Jewish soul mentioned in the Torah. She is also the only woman in the Torah whose lifespan is stated explicitly.[18] As Abraham's wife, Sarah is the mother of all of the generations that follow. In this context, God commanded Abraham, "Listen to her voice"[20] i.e., heed her advice to banish Ishmael from their household because of his chaotic behavior. Considering Sarah in this way, it is natural to place her above all of the seven men in the chart, in the place of *binah* (the *sefirah* of understanding), the mother figure.

Sarah passed away at the age of 127.[21]◄ Rashi explains that, at 100, Sarah was as pure of sin as she was at 20 and as beautiful as at the age of 7. Kabbalah and Chassidut teach us that this number of years is the optimal lifespan for a woman because 100 (10^2) years corresponds to the perfected inter-inclusion of the 10 *sefirot* in the *keter* (super-rational crown of the soul).[22] At the level below *keter*, the two *sefirot* of the intellect, *chochmah* and *binah*, are multiplied by 10, yielding the second part of her age, 20 years, but *da'at* is not included. The seven lower *sefirot* appear without inter-inclusion, as the remaining 7 years.

In the feminine figure, *da'at* is present, but initially it is at a lower level than in the masculine.◄◄ as the verse in Proverbs states, "A woman of valor is the crown of her husband."[25]

► Miriam also lived to the age of 127 years. Although her age is not mentioned explicitly in the Torah, it can easily be calculated. The number 127 is further associated with Queen Esther, who ruled over 127 states.[19]

Adding Sarah's age to the sum of the ages of the seven men brings the total ages to 1,156, the perfect square of 34, which is the prime root of 137.

►► The first spark of feminine *da'at* is apparent in God's words to Abraham, "Listen to her [Sarah's] voice." "Her voice" (בְּקֹלָהּ) is a permutation of Kabbalah (קַבָּלָה) and has a numerical value of 137, i.e., 127 with the additional 10 interincluded *sefirot* of *da'at*. Eventually, feminine *da'at* will rise above that of her masculine counterpart. ◄

The Talmud states, "Women's *da'at* is easily upset" (נָשִׁים דַּעְתָּן קַלָּה עֲלֵיהֶן).[23] Alternatively, this can mean that their *da'at* is agile, as in "Be ... as agile as an eagle" (הֱוֵי ... קַל כַּנֶּשֶׁר).[24]

Lifespan	Biblical characters	Sefirot included in number	Other information
127	Sarah	*Keter* (100); *chochmah binah* (20); seven emotive *sefirot* (7)	associated with the feminine faculties
137	Ishmael, Levi, Amram	*Keter* (100); *chochmah binah da'at* (30); seven emotive *sefirot* (7)	most common explicit lifespan in the Torah▶
147	Jacob	*Keter* (100); *chochmah binah* and two inter-included halves of *da'at* (40); seven emotive *sefirot* (7)	average lifespan of the seven generations from Abraham to Moses

▶ The average of the three lifespans 127 (Sarah), 137 (Ishmael, Amram and Levi), and 147 (Jacob) is 137.

Jacob's Lifespan

On the one hand, a scientist who believes in God is a rare phenomenon. (Most scientists are skeptics, while others perceive religious faith as a crutch that we must hold on to until we discover a rational reason for why things are as they are.) On the other hand, a scientifically-minded man of religion is usually the exception rather than the rule.

In order to facilitate the union of Torah and science, the Torah must descend to discover the precision of science, and science must ascend to develop the aspect of faith that lies dormant in it. As mentioned in the Introduction, Torah reasoning reflects loving-kindness, and scientific reasoning is a manifestation of judgment. Both types of reasoning are properties of *da'at* (the *sefirah* of knowledge). Like all of the *sefirot*, the *partzuf* of *da'at* includes ten *sefirot*, however these are divided into two "crowns," which contain five loving-kindnesses and five judgments. [These two sets of five extend from the mind to permeate the first five emotive and behavioral characteristics of the heart (from *chesed* to *hod*). Each of these five characteristics becomes energized with the vitality necessary to express itself in reality by one "kindness" of *da'at* (positive, attractive force) and one "judgment" of *da'at* (negative, repellent force)[26]]. When each type of reasoning is inter-included with its counterpart, they double themselves. Thus, the two crowns of *da'at* become two complete faculties, each inter-included with ten *sefirot*. Jacob's faculty of *da'at* was rectified in this way to include both types of reasoning. His love for Rachel stemmed from the loving-kindnesses of *da'at*, while her sister Leah's sense that she was hated stemmed from the judgments of *da'at*. The inter-inclusion of these two aspects of *da'at* manifested when Jacob married Bilhah and Zilpah, Rachel and Leah's maidservants. Then his two wives became four, thus reflecting his perfected faculty of *da'at*. This offers one explanation why Jacob achieved a lifespan of 147 years and why the average value of all seven lifespans from Abraham to Moses is 147.▼

▶ In its fully articulated form, *binah* possesses two *partzufim*: the higher of these is *Ima Ila'ah* ("the Higher Mother"), the lower is *Tevunah* ("Comprehension"). These two *partzufim* are referred to jointly as *Ima* ("the Mother").

Leah and Rachel correspond to *tevunah* and to *malchut* (the *sefirah* of kingdom), respectively. The sum of the numerical values of *tevunah* (תְּבוּנָה) and *malchut* (מַלְכוּת) is 959, which equals 7 times 137. The combination of these two alludes to the Kabbalistic secret of the "greater Rachel"; i.e., the inclusion of Rachel's older sister, Leah, within Rachel. This union first occurred with Rachel's passing, as the Torah verse states, "And it came to pass when her soul left her" (וַיְהִי בְּצֵאת נַפְשָׁהּ).[27] The numerical value of this phrase is also 959.

During their lives, Leah and Rachel became mothers of eleven sons outside of the Land of Israel. The numerical value of their names; Reuben (רְאוּבֵן), Simeon (שִׁמְעוֹן), Levi (לֵוִי), Judah (יְהוּדָה), Dan (דָּן), Naftali (נַפְתָּלִי), Gad (גָּד), Asher (אָשֵׁר), Issachar (יִשָּׂשכָר), Zebulun (זְבוּלֻן), Joseph (יוֹסֵף) is 3,014, which equals 11 times 274 [the sum of the numerical values of Leah (לֵאָה) and Rachel (רָחֵל)], or 22 times 137.

A verse in Isaiah states, "The beauty of man dwells at home."[28] The Midrash[29] explains that this verse refers to Eve and relates to *tiferet* (the *sefirah* of beauty) and to feminine beauty in particular.

Hod (the *sefirah* of thanksgiving) is another inherently feminine *sefirah*.

The literal meaning of *hod* (הוֹד) is "splendor." It often accompanies expressions of royalty; i.e., the feminine *malchut* (the *sefirah* of kingdom).

These two *sefirot* (*tiferet* and *hod*) are two of eight synonyms for "beauty" in the Hebrew Bible. In particular, they refer to beauty associated with womanhood.[30] The sum of the numerical values of *tiferet* (תִּפְאֶרֶת) and *hod* (הוֹד) is 1,096, the numerical value of the phrase "an intelligent woman" (אִשָּׁה מַשְׂכָּלֶת), which will be discussed in more detail in Chapter 8. Like the words of this phrase, the names of these two *sefirot* also contain 8 letters, thus the average numerical value of each letter is 137.

The innate intelligence of a woman is her awe/fear of God, as the verse in Proverbs states, "Charm is false and beauty is futile; a God-fearing woman is to be praised."[31] At first glance, this verse invalidates all reference to beauty in a woman. However, as stated in another verse from Proverbs, "A woman of charm supports honor,"[32] meaning that a God-fearing woman is to be praised, not only for her fear of God, but also for her charm and beauty.[33] The numerical value of, "she is a God-fearing woman" (אִשָּׁה יִרְאַת הוי' הִיא) is 959 (the sum of *tevunah* and *malchut*, as mentioned), which equals 7 times 137. The difference between the values of "intelligent" (מַשְׂכָּלֶת) and "She is God-fearing" (יִרְאַת הוי' הִיא) is 137. Their sum is 1,443, which equals 39 [the numerical value of "God is one" (הוי' אֶחָד)] times 37, the companion of 137.

More About Lifespans and 137

> *There were ten generations from Adam to Noah. This is to teach us the extent of God's tolerance; for all these generations angered Him, until He brought upon them the waters of the Flood. There were ten generations from Noah to Abraham. This is to teach us the extent of God's tolerance, for all these generations angered Him, until Abraham came and reaped the reward for them all.*[34]

The above quotation from the *Ethics of the Fathers* points to three significant milestones of humanity: Adam, Noah and Abraham. Adam[35] lived to the age of 930; Noah[36] lived 950 years and Abraham[37] lived 175 years.▸ The sum of these three ages is 2,055, which equals 15 times 137. The average lifespan of these three is thus 685, which equals 5 times 137.▸▸

Abraham lived to 175, surpassing 137 by 38 years.▸▸▸ In a certain respect, his life reached a pinnacle when he withstood the tenth and final trial, when he was asked to bind Isaac as a sacrifice to God. The sages teach us that Sarah died as a result of being told of this test,[43] and we know that she lived to the age of 127. We also know that, when Isaac was born, Abraham was 100 years old[44] and Sarah was 90, meaning that Abraham was 10 years older than Sarah.[45] How old then was Abraham at the Binding of Isaac? He was 137 years old and Isaac was 37 years old. Accordingly, the Zohar[46] teaches us that 37 is the numerical value of the word, "and there were" (וַיִּהְיוּ), which begins the Torah verse relating Sarah's lifespan. This alludes to the best days of her life, from the day she gave birth to Isaac when she was 90 until she passed away at the age of 127, when Isaac was 37 years old.

Action from a Distance

Isaac lived for 180 years, which was the maximum lifespan of the

▸ Adam should have died on the day he ate from the Tree of Knowledge.[38] However, God granted him one of His days of 1,000 years. Adam donated 70 years of his life to King David, who he saw was destined to die at birth.[39] Thus, Adam lived only 930 years.

The Torah summarizes Adam's life with two verses that begin, "And Adam lived..." (וַיְחִי אָדָם)[40] and conclude, "and he died" (וַיָּמֹת).[41] In these two verses there are 37 words and 137 letters. The numerical value of "And [Adam] lived" (וַיְחִי) is 34. We thus see here a combination of 37 (the companion of 137), 137 itself and 34, (the prime "root" of 137), all in the context of the lifespan of the first man.

▸▸ The average lifespan of Adam, Abraham and Noah is 685, the 19th inspirational (interface) number, represented algebraically by the formula $n^2 + (n - 1)^2$. In this case, 685 equals $19^2 + 18^2$. 19 is the numerical value of Eve (חַוָּה), who is called "the mother of all life"[42] and 18 is the numerical value of "life" (חַי). The sum of these two numbers is 37, the companion number of 137.

▸▸▸ Each of the four basic words in Hebrew that have a numerical value of 137 (קַבָּלָה, אוֹפָן, מוֹצָא, מְצֻבָּה) has an ordinal value of 38. The sum of the normative value (137) and the ordinal value (38) of each word is thus 175, which is the lifespan of Abraham.

Patriarchs. Let's take a look at what happened when he was 137 years old.

Isaac married Rebecca when he was 40 years old, and she bore his twin sons, Jacob and Esau, when he was 60.[47] When Isaac blessed Jacob, Jacob was 63 years old. Isaac was then 123 years old. Immediately after receiving the blessing, at his parents' command, Jacob left home to find his soul-mate. For the next 14 years, he studied Torah in the yeshivah of Ever.[48] This brings Jacob's age to 77 years old when he reached Charan and saw Rachel for the first time. At that very time, Isaac was 137 years old.

By delving into the story of Naomi and Ruth (as told in the Book of Ruth), we can conjecture that Isaac was involved from afar in Jacob's falling in love with Rachel at first sight.

Ruth, a Moabite woman, pledged her faithfulness to Naomi and the Jewish people, and their souls became eternally entangled. Naomi and Ruth became inseparable and to a certain extent, they are indistinguishable from one another. Naomi sent Ruth to Boaz, giving her explicit directives how to behave. In the verses that describe Ruth's actions, all of the verbs suggest that Naomi was acting through Ruth, who followed her directives exactly. Like a parent overseeing their child's behavior, Naomi oversaw Ruth's actions and lived them out through Ruth. In the same way, Isaac directed Jacob where to go and who to marry. Then it was as if the soul of Isaac at the age of 137, was living out Jacob's actions and emotions as he met his soulmate.

We have seen that the age of 137 is explicitly significant in the three Biblical personalities of Ishmael, Levi and Amram, and that it is implicitly significant for Abraham and Isaac. The last of the first seven generations charted earlier is Moses, who only lived to the age of 120. Is the number 137 somehow implicit in his lifespan?

God punished Moses for hitting the rock instead of speaking to it.[49] Rashi explains that had he not hit the rock, his time would not have come to die. From this we learn that Moses was destined to live longer than his lifespan of 120 years.

There are good reasons why we can postulate that Moses should have lived another seventeen years, bringing his lifespan to a total of 137.

- First, every child should expect to live to the age of their parents.[50] For this reason, it is customary to make a special thanksgiving meal when one passes the age of one's parents. Thus, Moses could expect to live to the age of his father, Amram, who lived to the age of 137.

- The second reason is far more profound. In the chart, Moses appears below his father, Amram, who is directly beneath Jacob in the same column. In Kabbalah we find that, "Jacob is from without and Moses from within,"[51] meaning that Moses reflects Jacob's inner soul. At first glance, we might infer from here that Moses' lifespan should be equated with Jacob's 147 years, and Moses should have lived another 27 years. But, at a deeper level, the lives of Jacob and Moses parallel and complement one another. Jacob lived 110 years in the Land of Israel,[52] 20 years in Charan and another 17 years in Egypt. In contrast, Moses spent the entire 120 years of his life outside of the Holy Land, in Egypt and in the desert. His life's goal was to bring the entire Jewish People to the Promised Land. This suggests that Moses should complement Jacob's 17 years of living in Egypt by living in the Land of Israel for the final 17 years of his life. Had Moses lived for another 17 years, he would have been 137 years old at his passing.

- Another allusion to Moses' inherent connection with the number 17 is his relationship with goodness. At his birth, his mother saw "that he is good."[53] One interpretation of this verse is that she called him Good (טוב),[54] which has a numerical value of 17. The Talmud[55] states, "Good shall come and receive goodness from the Good, for those who are good." This riddle refers to Moses, who received the Torah (which is called "good teachings") from God, who is the source of all good, "for those who are good."▾

(Curiosity Killed) Schrödinger's Cat

The theory of quantum mechanics contains many aspects that defy human logic. At the sub-atomic level, these phenomena suggest that the determining factor in quantum experiments is the choice of a conscious observer. However, quantum effects are being demonstrated experimentally in larger and larger objects. In principle, there is no limit to the size of objects that can display enigmatic quantum phenomena. This is true for atoms, molecules, viruses and even animals, as Rosenblum and Kuttner have observed:

> *As the experimental objects become more macroscopic, the bafflement arising directly from experiment becomes more compelling. We can no longer relegate paradoxical quantum phenomena to objects so small as to be considered mere "models," qualitatively different from objective physical entities like chairs and cats. Physics is increasingly pressed to confront the issue of observation, or at least acknowledge it as a hint that we have not yet told the whole story. If*

▸ When filled with its letters, the numerical value of "that he is good" (כף יוד טית וו בית הה וו אלף) is 1,096, which equals 8 times 137. Thus, the average numerical value of each of the 8 letters of "for he is good" (כִּי טוֹב הוּא) is 137. The first appearance of the phrase "that it is good" (כִּי טוֹב) is in God's original assessment that light is good. The numerical value this phrase when filled (כף יוד טית וו בית) is 963, which is the numerical value of "and God saw the light that it was good"[56] (וַיַּרְא אֱלֹהִים אֶת הָאוֹר כִּי טוֹב).

Another verse that includes the phrase "that it is good" (כִּי טוֹב) is "taste and see that God is good"[57] (טַעֲמוּ וּרְאוּ כִּי טוֹב הוי). The numerical value of this phrase is 411, the same as the numerical value of "chaos" (תהו), which equals 3 times 137.

There are 17 letters in this phrase, relating to the numerical value of "good" (טוב), which is 17. Dividing the phrase into three groups with 9, 6 and 2 letters according to the numerical value of each of the three letters of "good" (טוב), we find that the numerical values of the three

groups are 358 (טַעֲמוּ וּרְאוּ כ), 42 (י טוֹב יה) and 11 (וה), respectively. 358 is the numerical value of Mashiach (מָשִׁיחַ); the group of six letters (י טוֹב יה) spells "Let there be... good" (יְהִי...טוֹב), which are the key words in creation, as explained in Kabbalah with reference to the 42-lettered Name of God, with which He created the world. The sum of these two groups is 400, the numerical value of the letter *tav* (ת), the first letter of "chaos" (תהו). Finally, the last two letters of the phrase (וה) are the last two letters of "chaos" (הו), thus completing the word.

Let's meditate further on these three numbers by continuing the mathematical series that they form.

$$11 \quad 42 \quad 358 \quad 959$$
$$31 \quad 316 \quad 601$$
$$285 \quad 285$$

The fourth number in the series is 959, which equals 7 times 137. Together, the sum of the first four numbers is 1,370, which equals 10 times 137.

quantum mechanics is completely correct, the notion that we make free choices and the notion that a physical world exists independent of those choices confront each other in the quantum experiment to produce a measurement problem intimately involving the observer that arises independently of the quantum theory.[58]

In view of the bizarre ramifications of his own calculations, early quantum physicist Erwin Schrödinger believed that quantum mechanics would eventually be proven incorrect. Einstein and other critics of quantum theory agreed that the whole concept was completely absurd. Einstein labeled another strange phenomenon of the quantum theory, now called quantum entanglement, "spooky action at a distance." He didn't believe it could be true. In order to illustrate the absurdity of the theory, Schrödinger developed what has become one of the most famous quantum thought experiments.

This imaginary experiment involves measuring the lifespan of a subatomic particle by placing it in a closed box together with a live cat (since the cat is also composed of atoms etc., it must also experience the effects of the quantum world). If the particle decays, it sets off a mechanism that shatters a flask of poison, which instantly causes the cat to die. The only way that one can discover what has happened to the particle is by opening the box to see whether the cat is alive or dead. As long as the observer has not opened the box to see how the cat has fared, he cannot know if the particle has decayed or not. The experiment yields the paradoxical assumption that before opening the box, the particle is in superposition and, therefore, the cat is both alive and dead simultaneously. Analysis of this assumption with simple logic indicates that the whole theory of quantum mechanics must be incorrect. However, for the past century, the theory has been put to the test by rigorous laboratory experimentation (albeit without an actual cat), and no prediction made by quantum theory has ever been proven wrong. In fact, every experiment to date has proven unequivocally that the paradox always holds true.

Schrödinger's Cat reveals that the observer's freedom of choice (whether to open the box or not), and subsequently seeing the cat in the open box is the factor that determines the objective fate of the cat (whether the cat is alive or dead).

Schrödinger's cat experiment involves the lifespan of an elementary particle (or the lifespan of the accompanying cat). A similar thought experiment named "quantum suicide," involves a scientist setting up a gun that will shoot him depending on the spin of a particle. In this thought experiment, the scientist himself replaces the cat in Schrödinger's box and records his own observations. This version is used to describe the effects of the many-worlds theory of quantum mechanics (see Chapter 6), which holds that after the experiment has been carried out once, a superposition of the experimenter exists and he is both alive and dead at one and the same time.

The Torah upholds the fact that conscious observation determines one's fate. This is clearly stated in Deuteronomy, "See, life and death I have placed before you … choose life."[59] If we see life and choose life, then we will certainly be alive. When a good person dies, his soul lives on. By contrast, an evil individual may be alive, but, spiritually, he has already died.[60] So, ultimately, if you choose good, you are alive, but if you choose evil you are dead. Until you make your choice, you are both alive and dead, simultaneously! ◄

In the *Ethics of the Fathers*, we read, "At the age of 100, it is as if he has died and has passed on and is nullified from this world."[63] How much more so is this true at the age of 137! Yet, Abraham reached the apex of his life at 137 and continued to live. Significantly, Ishmael, his son, who did not live a righteous life, died when he reached the age of 137.

The lifetime of a charged particle depends on the probability that it will decay, which is proportional to the fine-structure constant. Thus, both physically and spiritually, 137 controls the life of Schrödinger's Cat.

► There is one verse in the Hebrew Bible where the idiom, "living man," (אָדָם חַי; 63)[61] appears. Similarly, there is one verse where the idiom, "dead man" (מֵת אָדָם; 485)[62] appears. The sum of these four words is 548, or 4 times 137, i.e., the average value of each of the four words is exactly 137.

In the Shadow of Life

There is a verse in Ecclesiastes[64] that concisely summarizes these ideas:

For who knows what is good for man in life, in the numbered days of his life of folly, which he lives out like a shadow, for who can tell a man what will be after him under the sun?	כִּי מִי יוֹדֵעַ מַה טּוֹב לָאָדָם בַּחַיִּים מִסְפַּר יְמֵי חַיֵּי הֶבְלוֹ וְיַעֲשֵׂם כַּצֵּל אֲשֶׁר מִי יַגִּיד לָאָדָם מַה יִּהְיֶה אַחֲרָיו תַּחַת הַשָּׁמֶשׁ:

Let us examine the wording of this verse:

♦ "For who knows what is good for man in life?" Who can claim to know which choices one makes are good? [Note that the numerical value of "what is good for man" (מה טוב לָאָדָם) is 137.]

♦ "Like a shadow" (כַּצֵּל) appears numerous times in the Hebrew Bible, generally relating to the human lifespan.▸ As such, it does not refer to the shadow of a fixed object such as a tree or a wall, but to the fleeting shadow of a bird flying overhead.[65]

♦ "Who will tell a man what will be after him under the sun?" Who knows how long anyone will live? For practical purposes, the manifestation of a lengthy lifespan is old age, which begins at 60, as mentioned. This is also the age when one becomes wise, as Rashi states, "An elder is one who has acquired wisdom."[66] It is then that one is able to correctly infer the outcome of events that take place, as we read in the *Ethics of the Fathers*, "Who is wise? One who perceives the outcome [lit.: 'that which will be born']."[67] We might say that the ability to perceive the outcome of an event relates to conducting a thought experiment that results in a paradox.

From 60 to 137 remain another 77 years.[68] The numerical value of "mighty" (עז) is 77.▸▸ "Mighty" (עז) implies the ability to simultaneously bear two opposite states, i.e., to be in a state of superposition.

▸ The numerical value of "shadow" (צֵל) alone (without the prefix) is 120, the lifespan of Moses, which is accepted as the epitome of longevity.

▸▸ This suggests that living from 60 to 137 years old entails being "bold" (עז) as we learn from, "Be as bold as a tiger... to do the will of your Father in Heaven."[69]

▶ 137 is either 60 plus 77 [the numerical value of *mazal* (מַזָּל)], or it is 2 times 60 (Moses' lifespan of 120) plus 17, the numerical value of "good" (טוֹב). Together, these two additions (17 plus 77) form the common blessing of congratulations, "*mazal tov*!" (מַזָּל טוֹב). Isaac's lifespan of 180 years plus *mazal tov* (94) equals 274, which equals 2 times 137.

Its numerical value also alludes to the revelation of one's "fortune" or *mazal* (מַזָּל), the paradoxical, super-conscious source of the soul. We shall discuss this division into 60 and 77 further in Chapter 15 with reference to the Biblical expression, "a mighty instrument" (כְּלִי עֹז).[70]◀

REFERENCE NOTES FOR CHAPTER 7

1. See also, *The Twinkle in Your Eye*, pp. 109-110.

2. Genesis 25:17.

3. Exodus 6:16.

4. Ibid 6:20.

5. Genesis 16:12; see also Rashi on Genesis 21:9.

6. Ibid 21:11.

7. Ibid, v. 12.

8. Ibid 25:17.

9. Exodus 6:16; 6:20.

10. See our books in Hebrew, *Mudaut Tiveet* p. 72-73 and *Teshuvat Hashanah* p. 289-295.

11. Rashi on Genesis 25:9. Ishmael's offspring, who were born before his repentance, do not have the same merit.

12. Genesis 1:3.

13. Ibid 1:4.

14. Ibid 1:26.

15. See our book, *From Darkness to Light*.

16. *Vayikra Rabah, Parashat Emor* 29.

17. See also our book in Hebrew, *Einayich Breichot Becheshbon* pp. 131-133.

18. See *Zohar Bereishit* 122a.

19. *Esther* 1:1.

20. Genesis 21:12.

21. Genesis 23:1.

22. *Likutei Torah, Vezot Haberachah* 93.

23. *Shabbat* 33a.

24. *Avot* 5:20.

25. Proverbs 12:4. For more on how female consciousness rises above male consciousness, see *The Twinkle in Your Eye*, Chapter 1, esp. in endnotes.

26. See our book, *Body, Mind and Soul*, p. 41 ff.

27. Genesis 35:18.

28. Isaiah 44:13.

29. *Bereishit Rabah* 21:2.

30. See *Ketubot* 59b.

31. Proverbs 31:30.

32. Ibid 11:16. The numerical value of this phrase is 1,651, which equals 13 times 127, the lifespan of Sarah.

33. See *Chidushei Maharam Shif* on *Ketubot* 17a; *Avodat Yisrael, Parashat Vayetze*. See also our book in Hebrew, *Yayin Mesameiach*, Vol. 1, pp. 22, 28.

34. *Avot* 5:2.

35. Genesis 5:5.

36. Ibid 29:9.

37. Ibid 25:7.

38. *Shemot Rabah* 22.

39. *Zohar Shemot* 235a.

40. Genesis 5:3.

41. Ibid 5:5.

42. Ibid 3:20.

43. Rashi on Genesis 23:2.

44. Genesis 21:5.

45. Ibid 17:17.

46. *Zohar Bereishit* 123a.

47. Genesis 25:26.
48. Rashi on Genesis 25:17.
49. Numbers 20:12.
50. See Rashi on Genesis 27:2.
51. See *Tikunei Zohar, Tikun* 13 29a.
52. Genesis 47:9.
53. Exodus 2:2.
54. *Sotah* 12a; *Shemot Rabah* 1:16.
55. Ibid 1:20.
56. Genesis 1:4.
57. Psalms 34:9.
58. See *The Observer in the Quantum Experiment* by Bruce Rosenblum and Fred Kuttner. Or in the book by the same authors, *Quantum Enigma*.
59. Deuteronomy 30:19.
60. *Berachot* 18a, b.
61. Lamentations 3:39.
62. Ezekiel 44:25.
63. *Avot* 5:21.
64. Ecclesiastes 6:12.
65. Ibid, Rashi ad loc.
66. *Midrash Agadah Bereishit* 24:1.
67. *Avot* 4:1.
68. See also, *The Twinkle in Your Eye,* Chapter Thirteen.
69. *Avot* 5:20.
70. II Chronicles 30:21

THE ROLE OF WOMAN

A Feminine Number

There is one combination of three letters in the Hebrew alphabet that, when spelled in full, has a numerical value of 137. These are the three letters *alef*, *hei* and *yud* (אלף הא יוד).▸

Each of the three letters *alef* (א), *hei* (ה) and *yud* (י)▸▸ appears for the first time in the Torah in the very first phrase of creation, "In the beginning God created" (בְּרֵאשִׁית בָּרָא אֱ-לֹהִים). Their order there spells out the word "where" (אַיֵּה), which appears in the rhetorical question asked by the angels (recited in Shabbat liturgy)[1], "Where is the place of His glory?" (אַיֵּה מְקוֹם כְּבוֹדוֹ).▸▸▸

The next consecutive appearance of these letters in the Torah is in the phrase, "'Let there be light!' and there was light" (יְהִי אוֹר וַיְהִי אוֹר.). Here they appear twice and each time spell out the word "she" (הִיא). In both these cases, the letters appear without any other letters in between them. The letter order of "she" (הִיא) is the reverse of the original order "where" (אַיֵּה). "She" is a reflection, or the response to the original question of "where," i.e., God's glory resides in the Divine feminine figure.▸▸▸▸

As mentioned above, one method of calculating the numerical value of a letter is by "impregnating" it, i.e., filling it with the letters that spell its Hebrew name. The letters *alef*, *hei* or *yud* are interchangeable as the pregnant letters of three particular letters [*hei* (ה), *vav* (ו) and

▸ The three letters *alef*, *yud*, and *hei* are the 3rd, 5th, and 12th letters of the Torah, respectively. We can continue this sequence in a quadratic series:

3	**5**	**12**	24	41	63	90
	2	**7**	12	17	22	27
		5	5	5	5	5

The sum of the first 7 numbers is 238, the numerical value of Rachel (רָחֵל).

The first three letters of *Elokim* (אֱ-לֹהִים) are a permutation of "Leah" (לֵאָה), Rachel's older sister. The phrase can therefore be read, "In the beginning, God created Leah" (בְּרֵאשִׁית בָּרָא לֵאָה). The average numerical value of Rachel (רָחֵל) and Leah (לֵאָה) is 137.

▸▸ In Chapter 1, we saw that the filled numerical value of the phrase "'Let there be light!' and there was light" is 1,370. The filled value of "she" (הִיא), which appears twice in this phrase, is 137. The numerical value of the remaining letters in the phrase is thus 1,096, which equals 8 times 137, the value of "an intelligent woman" (אִשָּׁה מַשְׂכֶּלֶת).[2]

▸▸▸ The numerical value of the "back" of "where" (א אי איה) is 28, the triangle of 7. The numerical value of the "back" of "she" (ה הי היא) is 36, the triangle of 8. The mathematical rule is that the sum of two consecutive triangles is the perfect square of the larger number. In this case, 28 + 36 = 64, which equals 8². 64 also equals 4 times 16, the original numerical value of both words.

▸▸▸▸ Alef, *hei* and *yud* are three of the four Hebrew letters that can serve as either a vowel or a consonant (אהוי). The fourth letter, *vav* (ו) is discussed in detail in Chapter 9. The number of consonants in the *alef-bet* is 22, adding these four letters in their vowel form brings the total number of Hebrew letters to 26, the numerical value of the essential Name of God, *Havayah*.

▶ The numerical value of "parallel" (מַקְבִּילֹת), is 582. The average numerical value of the six letters is 97, which is the numerical value of Meheitavel (מְהֵיטַבְאֵל), the archetype of rectified femininity in the Torah (see Chapter 1). The four letters *mem* (מ), *yud* (י), *bet* (ב) and *lamed* (ל) are common to both words.

▶▶▶ Another word with the same letters is the word, "in the congregation" (בְּקָהָל). This word appears exactly 23 times in the Bible, which is the numerical value of the name Chayah (חַיָה), the rectified name of Eve, the first woman. The six appearances of "in the congregation" (בְּקָהָל) in the Torah all appear in conjunction with God's Name, as the phrase "in the congregation of *Havayah*" (בִּקְהַל הוי').[6] This phrase appears a seventh time in Micah.[7]

This teaches us that the 137 principle is significantly related to the secret of God's Name, *Havayah*.

pei (פ)], which affects the numerical values of their Hebrew names when spelled in full. This quality of "pregnancy" further alludes to the innate femininity of the three letters *alef*, *hei* and *yud*.

Our meditation on the number 137 began by identifying it as the numerical value of Kabbalah (קַבָּלָה). In the Bible, the root of *Kabbalah* (ק-ב-ל) is seen in words meaning "parallel" or "corresponding." ◀ Kabbalah deals with correspondences and associations between different concepts. A secondary meaning of Kabbalah is "reception," from the root "receive" (ק-ב-ל). Indeed, in order to receive something, the receiver must stand parallel to the giver. Reception is an innately feminine quality. In the Torah, this root appears twice, both times in the feminine plural (מַקְבִּילֹת). The first appearance is in the phrase, "parallel were the loops [like] woman to her sister"[3] and the second time, "parallel were the loops one to the other."[4] ▼▼

We continued to analyze the number 137 according to the *sefirot*, showing how the deciding factor is *da'at*. The first appearance of the letters of Kabbalah in the Torah as a single word (albeit in a different order, but retaining their numerical value) is in God's words to Abraham, "[All that Sarah tells you, listen] to her voice" (בְּקֹלָה).[5] ◀◀

Sarah's lifespan of 127 alludes to the *sefirot* without including *da'at*, Yet, God refers to her opinion with a unique expression that equals 137. This indicates that the rectification of the *sefirah* of *da'at* in the female figure was initiated in Sarah, the first Jewish woman.[8] Indeed, although Abraham was a prophet in his own right, Rashi[9] teaches that Sarah was greater than Abraham in prophecy.

▶▶ The sum of the two phrases "parallel were the loops [like] woman to her sister" (מַקְבִּילֹת הַלֻּלָאֹת אִשָּׁה אֶל אֲחֹתָהּ) and "parallel were the loops one to the other," (מַקְבִּילֹת הַלֻּלָאֹת אַחַת אֶל אֶחָת) is 3,696, which equals 7 times 528, which is the sum of all four fillings (without the root letters) of the Goodly Name, *Akvah* (אהויה), associated with *da'at* (the *sefirah* of knowledge); see Chapter 1. The numerical value of one of the fillings equals exactly 137, the numerical value of another equals 119. In both of these fillings, the vav is spelled *vav-alef-vav*. As we shall discuss in depth later (Chapter 9), this spelling is most significant in our quest for *alpha*, the electromagnetic coupling

constant. The sum of these two fillings is 256, the square of 16 [the numerical value of "she" (הִיא)]. This closely relates to Eddington's caculations, as discussed above. The two remaining fillings of the Goodly Name (without the root letters) are 126 and 146, which total 272, the diamond form of 16.

Fillings of Akvah	Numerical value	Filling alone	Numerical value of filling alone	Corresponding filling of Havayah
אלף הי ויו הי	163	לף יויו	146	Ab
אלף הי ואו הי	154	לף יאו י	137	Sag
אלף הא ואו הא	136	לף א או א	119	Mah
אלף ההה וו הה	143	לף ה וה הה	126	Ban

Jacob's two wives, Rachel and Leah, were sisters. Kabbalah teaches us that they are the archetypal pair of women in the Torah. The numerical value of Leah (לֵאָה) is 36, and Rachel (רָחֵל) is 238. The sum of the numerical values of their two names is thus 274, and the average value of their two names is exactly 137.▼ They correspond to the above-mentioned two appearances of the word "she" (הִיא) in the verse referring to the creation of light.

Coupling Heaven and Earth

In the Torah in general, the feminine principle corresponds to speech, language and all of their components, including words and letters. The Talmud teaches that women are more attuned to speech than men, "Ten measures of speech descended to earth: women took nine and the rest of the world took one."[16] In Kabbalah, Leah represents speech in thought, while Rachel represents articulated speech. Speech in thought relates to the heavens and the hidden aspect of creation. Spoken words correspond to the earth and the revealed aspects of creation.

The Ba'al Shem Tov taught that when praying, studying Torah, and

▶ The difference (the "wing") between the values of their two names and 137 is 101. As mentioned above with reference to the source of Divine blessing, 101 is the numerical value of "from nothing" (מֵאַיִן). We can continue adding 101 to create a linear series that begins with these three numbers (36, 137, 238).

36 137 238 339 440 541
101 101 101 101 101

The fifth number in the series (440) is the numerical value of "sincere" (תָּם), which the Torah uses to describe Jacob in his youth, "And Jacob was a sincere man, who dwelt in tents."[10] Jacob's "tents" refer prophetically to Leah's and Rachel's tents.[11] The sixth number of the series is 541, the numerical value of Jacob's alternate name, Israel (יִשְׂרָאֵל).

The sum of the first five numbers of this series is 1,190, which equals 5 times 238 (the average of the five numbers), the numerical value of Rachel (רָחֵל), who is considered the mainstay of Jacob's home. 1,190 is also the diamond form[12] of 34, the prime "root" of 137.

The feminine *malchut* (the *sefirah* of kingdom) is often represented by Rachel. The ultimate revelation of this *sefirah* is Mashiach. One verse that explicitly refers to Mashiach in the Bible is, "Behold my servant will become wise, he will rise and be elevated and very high" (הִנֵּה יַשְׂכִּיל עַבְדִּי יָרוּם וְנִשָּׂא וְגָבַהּ מְאֹד).[13] The numerical value of this verse is also 1,190, 5 times the numerical value of Rachel (רָחֵל). The connection between Rachel and Mashiach is also alluded to in the phrase, "and the spirit of God hovers over the face of the water" (וְרוּחַ אֱ-לֹהִים מְרַחֶפֶת עַל פְּנֵי הַמָּיִם).[14] The Zohar[15] states that this refers to the spirit of Mashiach. The second, fourth and sixth letters from the beginning of the phrase spell Rachel (רָחֵל). The numerical value of the alternate words in the verse that refers to Mashiach (הִנֵּה עַבְדִּי וְנִשָּׂא מְאֹד) is 548, which equals 4 times 137, i.e., the average numerical value of each word is 137. The sum of the numerical values of the three intermediate words (הִנֵּה יַשְׂכִּיל יָרוּם וְגָבַהּ) is 642. The average numerical value of the three words is thus 214, the numerical value of "spirit" (רוּחַ), referring to the above-mentioned phrase, "and the spirit of God hovers over the face of the water," etc.

▸ The Ba'al Shem Tov's wife was called Leah Rachel (לֵאָה רָחֵל). The numerical value of her name is thus 2 times 137, as we have seen. The Ba'al Shem Tov's own given name is Israel, the name God gave to Jacob, the husband of Leah and Rachel, as we saw above. The Ba'al Shem Tov had two children, a daughter named Odel, and a son named Tzvi. The sum of the numerical values of Odel (35 ;אָדְל) and Tzvi (102 ;צְבִי) is also 137.

The numerical value of the initial letters of the names of the Ba'al Shem Tov's children (צא) is 91. The remaining letters have a numerical value of 46, which is the midpoint of 91. This is a significant ratio that is discussed in detail with reference to the word "origin" (מוֹצָא); see Chapter 10.

even when speaking with another individual, we should unite these two aspects of speech. The Ba'al Shem Tov's name literally means, "Master of the Goodly Name." This refers to the Divine Name, *Akvah* (אהוה).[17]◂ As we saw in Chapter 1, this Divine Name represents the primordial light of creation, which God saw was good, but then concealed. The light that God articulated into existence will be revealed to the righteous in the future.[18]

Like the concealed primordial light that it represents, this Divine Name does not appear explicitly in the Hebrew Bible. Its four letters are encoded in the initials of many Biblical phrases. The first and most fundamental instance is in the last four words of the opening verse of the Torah, "[In the beginning, God created] the heavens and the earth" (אֵת הַשָּׁמַיִם וְאֵת הָאָרֶץ).[19] God's purpose in creation is to manifest His infinite goodness throughout reality by unifying "the heavens" and "the earth" i.e., the spiritual and the physical realms.

The feminine nature of 137 relates to the union of Torah and science, which can be achieved if science, the feminine aspect of God's wisdom, stands opposite the Torah to receive and reflect it.

Two Eves

Leah and Rachel represent the two poles of womanhood.[20] This dichotomy has its foundations in the creation of the first woman, Eve. Although it may not be apparent from the explicit reading of the text, the sages teach us that a "first Eve" preceded Eve as we know her.[21] The first Eve either fled,[22] or "returned to the dust."[23] Her disappearance occurred either before or after the primordial sin. Whichever the case may be, a second Eve was created. The Arizal explains that these two Eves are the origin of Leah and Rachel, Jacob's two wives. Jacob is said to have resembled Adam.[24] Similarly, his two wives correspond to Adam's two Eves.

This concept becomes more significant when we consider the ordinal value of *Kabbalah* (קַבָּלָה). We first note that the letter *kuf* (ק)

has an ordinal value of 19, which is the numerical value of "Eve" (חַוָּה) by both the regular method of calculation and the ordinal method. The letters of *Kabbalah* that remain are *bet* (ב), *lamed* (ל) and *hei* (ה), which have ordinal values of 2, 12 and 5, respectively. The sum of these three numbers is also 19. The total ordinal value of "Kabbalah" (קַבָּלָה) is therefore 38, or 2 times "Eve" (חַוָּה). This reinforces our understanding of Kabbalah as a feminine concept that relates to the two aspects of Eve, and to their subsequent reincarnations as Rachel (רָחֵל) and Leah (לֵאָה), whose average value is 137, the regular numerical value of Kabbalah (קַבָּלָה), as mentioned previously.

The fact that there were two Eves is reflected in the Torah. Although Adam's name appears many times, the name Eve only appears twice. The first appearance of her name is after the primordial sin, when Adam named her "mother of all life."[25] The second time her name appears is several verses later, "and Adam knew his wife Eve, and she gave birth to Cain … and Abel."[26] When she had her third child, Seth, her name is not mentioned (she is just referred to as Adam's wife). There are exactly 256 letters between the first appearance of Eve's name (corresponding to Leah) and the second appearance of her name (corresponding to Rachel).[27]

Since 256 is equal to 16^2, we can write all the letters in the form of a 16 x 16 matrix:

▸ The word preceding the first appearance of Eve's name is "his wife" (אִשְׁתּוֹ) and the word following the second appearance is also "his wife" (אִשְׁתּוֹ). The numerical value of this word is 707, which equals 7 times 101. 101 is the numerical value of "from nothing" (מֵאַיִן), which is the "wing" of Rachel and Leah (רָחֵל לֵאָה), as mentioned.

```
ח ו ה א ה ו ה כ י ת ה א מ כ ל
ח ו י ע י ש ה ו ה א ל ה י ם ס ל
א ד מ ו ל א ש ת ו נ כ ת נ ו ת ע ו
ר ו י ל ב ש ם ו י א מ ר י ה ו ה
א ל ה י ם ה ד ה ה י ה כ א
ח ד מ מ נ ו ל ד ע ת ט ו ב ו ר ע
ו ע ת ה פ ן י ש ל ח י ד ו ו ל ק
ח ג ם מ ע ץ ה ח י י ם ו א כ ל ו
ח י ל ע ל ם ו י ש ל ח ה ו י ה
ה א ל ה י ם ס ג נ מ ג ע ל ע ב ד
א ת ה א ד מ ה א ש ר ל ק ח מ ש ם
ו י ג ר ש א ת ה א ד ם ו י ש כ ן
מ ק ד ם ל ג ן ע ד ן א ת ה כ ר ב
י מ ו א ת ל ה ט ה ח ר ב ה מ ת ה
פ כ ת ל ש מ ר א ת ד ר ך ע ץ ה ח
ה ו ת ח ד א ע ד י מ ס ה ו ה ו י
```

As soon as we have such a structure, we can examine many different variables.▶ Of particular interest to us is that the total numerical value of the four corners and the two middle letters of each side of the square, twelve letters in all (חאהלחחויימיה) is 137.

Feminine Intelligence

A unique idiom in the Bible describes feminine intelligence. "House and wealth are the inheritance of fathers but from *Havayah* [comes] an intelligent woman."[31]▶▶

Intelligence is a function of *binah* (the *sefirah* of understanding), the mother principle. The sages learn from the Torah verse describing how God formed Eve that a woman is endowed with greater understanding than a man.[35] In contrast, wisdom is essentially a male faculty, associated with the father principle. The intelligence of an intelligent woman is not merely a function of her inherent feminine insight and understanding, but also a function of her own faculty of wisdom.◀◀◀ This relates to the union of the father and mother principles in the faculty of *da'at*.

▶▶▶ The numerical value of the complete phrase, "But from *Havayah* [comes] an intelligent woman," (וּמֵהוי' אִשָּׁה מַשְׂכָּלֶת) is 1,168, which equals 16 times 73, the numerical value of "wisdom" (חָכְמָה).

The numerical value of "But from *Havayah*" (וּמֵהוי') is 72, which is the numerical value of the highest of the four primary fillings of the essential Name of God, *Havayah*, referred to as *Ab*. The numerical value of the filling letters of this Name alone is 46, which is the numerical value of the two prefix letters in this word. *Ab* corresponds to the ultimate intelligence and wisdom of God. This verse thus indicates that an intelligent woman comes from the highest level of God's Name.

▸ For example, the numerical value of the sixteen letters in the right-left diagonal is 613, which is the total number of commandments in the Torah. In Kabbalah, each commandment is a spiritual marriage of male and female Divine energy.[28] The total of the left-right diagonal is 613 ⊥ 365. 613 is the total number of commandments in the Torah, as stated, while 365 is the number of the prohibitive commandments alone.

The sum of the numberical values of the letters in the top line of the square is 572, which equals 22 (the number of letters in the *alef-bet*) times 26 (the numerical value of God's Name). With regards to marriage, 572 is also the numerical value of the phrase, "and they became one flesh" (וְהָיוּ לְבָשָׂר אֶחָד).[29]

The sum of the numbers in the bottom line of the square is 620, which is the total number of *mitzvot* when the 7 rabbinical commandments are included with the 613 commandments of the Torah.[30] The numerical value of "crown" (כֶּתֶר) is also 620. *Keter* (the *sefirah* of crown) is the highest of the ten *sefirot*.

The numerical value of the entire array is 14,900, which equals $100^2 ⊥ 70^2$. This number is represented by "forever" (לְעֹלָם), which appears in the array. The numerical value of the first two letters (לע) is 100 and the numerical value of the last two letters (לם) is 70. Dividing the word in half in this way yields a dot product of 3,700 (30·30 ⊥ 70·40), which equals 100 times 37, the companion number of 137.

The numerical value of the filling of "forever" (למד עין למד מם) is 358, which is the numerical value of Mashiach (מָשִׁיחַ). The numerical value of the second filling of the letters (למד מם דלת עין יוד נון למד מם דלת מם מם) is 1,592. The sum of the regular numerical value (170) and the filled value (358) and the filling of the filling is 2,120, which equals 5 times 424, the numerical value of, "Mashiach the son of David" (מָשִׁיחַ בֶּן דָּוִד).

The primordial value of "forever" (לְעֹלָם) is 680, which is equal to 4 times 170, the regular numerical value of the word, i.e., the average value of each letter in primordial numbering is equal to the regular numerical value of the entire word.

▸▸ The numerical value of "an intelligent woman" (אִשָּׁה מַשְׂכָּלֶת) is 1,096, which equals 8 times 137. This phrase has eight letters; thus, the average value of each letter is 137.

The number 1,096 also equals 4 times 274, which represents the union of Leah (לֵאָה) and Rachel (רָחֵל) in each of the four spiritual Worlds, *Asiyah*, *Yetzirah*, *Beriah* and *Atzilut*.

There are certain letters that have two or more possible spellings. Each alternative spelling of the same letter relates to a different *sefirah*. Filling the letter *tav* (usually spelled תו) with a *yud* (תיו) is associated with *chochmah* (the *sefirah* of wisdom). When the phrase "an intelligent woman" (אִשָּׁה מַשְׂכָּלֶת) is spelled in full using this spelling (אלף שין הא מם שין כף למד תיו), its numerical value is 1,507, which is equals 11 times 137. When calculated alone, the numerical value of this filling (לף ין א ם ין ף מד יו) is 411, the numerical value of "something from nothing" (יֵשׁ מֵאַיִן). Thus, the ratio between the numerical value of the root of the phrase (1,096) and the filling is precisely 8:3.

The 8:3 ratio corresponds to the letters of "festival" (חַג), the numerical values of which are 8 (ח) and 3 (ג). The commandment to celebrate the three prilgrimage festivals appears in the phrase, "Three pilgrimage festivals you shall celebrate for Me each year" (שָׁלֹשׁ רְגָלִים תָּחֹג לִי בַּשָּׁנָה).

The numerical value of "you shall celebrate" (תָּחֹג) is 411, and the average numerical value of each letter is 137.

The numerical value of "woman" (אִשָּׁה) is 306. The average numerical value of each letter is 102, which is the numerical value of "faith" (אֱמוּנָה), implying that a woman is a pillar of faith. The initial letters of "an intelligent woman" (אִשָּׁה מַשְׂכָּלֶת) are *alef-mem*, which are the initial letters of "faith" (אֱמוּנָה). They spell "mother" (אֵם). Another manifestation of the 3:8 ratio is between the numerical value of "woman" (אִשָּׁה), which is 3 times 102, and "lowliness" (שְׁפָלוּת), which is 816, or 8 times 102.

One possible spelling of the third level of filling of "an intelligent woman" (אלף למד פה שין יוד נון הא אלף מם מם שין יוד נון כף פה למד מם דלת תיו יוד וו). The numerical value of this spelling is 2,740, which equals 20 times 137.

The numerical value of the verse in Proverbs,[32] "It shall be a cure for your body and a tonic for your bones" (רִפְאוּת תְּהִי לְשָׁרֶּךָ וְשִׁקּוּי לְעַצְמוֹתֶיךָ) is 2,740. This verse refers to the Torah, the true "woman of valor"[33] and the ultimate wisdom of an "intelligent woman."[34]

When the entire phrase is spelled in full (1,507) and the root (1,096) is added, the result is 19 [the numerical value of Eve (חַוָּה)] times 137.

"Who is a wise person?" The literal meaning of the sages' response to this question is "one who sees what will be born."[36] This implies that the father principle of wisdom is complete when it can envision the final result, i.e., when it incorporates the mother principle of giving birth to a child.

Rebbe Shneur Zalman of Liadi, author of the *Tanya*, explains that the epitome of wisdom is the ability to experience the continual re-creation of the world together with the continual re-creation of oneself.[37] An intelligent woman is innately attuned to experience such continual re-creation.

The renewal of the feminine potential for fertility every month and her ability to give birth to new life are related concepts.▼ They become apparent in the male when he consciously develops feminine sensitivity in his psyche.

Three Components of a Happy Home

In his commentary on the verse, "House and wealth are the inheritance of fathers but from *Havayah* [comes] an intelligent woman,"[39] Rabbi Eliyahu of Vilna, the Vilna Gaon, explains that the three elements mentioned—house, wealth and inheritance—represent three ways of serving God.

Keeping Jewish tradition—that is, observing the Torah and its

▼ The Hebrew word for "month" (חֹדֶשׁ) pronounced with a different vowelization, means "new" (חָדָשׁ).

The numerical value of "month" (חֹדֶשׁ) is 312, which equals 12 times 26, the numerical value of *Havayah*, or two times Joseph (יוֹסֵף, 156). When spelled in full, the numerical value of "month" (חית דלת שין) is 1,212, which equals 12 times 101 [the numerical value of "from nothing" (מֵאַיִן) as seen above with reference to Rachel and Leah]. The numerical value of the second filling (חית יוד תו דלת למד תו שין יוד נון) is 2,244, which equals 12 times 187 (the probability that a word and its filling and the filling of its filling will all be divisible by 12, is 1:1,728). The number 2,244 also equals 22 times 102 (אֱמוּנָה), the average of each letter of "woman" (אִשָּׁה). All this relates to the feminine ability to

renew herself "from nothing" (the 22 letters) through the power of her faith.

The numerical value of the filling alone (ית לת ין) is 900, which equals 30², corresponding to the number of days in a full lunar month.

900 is also the numerical value of the filling of "'Let there be light!' And there was light." The number 30² represents the Jewish heart, which manifests in the Jewish woman in particular.[38]

The numerical value of "month" (חֹדֶשׁ, 312) plus its filling (1,212) plus the filling of the filling (2,244) is 3,768, which equals 12 times 314, the numerical value of the Divine Name, *Shakai* (שׁדּי). The interpretation of this Name is, "He who said to His world, 'Enough!'" According to the Zohar, God said "Enough!" when He created woman.

commandments—corresponds to a house. Torah observance provides the context in which we live, and in that respect, it is considered our home. Torah knowledge is symbolized by great wealth and is a direct inheritance from our parents who educate us.

Practicing Judaism and studying Torah seem to cover all aspects of life. Yet, the Vilna Gaon makes it clear that the way an intelligent woman builds her home and perfects her marriage is not only through Torah and its commandments. Her innate intelligence is the ability to serve God in every facet of life, as the verse in Proverbs states, "in all your ways, know Him."[40] Seeing and experiencing Godliness even in the most mundane activities renders them as holy as a *mitzvah*. An intelligent woman has the rare talent to translate the secular word into a world of sanctity.[41] Marriage contains an abundance of mundane activities, but an intelligent woman makes every facet of her marriage sacred.

In particular, this relates to the marriage of science and Torah wisdom. The ability to perceive how "God is all and all is God," as the Chassidic adage states, ultimately stems from perfected female intelligence. An intelligent woman is one who has been graced by God with the wisdom to understand the relationship between the mysteries of science and the hidden secrets of Kabbalah.▸

One example of a particularly intelligent woman is one of the daughters of Rabbi Shneur Zalman of Liadi:

Rabbi Shneur Zalman of Liadi (the Alter Rebbe) founded the Chabad Movement. His son, Rebbe Dov Ber (the Mittler Rebbe) was his successor. However, there were certain Chassidic discourses that the Alter Rebbe never taught his son. Instead, he taught them to one of his daughters, Freida. Her brother pleaded that she teach them to him. Thus, to some extent, Freida was the Alter Rebbe's true successor. Before Freida passed away, she atypically requested that she be buried next to her father. After some debate, and some uncanny Divine intervention, she was buried next to her father in her mother's—Sterna's—grave.▸▸

▸ In the previous chapter, we saw that a wise individual should be old, i.e., over 60 years old. The phrase in the Song of Songs, "There are sixty queens and eighty concubines, and innumerable maidens,"[42] is interpreted allegorically. "Sixty queens" refers to the three generations of Abraham, his sons and grandsons. "Eighty concubines" refers to the descendants of Noah until Abraham. The sum of the numerical values of "sixty" (שִׁשִּׁים) and "and eighty" (וּשְׁמֹנִים) is 1,096, which equals 8 times 137, the numerical value of "an intelligent woman" (אִשָּׁה מַשְׂכָּלֶת). This once again strongly links the number 137 to femininity. The numerical value of the entire verse "There are sixty queens and eighty concubines, and innumerable maidens" (שִׁשִּׁים הֵמָּה מְלָכוֹת וּשְׁמֹנִים פִּילַגְשִׁים וַעֲלָמוֹת אֵין מִסְפָּר) is 3,108, which equals 84 times 37, the companion number of 137.

▸▸ The sum of the numerical values of Freida (פְרִידָא) and Sterna (סְטֶערְנָא) is 685, which equals 5 times 137, the numerical value of "her voice" (בְּקֹלָהּ). 685 is also the 19th inspirational number and 19 is the numerical value of Eve (חַוָּה).

211

Uniting Two Halves into a Whole

The Torah commands us to collect one half-shekel coin from each individual in order to count the people during a census.[43] The great 16th century Biblical commentator, Rabbi Moshe Alshich, explains[44] that a half-shekel (and not a whole shekel) was collected to remind us all that we are not complete until we rectify our bond with the Almighty. An intelligent woman has the good sense to know how to unite the two halves to become one whole.[45]◄ An alternative interpretation is that we are not a whole person until we connect ourselves to another Jew. This is especially true of marriage, in which each spouse is married to his or her "other half."

One of the seven wedding blessings begins with the phrase, "Rejoice, You shall surely cause them to rejoice—the loving companions." The numerical value of the first two words of this phrase "Rejoice, You shall surely cause them to rejoice" (שַׂמֵּחַ תְּשַׂמַּח) is 1,096. Another classic blessing for a newlywed couple is that they should build "a faithful home within the Nation of Israel" (בַּיִת נֶאֱמָן בְּיִשְׂרָאֵל).◄◄ The numerical value of this phrase is also 1,096.

The first two letters of "woman" (אשה) are the two letters of "fire" (אש), which also appear in "man" (איש). This alludes to the holy fire that unites a married couple who merit it.◄◄◄ The remaining two letters in these two nouns form the Divine Name, Kah (יה), which is the Divine Name that corresponds to chochmah (the sefirah of wisdom). The special wisdom of the intelligent woman is to know how to create holy fire.[47] The primordial (or "cumulative") value (mispar kidmi) of "fire" (אש), is also 1,096.◄◄◄◄ This she does with the help of her innate feminine beauty.

Angel of Beauty

One of the angels named in the Zohar is Yofiel. The numerical value of Yofiel (יוֹפִיאֵ-ל) is 137, which is also the numerical value of Kabbalah (קַבָּלָה). Indeed, Yofiel is the angel of Kabbalistic wisdom.[48] As his name

---

▶ The numerical value of "half" (מֶחֱצִית) as in the phrase, "a half-shekel" (מֶחֱצִית הַשֶּׁקֶל) is 548. 548 equals 4 times 137 and is exactly half of 1,096, which is the numerical value of "an intelligent woman" (אִשָּׁה מַשְׂכֶּלֶת), etc.

This further alludes to the secret of the "trumpet" (חֲצֹצְרָה) that states that a person is only half a form (חֲצִי צוּרָה). In the Torah, the numerical value of one synonym for "trumpet" (כְּלִי עֹז) is 137 (See Chapter 15).

▶▶ In the contexts in which it appears in the Bible,[46] a "faithful home" (בַּיִת נֶאֱמָן) is a clear reference to malchut (the sefirah of kingdom), the sefirah most associated with femininity.

▶▶▶ One couple who did not merit this holy fire is Haman (הָמָן) and Zeresh (זֶרֶשׁ), the villains of the Book of Esther. The sum of the numerical values of their names (95 and 507, respectively) is 602, 2 times "fire" (אש), which is the average value of their two names.

▶▶▶▶ The primordial value of the letter shin (ש) alone is 1,095, which equals 15 [the numerical value of Kah (יה), the Divine Name that corresponds to chochmah (the sefirah of wisdom)] times 73 (the numerical value of ["wisdom" (חָכְמָה)]. The three main forms of wisdom are Torah, music and mathematics, the sum of the numerical values of these three (תּוֹרָה נְגִינָה חֶשְׁבּוֹן) is also 1,095.

The numerical value of, "fire of the Holy Tongue" (אֵשׁ לְשׁוֹן הַקֹּדֶשׁ) is also 1,096, the numerical value of "an intelligent woman" (אִשָּׁה מַשְׂכֶּלֶת). This refers to the rectified power of speech of an intelligent woman.

suggests, Yofiel is also the angel of beauty (יְפִי),[49] relating to Noah's youngest son, Japheth (יֶפֶת) and also to Shem, Noah's eldest son, who studied the inner secrets of the Torah from Yofiel.[50]

The phrase in Genesis that names both Japheth and Shem is "God beautifies Japheth and He dwells in the tents of Shem."[51] The numerical value of the first part of this phrase, "God beautifies Japheth" (אֱ-לֹהִים לְיֶפֶת) is 1,096, the numerical value of "an intelligent woman" (אִשָּׁה מַשְׂכָּלֶת), which equals 8 times 137, the numerical value of Yofiel (יוֹפִיאֵ-ל), the angel of beauty. Each of the Matriarchs was renowned for her unique style of beauty, as the Torah tells us.▸

- Sarah had "a beautiful appearance" (יְפַת מַרְאֶה).[53]
- Rebecca had "a very good appearance" (טֹבַת מַרְאֶה מְאֹד).[54] The numerical value of this phrase is 702, the numerical value of Shabbat (שַׁבָּת). 702 is the reflection of 207, the numerical value of "light" (אוֹר). True beauty is the result of light contained in a translucent vessel. Like *malchut* (the *sefirah* of kingdom), which reflects the six emotive powers of the heart, Shabbat reflects the light of the six days of the week. The sum of the numerical values of "light" (207, אוֹר) and "beauty" (100, יְפִי) equals Rebecca (רִבְקָה).
- Rachel had "beautiful form and beautiful appearance" (יְפַת תֹּאַר וִיפַת מַרְאֶה).[55] God gave her a beautiful appearance because her behavior proved that she was worthy of it.▾▾ She bore Joseph who inherited her double measure of beauty and is described with the same Hebrew words (יְפֵה תֹאַר וִיפֵה מַרְאֶה).[58]
- The Torah states that, "Leah's eyes were soft."[59]▸▸▸ The general opinion is that despite the fact that she was not physically beautiful, her inner qualities were beautiful.[60] Unkelus translates

▸ The phrase "God beautifies Japheth" (יַפְתְּ אֱ-לֹהִים לְיֶפֶת) alludes to three of the four Matriarchs, Sarah, Rachel and Leah:

The sum of the numerical values of the two words "beautifies Japheth" (יַפְתְּ... לְיֶפֶת), is 1,010, which is 2 times 505 (the average value of each word), the numerical value of "Sarah" (שָׂרָה), the first Matriarch. Rachel was "beautiful of looks and beautiful of countenance" (יְפַת תֹּאַר וִיפַת מַרְאֶה).[52] The numerical value of "God beautifies" (יַפְתְּ אֱ-לֹהִים) is 576, which equals 16 [the numerical value of "she" (הִיא)] times 36 [the numerical value of Leah (לֵאָה)].

The numerical value of Shem (340, שֵׁם) and Yofiel (137, יוֹפִיאֵל) is 477, which is the filled numerical value of "woman" (אלף שין הא).

The sum of the numerical values of the names of all four Matriarchs (שָׂרָה רִבְקָה רָחֵל לֵאָה) is 1,086. Sarah's original name was Sarai (שָׂרַי), but God replaced the *yud* of her name with the letter *hei*. Adding Sarah's missing *yud* brings the total to 1,096, the value of the phrase, "an intelligent woman" (אִשָּׁה מַשְׂכָּלֶת).

▸▸▸ The numerical value of "and Leah's eyes [were soft]" (וְעֵינֵי לֵאָה) equals 182, the numerical value of Jacob (יַעֲקֹב). This indicates that Jacob is reflected in particular through the eyes of Leah.

▾▾ God gives beauty to those who are inwardly beautiful n the same way that "He gives wisdom to those who are wise,"[56] as the following anecdote illustrates:

A matron once asked Rabbi Yossi bar Chalaftah, "Should the verse not have said, 'He gives wisdom to those who are stupid!'?" He answered her by way of a parable, "If two people came to borrow money from you, one was rich and one was poor, who would you loan the money to?" "To the rich man," she replied, "Why?" he asked. "Because if the rich man loses my money, he has from where to pay it back, but if the poor man loses the money, where will he find the means to pay it back to me?" He said, "Your ears should hear what your mouth is saying! Would God give wisdom to the stupid, they would sit around and philosophize in the lavatories and theatres and washhouses, rather, God gave wisdom to the wise and they sit and discuss it in the synagogues and study halls."[57]

"Leah's eyes were beautiful." According to a Talmudic rule[61] this means that no further proof of her beauty is needed.

A Majestic Balance

Above, we mentioned the Vilna Gaon's interpretation[62] to the verse "house and wealth are the inheritance of fathers but from *Havayah* [comes] an intelligent woman."[63] The Vilna Gaon finds a parallel to this verse in another well-known verse from Proverbs, "for a candle is a *mitzvah* [Jewish tradition] and the Torah is light [Torah knowledge], and the way of life is moral rebuke [the innate wisdom of an intelligent woman]."[64] ◄

▶ The Talmud teaches us that "way" (דֶּרֶךְ) refers to a woman.[65] Similarly, the numerical value of "moral" (מוּסָר), 306, is equal to the numerical value of "woman" (אִשָּׁה), i.e., a woman is the source of morality.

When the Ba'al Shem Tov visited a home, he would often request that more candles be lit. Every *mitzvah* is a candle, which once lit, illuminates the home with the great light of the Torah. As mentioned, in addition to the Torah and its commandments, one needs a way of life paved by an intelligent woman, who knows how to rebuke her husband in an acceptable way.

The verse in Genesis states that a woman is "a helpmate facing"[66] her husband. One interpretation is that the woman is "a helpmate to confront," her husband (when his behavior does not merit her direct support and assistance).[67] The subtlety of rectified feminine intelligence recognizes how to correctly judge her husband's ego level. She knows when to boost his morale when he is disheartened, and when to chastise him to balance his exaggerated self-image.

▶▶ The words, "low tide" (שֵׁפֶל) and "high tide" (גֵּאוּת) also mean "lowliness" and "pride," respectively. The two words share the same numerical value of 410, which equals 10 times 41, the numerical value of "mother" (אֵם).

The feminine talent for psychological balance stems from her own inner stablity. By correctly balancing her fluctuating emotions and feelings, she develops a heightened sensitivity to equilibrium in her environment. This relates in particular to coping with her inner monthly cycle. Every woman is like the moon, which governs the rise and fall of the ocean tides. The lower the low tide sinks, the higher the high tide can rise. ◄◄ The waxing and waning of the moon thus relates to the balance between humility and assertiveness in the psyche. The humbler one feels within, the more effective is one's assertiveness.

Assertiveness does not contradict one's inner sense of humility. It is worn like a majestic cloak, as stated with reference to the Almighty in Psalms, "God has ruled, He has donned majesty/assertiveness."[68]

The intelligent woman senses how to balance and regulate her household. In this way, a woman is like a king, who is sensitive to the ego of each of his subjects and knows when to elevate it and when to lower it, as the verse in Samuel states with reference to God, the King of Kings, "He promotes and demotes."[69]▶

Humility positively affects the level of joy in the soul, and joy is reflected in physical beauty. Rebbe Dovid of Lelov taught that Jacob loved Rachel because she was a happy woman. As we saw with reference to Leah, physical beauty manifests principally in the eyes,[70] which are balanced above the nose, as the two pans of a scale are balanced above its foot. Indeed, the Hebrew word for "eye" (עַיִן) is closely related to the Aramaic word "balance" (עַיִן) used in the Talmud.▶▶

"Balance" (עִיּוּן) also means deliberate contemplation, as in the observation necessary for scientific experimentation. We have already pointed out that this is the paradox of quantum mechanics. As mentioned, the solution to the enigma is to "think/observe good and it will be good." Fulfilling God's commandments brings increasingly more good energy into the world. Similarly, positive thinking and positive speech boost the positive spiritual energy in the world. This serves as a catalyst to bring about the definitive positive outcome of creation, i.e., the advent of *Mashiach* and the Final Redemption.

▶ An intelligent woman knows when to be silent and when to speak (see reference to *chashmal*). This is alluded to in the numerical value of "but from *Havayah* [comes] woman" (וּמֵהֹוִי׳ אִשָּׁה), which is 378, the numerical value of *chashmal* (חַשְׁמַל).

▶▶ In modern Hebrew, "balance" (אִזּוּן) is conjugate to "ears" (אָזְנַיִם), where the sense of balance is situated.

REFERENCE NOTES FOR CHAPTER 8

1. From the *Kedushah* in the *Musaf* prayer of Shabbat; *Zohar Shemot* 100b, etc.

2. Proverbs 19:14.

3. Exodus 26:5.

4. Ibid 36:12.

5. Genesis 21:12.

6. Deuteronomy 23:2-4, 23:9.

7. Micah 2:5.

8. This alludes to the elevation of the feminine figure (*malchut*) to the height of the masculine *Ze'ir Anpin*. This subject is discussed in detail in our book *The Twinkle in Your Eye*.

9. Genesis 21:12; Rashi ad loc.

10. Genesis 25:27.

11. See *Kli Yakar*, ibid.

12. See our book in Hebrew, *Einayich Breichot Becheshbon*, pp. 28-29 for an explanation of diamond numbers.

13. Isaiah 52:13.

14. Genesis 1:2.

15. *Zohar Chadash Bereishit* 31b.

16. *Kidushin* 49b.

17. Not to be confused with the similar Hebrew word "love" (אַהֲבָה).

18. *Chagigah* 12a.

19. Genesis 1:1.

20. *Etz Chayim* 38:2.

21. *Bereishit Rabah* 22:7; "Yehudah bar Amei said, 'They [Cain and Abel]

22. *Zohar Vayikra* 19a.

23. *Bereishit Rabah* 22:7

24. *Baba Metziah* 84a; *Baba Batra* 58a

25. Genesis 3:20.

26. Genesis 4:1.

27. This subject is covered in greater detail in our book in Hebrew, *Einayich Breichot Becheshbon*, pp. 45-49.

28. see *The Twinkle in Your Eye*, p. 25.

29. Genesis 2:24.

30. See our book in Hebrew, *Einayich Breichot Becheshbon*, Chapter 1

31. Proverbs 19:14.

32. Ibid 3:8.

33. Ibid 31:10.

34. Ibid 19:14.

35. *Niddah* 35b; *Sotah* 45b.

36. *Avot* 4:1.

37. *Tanya*, Chapter 43.

38. See Chapter 1, p. 38, note at bottom of page.

39. Proverbs 19:14.

40. Ibid 3:6.

41. See *Likutei Moharan* 19.

42. Song of Songs 6:8.

43. See Exodus 30:13; 30:15; 38:26.

44. *Torat Moshe, Parashat Ki Tisa*.

45. See Chapter 15 and also our book *The Twinkle in Your Eye* p. 107.

were arguing over the first Eve.' Rabbi Ayvu said, the first Eve returned to the dust.'"

46. I Samuel 2:35, 25:28; I Kings 11:38.

47. See *The Mystery of Marriage*, pp. 109-111. See also *Likutei Moharan* I 19.

48. *Zohar Shemot* 247b.

49. Ibid 70a.

50. Introduction of the Raavad to *Sefer Yetzirah*.

51. Genesis 9:27.

52. Ibid 29:17.

53. Ibid 12:11.

54. Ibid 24:16.

55. Ibid 29:17.

56. Daniel 2:21.

57. *Kohelet Rabah* 1:7.

58. Ibid 39:6.

59. Ibid 29:17.

60. See Rashi and *Siftei Chachamim* ad loc.

61. *Ta'anit* 24a.

62. Commentary of the Vilna Gaon on Proverbs 19:14.

63. Proverbs 19:14.

64. Ibid 6:23.

65. *Kidushin* 2b.

66. Genesis 2:18.

67. Rashi ibid.

68. Psalms 93:1.

69. I Samuel 2:7.

70. See *Ta'anit* 24a.

COMMUNICATION

QED

Quantum electro-dynamics (QED) relates to the interaction between light energy and subatomic particles. Photons are quanta of light energy; electrons are matter particles. An electron (and other particles of the same particle family) can absorb photons and become energized. *Alpha*, the fine-structure constant, is the coupling constant that defines the interaction between electrons and photons.

Electrons are subatomic elementary particles that participate in three of the four interactions discussed in Chapter 1: the gravitational (responsible for gravity), electromagnetic (responsible for light and electricity) and weak interactions (radioactivity). The passage of electrons through an element produces an electric current. Electrons are also a key factor in chemical interactions.

Electrons surround the nucleus of an atom in concentric shells. Beginning from the innermost shell, i.e., the shell closest to the nucleus, each electron shell is referred to by a number 1, 2, 3, and so on. Each shell can hold a certain maximum number of electrons. The distribution of electrons in the various shells is called the electronic arrangement (or electronic form or shape).

Until about 1925, the orbital of an electron around the nucleus of an atom was thought to be defined by only three different properties: 1) the principal quantum number, 2) the angular momentum (or secondary) quantum number, and 3) the magnetic quantum number. Quantum physicist Wolfgang Pauli later showed that the three-quantum number theory was not sufficient. Eventually, he proposed a fourth property, which he called "spin." Here is a brief summary of what these numbers mean:

▶ Physicists have stated that if an atom were the size of a football stadium, the nucleus would be like an ant in the middle of the playing field. On these scales, the electrons would be whizzing around the very edge of the stadium with a size at least 1,000 times smaller than the ant. All that lies between the nucleus and the electron is dark matter.

▶▶ According to the Bohr Model and the Dirac equation, the highest possible atomic number for an element is 137.

▶▶▶ The series of double square numbers [$2(n^2)$] plays a prominent role in Kabbalah.[1] Some of the most prominent double squares are the 32 ($2 \cdot 4^2$) pathways of wisdom, the 50 ($2 \cdot 5^2$) gates of understanding and the 72 ($2 \cdot 6^2$) bridges of knowledge.

1. The Principal Quantum Number (n): $n = 1, 2, 3, \ldots \infty$

This number specifies the energy level of an electron and the size of its orbital, which depends on its distance from the nucleus.◀ The farther away the electron is from the nucleus of the atom, the larger its orbital and the higher its energy level. In accordance with the law of entropy, each electron "prefers" to occupy the lowest possible energy level.

The principal quantum number is significant in chemical reactions, which rely on the exchange of electrons between atoms. Atoms with outer shells filled with the maximum number of electrons are poorly reactive, because they cannot accept any more electrons. By contrast, atoms with outer shells that lack electrons are reactive, because they attract electrons from other atoms. For example, in a hydrogen atom (which has 1 electron) the electron is in its ground state in the orbital closest to the nucleus ($n = 1$). In this case, the hydrogen atom is stable and will not easily relinquish its electron to another atom. If the electron is in the $n = 2$ orbital, it is in an excited state and is likely to be transmitted to another atom, which lacks an electron, and a chemical reaction will occur. The number of electrons in atoms is the underlying basis of the chemical periodic table.◀◀

Each shell can contain only a fixed number of electrons. The first shell can hold up to two electrons, the second shell can hold up to eight ($2 + 6$) electrons, the third shell can hold up to 18 ($2 + 6 + 10$) and so on. The general formula is that the nth shell can in principle hold up to $2(n^2)$ electrons.◀◀◀

2. Angular Momentum (Secondary) Quantum Number (l): $l = 0, \ldots n - 1$.

The angular momentum of an electron specifies the shape of its orbital within the electron shell. The energy of the resulting sub-shell increases slightly, depending on the shape of the orbital.

Subshell label	ℓ	Max. electrons	Shells containing it
s (sharp)	0	2	Every shell
p (principal)	1	6	2nd shell and higher
d (diffuse)	2	10	3rd shell and higher
f (fundamental)	3	14	4th shell and higher
G	4	18	5th shell and higher (theoretically)

3. Magnetic Quantum Number (m_l): $m_l = -l, \ldots 0, \ldots +l$.

The magnetic quantum number specifies the orientation in space of an orbital of a given energy (n) and shape (l). This number divides the sub-shell into individual orbitals that hold the electrons; there are $2l+1$ orbitals in each sub-shell. Thus, the s sub-shell has only one orbital, the p sub-shell has three orbitals, and so on.

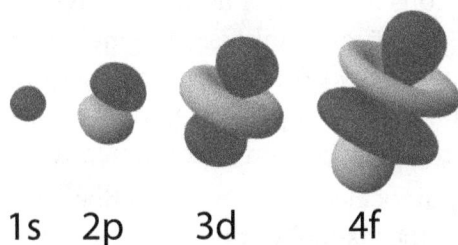

1s   2p   3d   4f

Possible shapes of various electron orbitals

4. Spin Quantum Number (m_s): $m_s = +\frac{1}{2}$ or $-\frac{1}{2}$.

In 1924, Wolfgang Pauli showed that the three-quantum properties could not explain why the electron orbit did not collapse in the presence of the proton in the core of the hydrogen atom. If there were only three quantum numbers, as thought previously, the positively charged proton should attract the negatively charged electron in orbit. This would draw it away from its orbital and into the atom's core.

Obviously, this never occurs. Pauli searched for a theory that would resolve this quandary. He eventually proposed a fourth quantum number for the electron that depends on one of two different charged states. Pauli called this state "spin." He gave it a value of either $+\frac{1}{2}$ or $-\frac{1}{2}$.

The spin quantum number specifies the orientation of the spin axis of an electron. Each electron spins in one of two directions (sometimes called up and down). The term "spin" is a symbolic representation. Technically, there is no difference between an electron with up spin and that with down spin. But, when lines of the hydrogen spectrum are examined at very high resolution, they are found to be closely spaced doublets. This was the first experimental evidence for electron spin. The fine structure of the visible spectrum is defined by the fine-structure constant, $\frac{1}{137}$.

Together with his discovery of spin, Pauli came to the realization that no two electrons in the same atom can have identical values for all four of their quantum numbers. This means that no more than two electrons can occupy the same orbital, and any two electrons in the same orbital must have opposite spins. The exclusion principle, as this is called, has been proven experimentally.

The fact that a pair of electrons—by definition—must have opposite spins revealed another "spooky" phenomenon of the quantum world: entanglement. Two particles that have been formed together as a result of a quantum reaction are so sensitive to one another that their communication is no longer subject to the limitation of the speed of light. In the case of entangled electrons with different spins, if they are separated and brought to opposite ends of the cosmos and an action is performed on one to change its spin, its partner accommodates by instantaneously changing its own spin state. This logical conclusion of the exclusion principle brought even more bewilderment to the scientific world, which prefers to avoid the psychological ramifications implied by such non-local communication. Einstein found the concept of non-local phenomena so intolerable that he said that he would rather become a cobbler than accept the idea as a physicist!

The more science develops, the clearer it becomes that non-local phenomena definitely exist. Quantum entanglement has been subjected to rigorous experimentation that has proven it time and again to be true. The possibility of instantaneous communication via quantum computers is quickly becoming reality.

Quantum Communication

As we have seen in earlier chapters, there are four spiritual Worlds, representing four levels of consciousness: *Atzilut* (the World of Emanation), *Beriah* (the World of Creation), *Yetzirah* (the World of Formation) and *Asiyah* (the World of Action).

The four quantum numbers of the electron orbit correspond to these four worlds. The primary quantum number has the most effect on the energy level of the atom. It corresponds to *Asiyah* (the World of Action). The secondary quantum number refers to the shape, or form of the electron orbit. It corresponds to *Yetzirah* (the World of Formation). The magnetic quantum number corresponds to *Beriah* (the World of Creation), and the fourth quantum number, referred to as "spin" corresponds to *Atzilut* (the World of Emanation), the highest of the four Worlds. In *Atzilut* there is total unity, even between entities that in the lower worlds are innately paradoxical, such as a pair of electrons in two different states of spin. ▸

Spiritual World	Quantum number	Effects:
Atzilut	spin	fine structure of spectral lines
Beriah	magnetic	energy changes in magnetic field
Yetzirah	angular momentum	changes in energy due to orbital shape
Asiyah	principal	potential electromagnetic energy

▸ The property of spin relates to the relative movement of the sun around the earth, "It goes to the south and goes around to the north: the will goes around and around and the will returns to its circuits."[2] This verse contains the letter *vav* 11 times. the sum of the remiaing letters is 2,140, which equals 10 times 214, the numerical value of "spirit" (רוח), which is the word used in this verse for "will."

Will Power, Telepathy and Speech

Each of the four spiritual Worlds is associated with a different form of communication. In *Atzilut*, communication is silent. It is referred to in Kabbalah as "light that illuminates itself"; this corresponds to the spin of an electron, which is relative only to the electron itself.

Communication with others begins in the first of the three lower Worlds, *Beriah*, in the initial thought that will later accompany the spoken word. This corresponds to the magnetic quantum number, which assigns the electron cloud (i.e., the atomic orbital) its orientation.

In *Yetzirah*, the next World down, communication is achieved through the intonation of speech, which need not relate to the spoken words. This corresponds to the shape of the electron cloud, which depends on the angular momentum of the electron.

In the lowest World, *Asiyah*, communication is effected through the medium of spoken language. This corresponds to the primary quantum number, which produces chemical interactions between atoms.

In terms of quantum mechanics, communication at the levels of speech and intonation are local phenomena. They can only occur with a time-lapse. The maximum speed at which verbal communication occurs is the speed of sound.

The power of communication through thought, like verbal communication, is also a local phenomenon.[3] Thought-transmission relies on electromagnetic radiation, whose maximum velocity is the speed of light. Sophisticated equipment allows individuals with impaired communicative powers to communicate through a brain-computer interface. Computer games have also been developed that are played through thought interaction.

As mentioned, a non-local phenomenon depends on the "entanglement" of two particles, such as two electrons from the same atom, one with an up-spin, and the other with a down-spin. Once they are separated, each particle retains an inherent "knowledge" of its partner's state such that, even if they are at different ends of the universe, each particle instantaneously adapts its spin to the changes in spin that occur in its partner. There is no physical explanation how particles communicate instantaneously, yet they do. This type of non-local communication corresponds to *Atzilut*. As discussed in Chapter 6, in the super-conscious crown of the soul, there are three different levels, corresponding to faith, pleasure and will. The power of will, the lowest of these three super-conscious powers of the soul, projects itself outwards and so has the ability to communicate. The light of will manifests in *Atzilut* as a lightening flash of insight or inspiration. Because of its lofty origin, willful communication pays no regard to the speed of light. Thus, two entangled souls can simultaneously become aware of the same insight.

The concept of entanglement is apparent in Kabbalah with reference to a married couple. In *Atzilut*, a couple is one "entangled" soul. At the lower levels, their one soul splits into two. There is no local communication between them until they meet. Nonetheless, they do communicate non-locally. At this stage, one partner's actions and choices immediately affect the other's, without them being aware of it. When they first meet, their entanglement begins to manifest in the lower spiritual realms and local communication becomes possible. Communication at the physical level manifests once they are married.

One Man, Two Wives

All types of communication must be utilized in order to create a better world. In order to achieve this, will—the most potent form of communication—must unite with, or "marry" the lower levels of communication.

According to Kabbalah, the three levels of *keter* (the super-conscious crown)—namely faith, pleasure and will—correspond to Abraham, Isaac and Jacob respectively.▸ Will, associated with Jacob, is a relatively masculine power. By comparison, the powers of thought and

▸ In general, the three forefathers correspond to the first three emotive qualities, *chesed* (lovingkineness), *gevurah* (might) and *tiferet* (beauty), respectively.[4]

speech are feminine. Indeed, as mentioned in the previous chapter, Jacob had two wives, Leah and Rachel.

The Torah juxtaposes the two sisters with these words, "Leah's eyes were soft, but Rachel had beautiful form and beautiful appearance."[5] Leah's eyes were soft because she cried so much. She knew that Rachel was destined to marry Jacob, and she concluded that she would fall into the hands of Jacob's older brother, the evil Esau.[6] These thoughts made her cry. Thus, she corresponds to communication through thought. A hidden mindset such as that of Leah is constantly in a state of tears. These might be tears of anxiety, or tears of joy that result from the revelation of Torah mysteries.

Jacob fell in love with Rachel not only because of her physical beauty, but because she had a practical and less philosophical mindset. Someone with an outlook on life such as Rachel's rejoices at every good deed they do and every word of Torah that they study. Obviously, men are attracted to happy girls, and Jacob was no exception.[7] In Kabbalah, Rachel is associated with speech.

Jacob's two wives correspond to the two local types of communication; Leah represents communication through thought, the concealed world. Rachel represents the two revealed methods of communication through intonation and speech. Jacob, as mentioned above, represents the male energy of communication through will.

The Power of Silence

As explained in Chapter 1, one of the words that appear in Ezekiel's most esoteric vision is "wheel" (אוֹפָן).[8] It has a numerical value of 137. Like the wheel of the Divine Chariot in Ezekiel's prophecy, electrons revolve around the nucleus of an atom in concentric shells. This is why we have suggested that *ofan* (אוֹפָן) be chosen as the official Hebrew word for "electron." The most esoteric word in Ezekiel's vision, and perhaps the most mysterious word in the Bible is *chashmal* (חַשְׁמַל).[9]

The literal reading suggests that *chashmal* is a hue of brilliant light.[10]

This is the visual aspect of *chashmal*. The more profound aspect of *chashmal* is thought transmission.▸

Electrons are the key factor to all chemical interactions. As their name implies, they relate to electricity. Their passage through an element produces an electric current.

There are three types of electricity, static electricity, direct current and alternating current. Industrial electricity began with direct current, which is a flow of electrons moving in one direction through the wire. In contrast, electrons in an alternating current move back and forth at a much slower speed than electrons in a direct current. Nonetheless, alternating current projects an electric field around the wire, which moves at the speed of light. It is thus far more efficient. This idea is reminiscent of another most important verse in the same chapter of Ezekiel: "and the living beings run and return like a flash of lightning."[11] Reading this verse with electricity in mind, we conlude that it is clearly referring to an alternating current. Chassidut explains that the living beings run and return, without advancing. The result of this constant movement is "like a flash of lightning," i.e., an electromagnetic field that moves at the speed of light, the type of radiation that exudes from alternating current. Appropriately, the modern Hebrew word for "electricity" is *chashmal* (חַשְׁמַל).

The three states of electricity are reflected in the word *chashmal* when the second syllable of the word is repeated, creating three syllables: *chash-mal-mal*, as taught by the Ba'al Shem Tov.[12] The first syllable, *chash* (חַשׁ) means "silence." This corresponds to static electricity. The first *mal* (מַל) relates to direct current and the second relates to alternating current. This division of *chashmal* into three syllables will be discussed in more detail later.

In the twenty-eight "times" that change in the world—as enumerated in Ecclesiastes—there is "a time to be silent and a time to speak."[13] In the consciousness of the three lower Worlds in which human consciousness functions, time is divided into discrete sections to allow for silence and speech to manifest at separate moments. In these

▸ The numerical value of *chashmal* (חַשְׁמַל) is 378. The Arizal referred to *chashmal* as "types of colors" (מִינֵי צְבְעוֹנִים), a phrase which also has a numerical value of 378. Appropriately, he taught that the visible spectrum of light contains 378 hues. The fact that *chashmal* refers to a brilliant light is reflected in its numerical value, 378, which is the triangle of 27 (3^3). 27 is the numerical value of "brilliant" (זַךְ). Similarly, the ordinal value of "light" (אוֹר) is also 27.

Chashmal appears three times in Ezekiel. These three appearances are the only times the word appears in the entire Bible. The first time *chashmal* appears is in the 4th verse of the book. The third time it appears is in the 142nd verse from the beginning of Ezekiel. Between the first and the last appearances there are exactly 137 verses.

three Worlds, we are unable to simultaneously tolerate two opposites. Although communication in these Worlds may sometimes be implicit in silence, speech cannot be explicit at those levels. Maimonides[14] explains that there are angels called *chashmalim* (חַשְׁמַלִים), that are present in the lower Worlds, who are "sometimes silent [חָשׁוֹת] and sometimes speak [מְמַלְלוֹת],"[15] but never both together.

In *Atzilut* (the World of Emanation), there are no such contradictions. In *Atzilut* everything is unified in harmony and all paradoxes, including the paradox of silence and speech, coexist in superposition. The consciousness of *Atzilut* incorporates communication through silence.[16]

The Letter of Communication

Of the entire Hebrew alphabet, the letter *vav* (ו) is the most communicative. In Chapter 8, we noted that it is one of the four letters (א ה ו י) that double as a consonant and a vowel. Its name, *vav* (וו), means "hook" or "link." It has many grammatical uses, also serving as two different vowels that aid in pronunciation, the *cholam* (ו) and the *shuruk* (ו).

In Hebrew, the word "and" (the most connective word) is represented by prefixing a *vav* to any word. Appropriately, this type of *vav* is referred to as "the connecting *vav*" (וו הַחִבּוּר). The very first *vav* that appears in the Torah is in the first verse, "In the beginning God created the heavens *and* the earth."[17] This indicates that the letter *vav* connects between the heavens and the earth, soul and body, energy and matter. The connective quality of the *vav* is its most common and most important function.

In the Torah, as a prefix to a verb, the *vav* often transforms it from one tense to another. It is then called a "transforming *vav*" (וו הַהִפּוּךְ). The first time the transforming *vav* appears in the Torah is in the phrase "and He said"[18] (וַיֹּאמֶר)—the phrase that precedes the creation of light. This is the first phrase in the Torah that explicitly relates to communication. Without the *vav*, the phrase would be in the future

▶ The numerical value of the phrase, "they are sometimes silent and sometimes they speak" (עתּים חָשׁות עתּים ממללות) is 2,300, which equals 100 times 23, the numerical value of "living one" (חַיָּה), the singular form of those angels who are called "living beings" (חַיּות).

▶▶ In *Atzilut*, time does not exist (see Chapter 13). There, *chashmal* refers to the ethereal light that garbed the Divine Chariot that Ezekiel saw in his prophecy. The numerical value of *chashmal* (חַשְׁמַל) is 378, which is also the numerical value of "garment" (מלבּוש).

▶▶▶ The *vav* in this verse is the 22nd letter from the beginning of the Torah. When the letter *vav* is spelled with a *yud* (ויו, the form associated with the *sefirah* of wisdom) its numerical value is 22, relating to the power of communication through the 22 letters of the *alef-bet*.

The author of *Haktav Vehakabbalah* writes that the letter *vav* is sometimes used comparatively. In this way, the phrase, "the heavens and the earth" means "the heavens like the earth." This alludes to an entirely new interpretation of the first verse of creation. God created reality in such a way that heaven and earth are equal; i.e., spirituality and materialism are parallel to one another.

tense ("He will say"; יֹאמַר). The prefix of the *vav* transforms the word into the past tense.

The connecting quality of the *vav* is particularly apparent in its place as the third letter of the essential Name of God, *Havayah*. The four letters of the Divine Name correspond to the ten *sefirot*. The *yud* (י) corresponds to *chochmah* (the *sefirah* of wisdom). The upper *hei* (ה) corresponds to *binah* (the *sefirah* of understanding). The letter *vav* of the Name (which has a numerical value of 6), corresponds to the six *sefirot* of the heart, or to *da'at* (the *sefirah* of knowledge). As mentioned, *da'at* is the connective faculty of the soul. The final *hei* (ה) of the Name corresponds to *malchut* (the *sefirah* of kingdom). The origin of the *vav* in the Divine Name is in the super-conscious crown, alluded to by the upper cusp of the initial *yud*. The *vav* is therefore the most significant letter of the essential Name of God.

The Connecting Letter and the Coupling Constant

The letter *vav* (ו), the letter of connection, is one of the most important allusions to the inverse of the electromagnetic coupling constant (137). The numerical value of the letter *vav* when it reaches its mature state of development is 137, as we shall see.

When contemplating any of the letters of the Hebrew *alef-bet*, one may do so through three stages: 1) the letter, 2) its filling (the letter's spelling) and 3) the filling of its filling (the spelling of the filling letters). Theoretically, this process of calculation could continue indefinitely. However, the Arizal[19] teaches that these three stages suffice to grasp the full meaning and implication of the letter. This three-stage process corresponds to the three syllables of *chash-mal-mal* and to the three stages of silence, separation and sweetening derived from them by the Ba'al Shem Tov.[20]

The letter alone (before any filling) corresponds to the stage of silence; it has no particular form or expression. The first filling corresponds to the stage of separation. It allows us to distinguish one letter from another through its name. The second filling corresponds to the stage of sweetening. It turns the letter into a complete form of expression.

Expanding the letter *vav* (when the letter is spelled with an *alef*, ואו) to the third level results in nine letters: *vav-alef-vav, alef-lamed-pei, vav-alef-vav* (ואו אלף ואו). The numerical value of these nine letters is 137.▾

Point, Line, Area

This three-stage process of developing a letter can also be described as an expansion from zero-dimension to two-dimensions.

Every letter begins as a zero-dimensional point.

The spelling of the letter's name (its first filling) corresponds to a one-dimensional line. [In the case of the letter *vav* (ו), its form is also a straight line, alluding to its filling].

With the next filling, the letter becomes two-dimensional, and it can be written in the form of a rectangle, or in our case as a square, as we shall see.

Each filling thus adds a new dimension. Mathematically, all three space dimensions can be projected onto a two-dimensional area, which is why the second filling of the letter suffices for most purposes.

▸ Some letters have two or more variant fillings, each of which is authentic. The *vav* is one of these letters. Its name can either be written *vav-vav* (וו), *vav-alef-vav* (ואו) or *vav-yud-vav* (ויו). In general, each possible filling has a different significance. In particular, three of the four letters of God's Name, *Havayah*, have three variant fillings. These are the two letters *hei* [which can be spelled *hei-alef* (הא), *hei-hei* (הה) or *hei-yud* (הי)] and the *vav*. Thus, there are 27 different ways of filling God's Name, with 27 different numerical values (some of which appear more than once). The four most significant fillings are referred to by their numerical values (72, 63, 45, 52), as spelled with Hebrew letters (עב, סג, מה, בן). Each of these values corresponds to one of the four letters of *Havayah*. The Name that corresponds to the letter *vav* is the Name *Mah*, which has a numerical value of 45 (מה). It is derived when the *vav* of God's Name *Havayah* is written *vav-alef-vav* (ואו).

As we saw previously, the *vav* of God's Name *Havayah* refers to the *sefirah* of *da'at*, the *sefirah* of connection, consciousness and communication. Here, we see that the letter *vav* in its full spelling equals the inverse of the electromagnetic coupling constant, 137.

Da'at contains two hemispheres, relating to love and fear. The third level of filling the *vav* (ואו אלף ואו) reflects this division. The sum of the numerical values of the first letter of the first word (ו), the second letter of the second word (ל) and the third letter of the third word (ו), is 42, alluding to the 42 times the root of the Hebew word for "love" appears in the Torah (see Chapter 12). The value of the remaining letters is 95, which relates to the 95 times that the root of the word for "fear" appears in the Torah.

Let us portray the point-line-area progression of the letter *vav*:

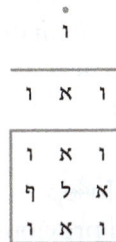

ׁו

ו	א	ו

ו	א	ו
א	ל	ף
ו	א	ו

The numerical value of *vav* (ו) is 6. The numerical value of the second stage of development (ואו) is 13, ▶ and the filling of the filling, the third stage of development (ואו אלף ואו) is 137.

The sum of the numerical values of the 13 letters contained in all three stages (6 ⊥ 13 ⊥ 137) of development is 156, which equals 6 (the initial numerical value of the letter *vav*) times 26 (the numerical value of the essential Name of God, *Havayah*). ▶▶ Here, the initial numerical value of the *vav* is multiplied by the Divine Name, which is another indication that the *vav* is the most significant letter of the essential Name, as explained previously. ▶▶▶

Vav and Joseph

In the Genesis account of creation, the verse[21] in which God blesses mankind with the power of procreation has 13 *vav*s. This is a self-reference to the filled value of the letter *vav* and the 13 letters of all three stages of filling. ▶▶▶▶

And God blessed them, and God said to them, "Be fruitful and multiply and fill the earth, and subdue it, and rule over the fish of the sea and over the birds of the sky and over all the beasts that tread upon the earth."

וַיְבָרֶךְ אֹתָם אֱ-לֹהִים
וַיֹּאמֶר לָהֶם אֱ-לֹהִים פְּרוּ
וּרְבוּ וּמִלְאוּ אֶת הָאָרֶץ
וְכִבְשֻׁהָ וּרְדוּ בִּדְגַת הַיָּם
וּבְעוֹף הַשָּׁמַיִם וּבְכָל חַיָּה
הָרֹמֶשֶׂת עַל הָאָרֶץ

▶ The *alef*, which lies in-between the two letters *vav*, represents Shabbat, which lies in-between the six days of preparation that precede it, and it blesses us for the six days until the next Shabbat.

In the phrase, "'Let there be light!' and there was light" (יְהִי אוֹר וַיְהִי אוֹר) there are 13 letters that are divided in the same way as the numerical value of the second level of spelling the letter *vav* (6-1-6, ואו). The first six letters of the phrase (יְהִי אוֹר), followed by the letter *vav* (ו) itself, and then the same six letters again (יְהִי אוֹר).

▶▶ 156 is equal to 12 times 13, where 13 is the one dimensional stage of development of the *vav* and 12 is the numerical value of *vav-vav* (וו), the simplest spelling-filling of the letter *vav*.

▶▶▶ This is further emphasized by the fact that there are exactly 26 *vav*s (with a total numerical value of 156) in the entire account of the first day of creation.

▶▶▶▶ The entire verse contains 22 words. The 13 *vav*s appear in nine words in the verse. This means that there are another 13 words that do not contain a *vav*. Once again, 13 is the numerical value of the *vav* when filled.

The sum of the numerical values of the 9 words with a *vav* is 1,849, which equals 43^2. The sum of the first 8 words in the verse without a *vav* is also 1,849.

The twenty-two words in this verse reflect the numerical value of the letter *vav* when filled with the letter *yud* (ויו).

As mentioned in Chapter 1, each of the *sefirot* has an inner motivating power. The inner power of *yesod* is truth—that is, the power that motivates procreation is self-verification. In the verse being discussed here, the first word that does not contain a *vav* is "them" (אֹתָם). The letters of this word permute to read "truth" (אֱמֶת). When the triangular value of each letter of the word *vav* (ואו) is calculated (i.e., the sum of all numbers to that number), the product (△6 times △1 times △6 = 21·1·21) is 441 (21²), which is the numerical value of "truth" (אֱמֶת).

This verse includes seven verbs, one of which is "And He said" (וַיֹּאמֶר), which is directly connected to communication. The sum of the numerical values of the other six verbs (וַיְבָרֶךְ פְּרוּ וּרְבוּ וּמִלְאוּ וְכִבְשֻׁהָ וּרְדוּ) is 1,370, which equals 10 times 137.

The numerical value of Ezekiel (יְחֶזְקֵאל) is 156, the same as the numerical value of Joseph (יוֹסֵף). It should also be noted that the first time a chariot is mentioned in the Torah is with reference to Joseph.[22]

In a somewhat lighthearted play on words, we find that the principal letters of Joseph's name (יוֹסֵף) are also the principal letters of the transliteration of "spin" (סְפִּין), for which a Hebrew equivalent has not yet been adopted. Similarly, "wheel" (אוֹפָן), mentioned in Ezekiel's vision of a chariot, has two letters in common with "spin" (סְפִּין). These two letters mean "face" (פַּן), which is mentioned in the plural (פָּנִים) with reference to the four faces of each figure that Ezekiel saw in his vision.[23]

The numerical value of Joseph (יוֹסֵף) is also 156. Joseph is the archetypal righteous soul of connection. He represents the sixth emotive attribute of the soul, *yesod* (the *sefirah* of foundation), associated with the letter *vav*, which has a numerical value of 6. In human physiology, this *sefirah* corresponds to the male procreative organ, and to the uterus in the female.

Of all the twelve tribes, only Joseph's soul derives directly from the World of *Atzilut*. As mentioned previously, *Atzilut* corresponds to the spin property of an electron. The fine-structure constant is related to spin and was first noted with reference to the splitting of spectral lines into two. Joseph too is particularly associated with splitting. Each of Joseph's two sons, Ephraim and Menasseh, became a tribe in their own right, bringing the total number of tribes to 13.

Not only did the tribe of Joseph split in two via his sons, Joseph himself had previously split himself into two. As a slave to Potiphar he detached himself from his clothing as he heroically fled from the clasp of his master's wife.[24] In the Torah verses that describe Joseph's flight, the key word "and he fled" (וַיָּנָס) appears four times.

When Joseph died, his coffin was buried in the Nile. Moses extricated the coffin before the Exodus.[25] Later, when the Children of Israel stood at the shores of the Red Sea, the waters split when Moses approached with Joseph's bones.[26] The Zohar[27] states that the verse in Psalms that refers to the splitting of the Red Sea, echoes Joseph's flight from the wife of Potiphar by using exactly the same word, "[the sea saw] and it fled" (וַיָּנָס).[28] The next phrase in this psalm states, "the [River] Jordan spun about face." The Midrash states that this was also in Joseph's merit.[29]

These verses demonstrate that spectral splitting—the Splitting of the Red Sea—is directly related to electron spin—the spin of the River Jordan—both of which occurred by merit of Joseph's righteousness.

More specifically, Joseph—the connecting force—relates to the full spin of force particles (bosons). The Torah states that "Abram travelled back and forth southwards."[30] This relates to the spin-half particle, which alternates between spin-up and spin-down. Abraham therefore corresponds to the half spin of matter particles (fermions). The numerical value of Abraham (אַבְרָהָם) is 248, which is the same as the word for physical matter (חֹמֶר).

Speech and Circumcision

Let's examine the development of the letter *vav* as a mathematical series beginning with the three numbers 6, 13 and 137. The base of the series is 117, which equals 9 times 13 (the second number in the series).▶

$$
\begin{array}{ccc}
6 & 13 & 137 \\
 & 7 & 124 \\
 & & 117
\end{array}
$$

Let's continue the series to another four places.

$$
\begin{array}{ccccccc}
\mathbf{6} & \mathbf{13} & \mathbf{137} & 378 & 736 & 1211 & 1803 \\
 & \mathbf{7} & \mathbf{124} & 241 & 358 & 475 & 592 \\
 & & \mathbf{117} & 117 & 117 & 117 & 117
\end{array}
$$

The sum of the first 7 numbers in the series is 4,284, which equals 7 times 612, the numerical value of "covenant" (בְּרִית)—i.e., 612 is the average numerical value of all seven numbers. In particular, "covenant" refers to the covenant of circumcision, which is carried out on the male procreative organ. This corresponds both to Abraham, who was the first to carry out this commandment, and Joseph, who represents that organ. The act of circumcision relates to the mysterious word *chashmal* (חַשְׁמַל), mentioned earlier, which has a numerical value of 378.▶▶ It appears here as the fourth number in the series, directly following 137.

Chashmal is a key concept in communication. The first syllable means "silence" (חַשׁ) and the second syllable means "speech" (מַל). As mentioned, the Ba'al Shem Tov taught that the three-stage process of

▶ The differences between the three top numbers are 7 and 124 and when added together with 117 they equal 248, the numerical value of Abraham (אַבְרָהָם). The sum of all six numbers together is 404, which is the numerical value of "sanctity" (קֹדֶשׁ), and the numerical value of "God is one and His Name is One" (הוי' אֶחָד וּשְׁמוֹ אֶחָד).[31]

The fourth difference is 358, the numerical value of Mashiach (מָשִׁיחַ).

The sum of the first 13 numbers is 34,086, which is equal to 6 (ו) times 13 times 19 (חַוָּה, the original name of the first woman) times 23 (חַיָּה, the rectified name of the first woman, as explained in Kabbalah).

There are 613 commandments in the Torah, which are divided into 248 active commandments and 365 prohibitive commandments. If we form a mathematical series from these three numbers, we find that the difference between the first two numbers (248, 365) is equal to 117 (the base of the series above), and the difference between the second two numbers is 248. The base number is therefore 131, the sum of the two differences in the second line of the series above.

▶▶ 378 is also the numerical value of "clothing" (מַלְבּוּשׁ), which Joseph left behind when he fled from Potiphar's wife.

233

submission, separation and sweetening is alluded to by revealing a third, hidden syllable that lies inbetween the two syllables. The three syllables are thus *chash-mal-mal* (חַשׁ־מַל־מַל). As noted, *chash* means "silence," the first *mal* relates to circumcision, and the second *mal* means "speech."

The covenant of circumcision is the source of a man's ability to connect with a woman in sanctity. The act of circumcision comprises three stages:

1. First, the foreskin is removed. This stage is called *milah* (מִילָה), circumcision.

2. Then the transparent membrane is torn away, in order to reveal the "crown" of the limb, *pri'ah* (פְּרִיעָה).

3. In the third and final stage of circumcision, the resulting blood is drawn out of the wound. This stage is called *metzitzah* (מְצִיצָה).

The three stages of circumcision correspond to the refining process of submission, separation and sweetening. This process is demonstrated in the three stages of rectified communication: silence (submission); deliberating on one's words to clarify them of any negative implications (separation); and finally, communication via refined speech (sweetening).◄

Back to the Future

The fantasy of time travel tantalizes humanity. If it became possible, people could return to a previous point in their lives and rectify all errors that affect their current state of affairs. A trip into the future would allow them to glimpse the implications of their present efforts before they enacted them.

As of now, we have no experience of things moving backward in time. Similarly, we have no proven way to travel into the future. Some physicists believe that a time machine that enables time travel is a possible invention of the future.◄◄

The letter *vav* incorporates the necessary features of a virtual

▶ The numerical value of *milah* (מִילָה) 85, which is the numerical value of "mouth" (פֶּה). The sum of the numerical values of the three stages of circumcision (מִילָה פְּרִיעָה מְצִיצָה) is 685, which equals 5 times 137, relating to the five origins of the mouth, via which verbal communication is conducted.

▶▶ Currently, time-machines remain in the realm of science-fiction. Nonetheless, the numerical value of "time machine" (מְכוֹנַת זְמָן) is 613, the totality of all of the commandments in the Torah. This indicates that when fulfilled in its entirety, the Torah allows for time travel. In fact, the spiritual power of keeping all the 613 commandments allows us to travel forwards or backwards in time with perfect freedom, as we shall see later, regarding the power of sincere *teshuvah* (returning to God in repentance).

time-machine. When it appears as a prefix to a verb, it can transform it from the past to the future tense, or from the future to the past tense, as mentioned.

This mysterious quality of the *vav* alludes to the power of *teshuvah* (return to God). Every soul has the power to return to God through love. Heartfelt remorse for our past misdemeanors and sincere return to God, transform all our sins into merits.[32] Like the *vav*, *teshuvah* transforms the past into the future.

Through the power of *teshuvah*, we can repair the past. However, Rebbe Nachman of Breslov writes that in order to recapture the time wasted by our past misdeeds, we need to move extremely fast.[33] This suggests that time travel to the past would only be feasible if we could travel at speeds faster than the speed of light, as modern physics suggests.

One way of affecting the future is through telepathy.[34] A *tzadik*, a truly righteous individual, has the power to change the natural course of history by telepathically projecting his will and thoughts into the future through prayer. He too can transform the future into the past.

The intermediate, who is neither a *ba'al teshuvah* nor a *tzadik*, progresses by communicating to others through charitable deeds. Giving charity is one of the most refined ways of communication. This relates to the connecting *vav*, which is the most common use of the *vav* as a prefix.

Light to the Nations

In Chapter 2, we saw how the Torah passage relating to the menorah in the Temple is related to 137. The Temple is where the most potent communication with God takes place. The light of the menorah that shone out of the Temple was one way of transmitting that communication at the speed of light.

Like the Temple, Noah's ark was a haven of peace and salvation. In Kabbalah, the ark is sometimes more significant than the Temple. The

▶ The sum of the numerical values of the two synonyms for "word" 407 (תֵּבָה) and 75 (מִלָּה) is 482. The average value of the two words is thus 241, which is the numerical value of the root "to say" (אימ״ר). This is the first root in the Torah that refers to communication, in the verse, "And God said, 'Let there be light'."[35]

▶▶ The numerical value of "Make a light source for the ark" (צֹהַר תַּעֲשֶׂה לַתֵּבָה) is 1,507, which equals 11 times 137. The average value of each letter is exactly 137, the cosmic number of communication.

The numerical value of "ark" (תֵּבָה) is 407, which equals 11 times 37 (the companion number of 137).

The 11 letters of this phrase further correspond to the 11 dimensions of M Theory, which unites 5 different versions of string theory (corresponding to the five origins of speech). Each of these versions contains either 10 dimensions (the number of *sefirot*) or 26 dimensions (the numerical value of the Name *Havayah*).

Hebrew word for "ark" (תֵּבָה) is also a synonym for "word."◀ Indeed, many aspects of Noah's ark allude to the laws of refined speech.[36] Thus, Noah's ark is also a symbol of communication.

One of the most important phrases in God's commandment to Noah to build the ark, is "make a light source for the ark" (צֹהַר תַּעֲשֶׂה לַתֵּבָה).◀◀[37] One opinion states that the source of light in the ark was a window or a skylight, via which light entered. Another opinion is that a precious stone radiated light into the ark. According to the opinion that the source of light was a window or skylight, any light that radiated from the ark would also illuminate the world—like the light that shone from the Temple menorah. This opinion relates to the revealed level of verbal communication through speech, corresponding to Rachel.

The opinion that the source of light in the ark was a mysterious precious stone, relates to the hidden level of communication through thought and telepathy, corresponding to Leah.

Other permutations of Noah's "light source" (צֹהַר), create Hebrew words that mean "make narrow" (הָצֵר) or "hardship" (צָרָה). When presented with the confining hardships of life, we can transform them into a source of light. The Ba'al Shem Tov explains[38] that this can be achieved via another permutation of the same letters to create the Hebrew word for "will" (רָצָה), the highest form of communication (via will-power). This permutation is also the root of, "satisfying one's will" (לִרְצוֹת). By consciously acknowledging our own re-creation at every moment, we have the power to rearrange the basic letters of communication and create a new reality.

Thus, the phrase "make a light source for the ark" directs us how to achieve perfect communication. God commanded Noah, "Come into the ark," which we can interpret as, "come into the word." By bringing our innermost selves into the words that we speak, uniting our thoughts with our speech as we project our will via our thoughts in spoken prayer to God, we are able to illuminate our own inner world and the outside world around us.

The Message Inherent in Creation

According to quantum mechanics, a force manifests as an inter-exchange of force particles. Force particles are generally massless. For example, a photon, the conveyor of the electromagnetic force, has zero rest mass. The existence of such "objects" that have no form teaches us that we can no longer think of the universe in purely physical terms. Instead, they suggest that we consider the word/message that every phenomenon in the world transmits. Whereas an amorphous, inanimate object—a "particle" if you will—may appear to be a meaningless "thing," the word "particle" relays specific content and conveys a message to the world.

We have identified two Hebrew synonyms related to communication through speech (מִלָּה, תֵּבָה). A third synonym for "word" (דָּבָר) is a generic word that also means "thing." This suggests that every "thing" has a "name/word," through which it sends a message that it communicates to—and even commands—the conscious world.[39]

Commanding Creation

As mentioned, the epitome of refined communication is when pure will is transmitted via thought to speech. This type of communication creates reality.[40] The numerical value of the three Hebrew words "will" (רָצוֹן), "thought" (מַחֲשָׁבָה) and "speech" (דִּבּוּר) are 346, 355 and 212, respectively. The sum of these three numbers is 913, which equals the numerical value of the opening word of the story of creation, "In the beginning" (בְּרֵאשִׁית).[41] Thus we see that refined communication manifests in the ability to experience re-creation at every moment.

The medieval Kabbalist, Rabbi Abraham Abulafya, coined a Hebrew phrase that means "three things/words together" (שְׁלֹשָׁה דְּבָרִים יַחַד) that also has a numerical value of 913. The initials letters of Abulafya's three-worded phrase spell out the Divine Name *Shakai* (שַׁדַּי), usually translated as "the Almighty." Physicists have an aversion to infinities in the created world, and one interpretation of this Divine Name is that God set a finite limit to His creation by saying, "Enough!" (דָי).[42] Since the moment that creation began, God has interacted with us by instilling His will in creation through His speech. He created all that we know through the Ten Utterances, starting with "In the beginning," and He has placed limits on creation by saying, "Enough!" so that it should not expand infinitely.

But that's not all, God has also set up a two-way channel of communication between Himself and mankind. He has given us the gift of prayer and the ability to reach the heights of prophecy and hear His word.

These two channels are reflected in two of Maimonides' thirteen principles of faith. One of the levels of prophecy that Maimonides describes[43] is prophecy via Divine messengers, such as angels. Physicists might prefer to call them "force particles."

Angels are mentioned many times in the Torah—for example, in Genesis we read about the cherubim, which bar the entrance to the Garden of Eden. Other angels' names are mentioned explicitly in the various books of the Bible.[44] However, most are identified only in esoteric Kabbalistic texts.

The angel Matatron is associated with the Divine Name *Shakai*, "He who said 'Enough'." The two share the same the numerical value of 314. This Name of God is associated with *yesod* (the *sefirah* of foundation), and thus to Joseph. The Zohar refers to Matatron as a type of angel called an *ofan* (אוֹפָן),[45] which has a numerical value of 137. In fact, Matatron is another name for Yofiel (יוֹפִיאֵל), the Angel of Beauty, whose name has a numerical value of 137. Matatron is also referred to as "Minister of the Interior" (שַׂר הַפְּנִים). The numerical value of this title is 685, which equals 5 times 137.◀

Besides handing out beauty, the Arizal taught that Matatron/Yofiel is the ministering angel who praises God with the phrase, "Let the glory of *Havayah* be forever" (יְהִי כְבוֹד הוי' לְעוֹלָם).[46] This is the opening phrase in the compilation of eighteen verses that we recite during daily morning prayers. The entire compilation contains exactly 137 words.

▶ The numerical value of "minister" (שַׂר) is 500, 5 times 100, and the numerical value of "interior" (הַפְּנִים) is 185, 5 times 37. This division into 100 and 37 is particularly significant when we recall that 37 is the companion number of 137.

It is almost impossible to communicate without labeling things with names. "Noun," or "name" (שֵׁם), is also the name of Noah's son, Shem. Yofiel is also the Minister of the Torah.[47] It was he who taught Shem the secrets of the Torah.[48] As noted above, Jacob, who represents the initial state of unadulterated will, first studied Torah wisdom in the tent of Shem. Jacob is referred to as the "Pillar of the Torah." Beauty is also particularly relevant to Jacob, who is associated with *tiferet* (the *sefirah* of beauty).▶

In various Kabbalistic texts Matatron/Yofiel is also identified as the angel by whom God's 42-lettered Name was revealed to the Jewish People. This Divine Name—with which the world was created—enables us to rise in our prayers from one state of consciousness to the next.

Yofiel is responsible for all the keys of wisdom. He is the angel who conveys Torah wisdom to mankind. He examines our souls during Torah study and rewards individuals for their proper use of wisdom by giving them more knowledge,[50] as the verse in Proverbs states, "Give to the wise and he will become wiser."[51]

So, we see that Yofiel is a Divine messenger, a conduit of prophecy who imparts wisdom and knowledge to humans. Most significantly, as mentioned in the Introduction, the total numerical value of "prophecy" (נְבוּאָה, 64) and "wisdom" (חָכְמָה, 73) is 137, the numerical value of Yofiel (יוֹפִיאֵל) and the numerical value of the Torah's inner wisdom, Kabbalah (קַבָּלָה). Wisdom and prophecy are communicated through silence and speech, the secret of the mysterious word, *chashmal*.

In the above-mentioned mathematical series generated from the three stages of development of the letter *vav* (6 13 137), the fourth number is 378, the numerical value of *chashmal* (חַשְׁמַל). The sum of the numerical values of Yofiel (יוֹפִיאֵל, 137) and *chashmal* (חַשְׁמַל, 378) is 515, which is the numerical value of "prayer" (תְּפִלָּה), implying the union of the wisdom of the Torah (originally transmitted to Moses through prophecy) with prayer.

▶ The inner motivating power of *tiferet* is truth, as in the phrase in Micah, "Give truth to Jacob."[49] The sum of the numerical values of "truth" (אֱמֶת, 441) and "beauty" (יפי, 100)—both of which are perfect squares—is 541. This is also the numerical value of Jacob's God-given name, Israel (יִשְׂרָאֵל).

The Twelve Tribes and the Twelve Months

In Kabbalah, there are twelve spiritual senses, which correspond to the twelve months of the year, the twelve signs of the Zodiac, and the twelve tribes.[52]

The letter *vav*, whose connection with communication has already been explained, is associated with the month of Iyar, the second of the twelve months of the Hebrew calendar. The Zodiac sign of Iyar is an ox. The ox is a symbol associated with Joseph.[53] Nonetheless, the tribe associated with Iyar is Issachar, the scholarly tribe of the Jewish People. The talent associated with the month of Iyar is thought. In Hebrew, thought implies contemplation and introspection. Issachar's innate nature is contemplative. This tribe served as counselors to the other tribes of Israel and to their royal leader, Judah.

The Sanhedrin, the great court of sages, was composed mainly of sages from the tribe of Issachar.[54] In particular, the tribe of Issachar is considered master of the secret of the Jewish calendar,[55] as we read in the Chronicles, "And from the sons of Issachar, those who know the understanding of times...."[56] This refers to the complex astronomical calculations necessary to compute the Hebrew calendar, the specialty of the tribe of Issachar. In our context, Issachar's understanding of time relates in particular to "a time to remain silent and a time to speak."[57]

In the Hebrew Bible, the month of Iyar is called the month of *Ziv* (זִיו), meaning "radiance." The name Iyar (אִיָּר) is also cognate to "light" (אוֹר). Thus, both names of this month allude to thought communication at the speed of light. This idea is expressed in the Talmudic idiom, "Open your mouth and let your words shine forth."[58]

The Tribes of Speech and Hearing

Refined communication relies on the obvious connection between two senses: speech and hearing. The sense of speech relates to the tribe of Judah and the Hebrew month of Nisan, which precedes Iyar. A king rules by his speech, as we are taught, "There is one spokesman (i.e.,

> ▶ Previously, we saw that in Hebrew "thinking" (חֹשֵׁב) is cognate to "sense" (חוּש) indicating that thought is the most sensitive sense of all. But, "thinking" (חֹשֵׁב) with a different vowelization means "calculation" (חֶשְׁב), from which the Hebrew word for "mathematics" (חֶשְׁבּוֹן) also stems. Counting is the most basic act of calculation. Appropriately, during the entire month of Iyar we count the *Omer*.

the king) for every generation,"[59] and his word is law. Appropriately, Judah is the royal tribe.

Nisan and Iyar, speech and thought, correspond to Rachel and Leah, the two wives of Jacob. Judah and Issachar were Leah's sons; however, Judah has an inherent bond with Rachel's son, Joseph. Joseph's two sons were Ephraim and Menasseh, who merited tribes of their own. In the future, the tribes of Judah and Ephraim will unite to form the ultimate royal family.[60] ▶

The sense of hearing corresponds to the month of Av and to the tribe of Shimon. The men of the tribe of Shimon were schoolteachers. A crucial concern in education is that every child and adult realize their purpose in life by listening to, becoming conscious of, and manifesting their inherent talents and qualities.[62] The tribe of Shimon turned their listening ears to recognizing their students' potential and helping them actualize it.▶▶

The first month of the year, Nisan, speaks, as it were, and the fifth month of the year, Av, listens. In the Torah, Nisan is called "the Month of Spring" (חֹדֶשׁ הָאָבִיב),[64] which shares a common root with the word *Av* (אָב). Note that the portion of the tribe of Shimon in the Land of Israel lies within the boundaries of Judah's inheritance.▶▶▶ This suggests that true listening requires hearing the innate expression of the speaker's soul within their words.[65]

Refined Communication

As we saw in Chapter 2, the Hebrew language symbolizes the most refined form of communication. Through the Holy Tongue we can communicate verbally, as well as through thought. In one of his books on Hebrew grammar, Rabbi David Kimchi, known as the Radak, explains that the Hebrew language is divided into three sets of words:[66]

1. names (שֵׁמוֹת), including proper nouns, adjectives and numbers;
2. verbs (פְּעָלִים) and their conjugations;

▶ Ephraim and Menasseh correspond to the months of Tishrei and Cheshvan, and to physical communication (touch) and the sense of smell, respectively. At first glance, the sense of smell does not seem to be related to communication. Nonetheless, in Kabbalah, the sense of smell is the highest level of communion between humans.[61] Scientists today are investigating genetic matching by choosing a potential marriage partner through smell.

▶▶ The numerical value of Shimon (שִׁמְעוֹן) is 466. The sum of 466 and 860 (the sum of the numerical values of Judah and Issachar) is 1,326, which equals 13 [the numerical value of the *vav* when spelled in full, "love" (אַהֲבָה), "one" (אֶחָד) etc.] times 102 [the numerical value of "faith" (אֱמוּנָה)], and it is the triangle of 51. This is the numerical value of the final verse of the Priestly Blessing, "May God raise His countenance toward you and grant you peace" (יִשָּׂא הוי' פָּנָיו אֵלֶיךָ וְיָשֵׂם לְךָ שָׁלוֹם).[63] The Priestly Blessing is the epitome of refined speech in the Temple. (See Chapter 3.)

▶▶▶ The sum of the numerical values of Judah (30, יְהוּדָה) and Shimon (466, שִׁמְעוֹן) is 496, which equals the numerical value of "kingdom" (מַלְכוּת). *Malchut* (the *sefirah* of kingdom) is associated with the mouth and with speech.

The sum of the numerical values of "thought" (מַחְשָׁבָה), "speech" (דִּבּוּר) and "hearing" (שְׁמִיעָה), is 992, which equals twice 496.

241

3. words (מלים), including other types of nouns, prepositions and pronouns.

These three sets correspond to the three spiritual worlds of *Beriah* (the World of Creation), *Yetzirah* (the World of Formation) and *Asiyah* (the World of Action). Names correspond to *Beriah* because they form the initial basis of language, without which communication cannot be achieved; verbs correspond to the *Yetzirah*; all other words correspond to *Asiyah*.

The sum of the numerical values of the categories of these three sets (שֵׁמוֹת פְּעָלִים מִלִים) is 1,096, which equals 8 times 137, the numerical value of "an intelligent woman" (אִשָּׁה מַשְׂכֶּלֶת), as discussed in Chapter 8.

Reference Notes for Chapter 9

1. See also *913: The Secret Wisdom of Genesis*, p. 4 (in side-bar).

2. Ecclesiastes 1:6.

3. See our article, "Kabbalah and Telepathy"
   http://www.torahscience.org/communicat/kabbalah_and_telepathy.htm.

4. See our book in Hebrew, *Hanefesh*, p. 71.

5. Genesis 29:17.

6. Rashi on Genesis 29:17.

7. Rebbe Dovid of Lelov.

8. Ezekiel 1:15, 16.

9. Ibid 1:4, 27; 8:2.

10. Rashi on Ezekiel 1:4.

11. Ezekiel 1:14.

12. *Keter Shem Tov* 28a.

13. Ecclesiastes 3:7.

14. Maimonides, *Hilchot Yesodei Hatorah* 2:7; from the word *chashmal* (חַשְׁמַל); see Chapter 1.

15. *Chagigah* 13b.

16. See *Lectures on Torah and Modern Physics*, Lecture 1. See also our article "Kabbalah and Telepathy"
   http://www.torahscience.org/communicat/kabbalah_and_telepathy.htm.

17. Genesis 1:1.

18. Ibid 1:3.

19. *Etz Chaim* end of Chapter 18 and Chapter 19.

20. Explained in detail in our book, *From Darkness to Light*.

21. Ibid 1:28.

22. Genesis 41:43.

23. Ezekiel 1:6.

24. Genesis 39:12.

25. Exodus 13:19.

26. *Bereishit Rabah, Parashat Vayeishev* 84:8.

27. *Zohar Shemot, Beshalach* 48b.

28. Psalms 114:3.

29. *Bereishit Rabah, Parashat Vayeishev* 84:5.

30. Genesis 12:9.

31. Zachariah 14:9.

32. *Yoma* 86b.

33. *Likutei Moharan* 49:7.

34. Telepathy is strongly related to prayer. See our article, "Kabbalah and Telepathy," http://www.torahscience.org/communicat/kabbalah_and_telepathy.htm.

35. Genesis 1:3.

36. See the article, "Noah's Ark and the Gift of Cosmic Speech," http://www.inner.org/audio/aid/E_039.htm.

37. Genesis 6:16.

38. Additions to *Keter Shem Tov* 8-9.

39. See *Likutei Moharan* 1:1.

40. This refers in particular to the process of *keviah retzonit*, described in our book *Lev Lada'at*, pp. 91-94.

41. See our book *913: The Secret Wisdom of Genesis*.
42. *Chagigah* 12a.
43. Maimonides' *Guide for the Perplexed*, Part 2 Chapter 41.
44. Daniel 8:16; 9:21; 10:13, 21; 12:1.
45. *Zohar Bereishit* 21a.
46. Psalms 104:31.
47. *Zohar Chadash, Tikun Kadma'ah*.
48. Ra'avad on *Sefer Yetzirah*.
49. Micah 7:20.
50. *Pardes Rimonim* 24:2.
51. Proverbs 9:9.
52. *Sefer Yetzirah* 4:7-9.
53. See Deuteronomy 33:17.
54. *Midrash Tanaim Devarim* 33.
55. *Bamidbar Rabah* 13:15.
56. I Chronicles 12:33.
57. Ecclesiastes 3:7.
58. *Berachot* 22a.
59. *Sanhedrin* 8a.
60. Ezekiel 37:19.
61. This is discussed elsewhere in our teachings.
62. See our book *The Art of Education*.
63. Numbers 6:26.
64. Exodus 13:4; 23:15; 34:18; Deuteronomy 16:1.
65. *Likutei Moharan* 173.
66. See Introduction to *Sefer Michlol*.

THE EXPANDING UNIVERSE

Solving the Universal Paradox

The ancient Greeks, with their penchant for binary logic, posed a pertinent question regarding the size of the universe. Is the universe finite or infinite? Accepting the fact that the universe is infinite was difficult enough to imagine, but if it is finite, they asked, what is on "the other side"? This question has continued to puzzle astronomers, and it has led to a profound paradox with a more sophisticated twist.

On the one hand, if the universe is infinite and contains stars throughout, then in any direction that you happen to look, your line-of-sight should eventually fall on the surface of a star. Although the apparent size of a star in the sky becomes smaller as the distance to the star increases, the brightness of this smaller surface remains constant. Therefore, if the universe is infinite, the whole surface of the night sky should be as bright as a star. Since dark areas exist in the sky, the universe must be finite.

On the other hand, gravity always attracts, therefore every object in the universe effects an attractive force on every other object. If the universe is truly finite, the attractive forces of all the objects in the universe should have caused the entire universe to collapse in on itself. Since this cosmic catastrophe has clearly not happened, the universe must be infinite.

This means that the universe must simultaneously be both finite and infinite!

When Einstein was working on general relativity, his calculations indicated that the universe should be either expanding or collapsing, but he couldn't work out which it was. Ignoring his calculations, he assumed the universe to be static, and he added the cosmological constant to his equations, which cancelled the effects of gravity. As he himself stated later, ignoring the results of his calculations and assuming that the world was static was the "greatest blunder" of his professional life. Within a decade and a half, astronomical observations proved that the universe is expanding. This discovery resolved the questions concerning the finite/infinite nature of the universe. The expanding universe is finite in both time and space. However, it is in a constant state of change.

By extrapolation, proof of the expanding universe led to the understanding that the universe must have begun at some stage in history as an infinitesimal point. This is the basis of the Big Bang Theory and the belief that the universe is billions of years old.

In chapter 4, we explained that the basis of any Grand Unified Theory (GUT) is that, at the instant of the Big Bang, the universe reached a tremendously high temperature, at which all four natural forces were united as one. ◀ As the universe began to cool, a symmetry break occurred that threw the universe into a state of constant diffusion at a tremendous speed. Philosophically speaking, the expansion of the universe represents a state of disharmony and contention, as each entity becomes more distant from all the others.

▶ The verse, "For *Havayah* your God is a consuming fire"[1] alludes to the forces of nature (and all spiritual reality), which unite by means of fire (heat). The four forces that are manifest in nature correspond to the four letters of the Name *Havayah* (see Chapter 4). Prior to their creation, at the initial stage, all four forces are one, and "God is One."

One of the most fundamental rules of physics is that every force has an opposing force. Following this rule, the expanding, distancing force must be balanced by an attracting force that counteracts the diffusive effect of the former. Current scientific analysis of an expanding universe includes gravity in its considerations. Cosmologists today are in dispute regarding which of the two forces—gravity or expansion—will triumph. If the total gravitational pull within the universe overrides the expansion, it will cause contraction and the universe will eventually collapse in the "Big Crunch." If expansion is the stronger force, the universe will continue to expand forever, but the energy of the universe will gradually dissipate, leading to the "Big Chill."

At this level of discussion, cosmology remains mostly a matter of speculation.

The View of *Sefer Yetzirah*

Sefer Yetzirah describes how the universe began as an infinitesimal cube with eight zero-dimensional vertices, twelve one-dimensional bars, which connect the vertices, and six two-dimensional surfaces.[2] The total number of these three is 26, the numerical value of God's Name, *Havayah*. The power of expansion in that infinitesimal "atom" is in the bars, which are capable of expanding into three-dimensional space.[3] The six sides of the twelve bars in three-dimensions total seventy-two sides. On each side is written one of the seventy-two Names of God[4] that correspond to the power of loving-kindness. This is one interpretation of the verse in Psalms, "A world of loving-kindness is built."[5] The numerical value of "loving-kindness" (חֶסֶד) is 72.

Loving-kindness relates to the ability to extend outwards, representing the expansion of the universe. However, it also represents the power of attraction.[6] *Chesed* (the *sefirah* of loving-kindness) is represented by the right arm[7] that embraces the beloved, as the Song of Songs states, "His left arm is beneath my head and His right arm embraces me."[8] Thus, our Kabbalistic analysis suggests that there is an attractive force that works hand-in-hand with the expansion of the universe to prevent it from expanding infinitely.

The idea of a universe that begins from an infinitesimal size—as *Sefer Yetzirah* describes—is more appealing to scientists than the idea that the universe was created ex-nihilo. As Brian Greene writes,

> *In a universe governed by string theory, one is free to measure distances using either one of two approaches. There is a monumental difference in difficulty between the two methods. To date, the method used in all cases measures the size of the universe by examining photons that have traveled across the cosmos and happen into the telescopes of astronomers. According to this method, the universe is large and expanding. In principle, astronomers using vastly different (and currently nonexistent) equipment, should be able to measure the extent of the heavens with heavy wound-string modes*

and produce the result that the universe is tiny and contracting. There is no contradiction here; instead we have two distinct but equally sensible definitions of distance.

There are some physicists who have made use of these ideas to suggest a rewriting of the laws of cosmology in which both the Big Bang and the possible Big Crunch do not involve a zero-size universe, but rather one that is Planck-length in all dimensions. This is certainly a very appealing proposal for avoiding the mathematical, physical, and logical conundrums of a universe that emanates from or collapses to an infinitely dense point. Although it is conceptually difficult to imagine the whole of the universe compressed together into a tiny Planck-sized nugget, it is truly beyond the pale to imagine it crushed to a point of no size at all.[9]

A positive state of uncompromising expanse is one in which the hard core of sanctity expands to incorporate the entire physical universe. Of the Jerusalem of the future, it is written:[10]

Jerusalem will be settled as a city without walls.... "And I," declares Havayah, *"will be a wall of fire around it, and I will be a glory inside it."*	פְּרָזוֹת תֵּשֵׁב יְרוּשָׁלַם... וַאֲנִי אֶהְיֶה לָהּ נְאֻם הוי' חוֹמַת אֵשׁ סָבִיב וּלְכָבוֹד אֶהְיֶה בְתוֹכָהּ

First, it appears that Jerusalem will be "a city without walls." No walls will be able to contain it; neither will it need walls for protection. Then, God declares that He Himself will be Jerusalem's "wall of fire." The teaching is clear: physical walls will not be necessary, because Jerusalem will become Divine space with God around it and God within it.

The Divine "wall of fire" is not static, but eternally alive, it continuously expands. In the future, the Temple will expand to embrace the whole of Jerusalem; Jerusalem will expand to embrace the entire

Land of Israel;[11] and the Land of Israel will expand to embrace the entire world.[12]

Divine space at once defines the boundaries of our consciousness while simultaneously causing it to expand until it contains the entire cosmos. The sages refer to this phenomenon as "a boundless inheritance"[13] based on God's promise to Jacob: "You shall break forth to the west, and to the east, and to the north, and to the south."[14]

Large and Small

Brian Greene (as quoted above) mentions two possible methods of measurement. These are also important in the development of M-Theory, which attempts to synthesize all five versions of string theory into one. There is compelling evidence that each version of string theory is dual. For example, when the coupling constant of a Type I string is large, the particular masses and charges are precisely equal to those of the Heterotic-O string when its coupling constant is small, and vice versa. Like the transmutation between water and ice, which at first appear to be two distinct entities, each transforms into the other. This is true of two couples of string theories. Detailed analysis persuasively suggests that the fifth theory (Type IIB) is self-dual and transforms into itself.[15]▶

This paradoxical discovery by string theorists, indicates that a universe with a radius of R is identical to that with a radius of 1/R. This means that the larger the universe becomes according to the regular method of measurement, the smaller it is by the alternative method (currently unavailable). Taking this idea further, in a universe with an infinite radius, the alternative method of measurement would yield an infinitely small universe. In the Introduction, we mentioned that the equation that defines the fine-structure constant combines the one pure dimensionless number that connects relative infinity (the speed of light) and relative zero (Planck's constant). Thus, the pure dimensionless number 137 should be of great significance in deciphering this riddle.

▶ The number of gauge fields and gaugions necessary for Type I string theory is exactly 496, which is the numerical value of "kingdom" (מלכות), alluding to the *sefirah* of *malchut*.

In mathematics, 496 is the triangle of 31, which is the numerical value of *Kel* (אֵ־ל), the Divine Name associated with *chesed* (the *sefirah* of loving-kindness). In addition, 496 is a "perfect number" (i.e., a number that equals the sum of all of its divisors). 496 is also equal to 2 times 248, the numerical value of Abraham (אַבְרָהָם), the archetypal soul associated with *chesed*.

The Talmudic dictum states, "Wherever you find God's might, there you find His humility."[16] This also relates to an important moral teaching in the Zohar, "One who is large is small, and one who is small is large."[17] Those who diminish their ego are large, but those who aggrandize themselves are small.

The Ba'al Shem Tov and his disciples throughout the generations strive to be conscious of the higher worlds of consciousness even as they are engrossed in the mundane world. This idea is compacted in the Chassidic idiom *in velt ois velt* ("In the world and out of this world"). Similarly, the Maggid of Mezeritch coined the Yiddish phrase, *Atzilus iz oich doh* ["*Atzilut* (the World of Emanation) is here too"]. In order to reach this level of dual consciousness, we must learn how to "measure" the world by both standards of measurement, and unify them.

Continous Re-Creation

The expanding universe is the result of the cooling process that began after the original Big Bang. An alternative to the Big Bang theory is "Continuous Creation." This theory suggests that dark matter or elementary particles are created ex-nihilo in the vacuum of outer space at every moment.

The innovation of the Ba'al Shem Tov in Chassidic sources also teaches us that there is continuous creation. However, this is essentially different from the creation of new particles suggested by physicists. The Chassidic teaching is that not only are new particles created, but in essence, every particle is continuously being created anew.

Based on this teaching, we can hypothesize that a continual "big bang" takes place at every moment at the mid-point of the universe. Like a tree that expands with a new ring each year, this constantly introduces new energy into the universe, causing it to expand. Thus, the creation of the world was not a unique event at the beginning of history but a continuous process of expansion.

Scientists measure reality relative to the speed of light. Similarly, Kabbalah teaches us that God created the world by "measuring" it with the ray of light that penetrates the vacuum. As we began to explain in Chapter 6, this infuses the vacuum with new energy. Initially the vacuum is an infinitesimal point, called the "residual point." The ray of light, in the form of the time coordinate, enters that point in space and the vacuum begins to expand. This results in a new circle being created at every moment within the older circles as the line continues to extend.

According to this theory, at the first moment of creation, space and time were one unified point. As the ray of Infinite Light descends, it generates circles. Each new circle then pushes out all of the previous circles. This process continues and represents ex-nihilo creation at every moment of time.

The ever-expanding rings are held together by an opposing force that "hugs" the universe from the outside, preventing it from exploding. Thus, even though the universe appears to be constantly expanding, with all matter seemingly diffusing, there is another, yet unidentified force that "hugs" it together from "outside." ▶

Since the discovery of the expanding universe, observations imply that the speed of the expansion is not constant but accelerating. This fact has tremendous implications for modern cosmology. One of these is that entire galaxies of stars will disappear from view as they cross the "event horizon." On the basis of this understanding of the accelerating universe, some scientists believe that eventually, the universe will be "ripped apart" by the influx of dark matter. This climactic end to the universe is called the "Big Rip."

An alternative hypothesis is that gravitation will eventually overcome all other forces. This will eventually result in the collapse of all matter in "the Big Crunch." However, these suppositions about the end of the universe are merely wild speculations that cannot be verified and are not expected to occur for the next few billion years.

The theory we propose is that there is new energy being pumped

▶ Psychologically, we can achieve a state of positive expanse by striving for wholeness. We learn from the Ethics of the Fathers: "Do not be like a servant who serves his master in order to receive a prize, rather be like a servant who serves his master not in order to receive a prize."[18] Chassidut offers a witty interpretation. "A prize" (פְּרָס), also means "a part," as in "a slice of cake." The Chassidic interpretation humorously suggests that we should not be like students who study in order to receive a slice of cake. Rather, we should be like students who study without compromising an inch to receive the whole cake!

251

into creation at every moment. Simultaneously, a parallel hugging force holds the universe together by its power of attraction.

From a Kabbalistic perspective, the influx of new energy corresponds to the relatively masculine force of *yesod* (the *sefirah* of foundation) acting within *chesed*. The restraining force corresponds to the relatively feminine power of *malchut* (the *sefirah* of kingdom) acting within *chesed*; this feminine power reigns by setting rules and regulations.

The connection between *yesod* and *malchut* is often referred to as a covenant (בְּרִית).◀

There are a number of key covenants that God made with mankind, all of which correspond to *yesod* (the *sefirah* of foundation). They all relate mathematically to a most important series of numbers.

- The first covenant, which God made with Abraham, is the covenant concerning the Land of Israel.[19] In the Torah verses which relate God's promise of the Holy Land to Abraham's offspring, the word "covenant" (בְּרִית) appears once.◀◀

- The same word appears three more times with reference to the covenant God made between the Jewish People and the Torah. [22]

- Earlier in the Torah, in the chapters relating to the covenant of the rainbow—the universal covenant that God made with all mankind promising never again to destroy the world by flood—"covenant" (בְּרִית) appears seven times.[23]

- The sages teach us that the covenant of circumcision is greater than the covenant of the Torah. "Covenant" (בְּרִית) appears thirteen times in the context of Abraham's circumcision. This covenant is explicitly related to the *sefirah* of *chesed*, represented by Abraham. It is simultaneously associated with the *sefirah* of *yesod*, which corresponds to Joseph and the procreative organ, on which the act of circumcision is performed, as discussed previously.

The number of times that the word "covenant" is mentioned in these four places reveals an ascending progression from 1 to 3 to 7 to 13.

These are the first four numbers of the "covenant series." By calculating the finite differences between the numbers of the series, we can calculate the next numbers in the series and produce an algebraic equation that will define it.

$$1 \quad 3 \quad 7 \quad 13 \quad 21 \quad 31 \quad 43 \quad 57 \quad 73 \quad 91$$
$$2 \quad 4 \quad 6 \quad 8 \quad 10 \quad 12 \quad 14 \quad 16 \quad 18$$
$$2 \quad 2 \quad 2 \quad 2 \quad 2 \quad 2 \quad 2 \quad 2$$

The first ten numbers in the sequence of Covenant Numbers▸

▸ The tenth number of the series ("All tenths are holy to God") is 91, which is the sum of the numerical value of the Name of God as it is written (הו"י, 26) and the Name as it is read (אדנ"י, 65).

Although the original series contains four numbers and we would therefore expect it to generate a cubic function (in which the higest power of n in the equation is n^3), the second differences (the differences between the first differences of 2, 4, 6…) both equal 2. Thus, these four numbers generate a quadratic series (in which the highest power of n is n^2).

The algebraic formula for this series is

$$f(n) = n^2 \perp n \perp 1$$

Mathematically speaking, when we discuss the motion of an object, acceleration is expressed by a square number (n^2); constant velocity is described by n (n^1), and static position by a single unit (n^0; which always equals 1). Thus, in order to "measure" the present position of the universe, the equation: $f(n) = n^2 \perp n \perp 1$ is a useful one indeed.

Quality versus Quantity

Every algebraic statement has a geometric form. For example, as explained in Chapter 1, the algebraic equation

$$f(n) = n(n \perp 1)/2$$

is represented by a triangle. The equation for two triangles is therefore

$$n(n \perp 1) = n^2 \perp n$$

Our original equation of $f(n) = n^2 \perp n \perp 1$ is thus equal to two triangles $\perp 1$. The series of covenant numbers can therefore be represented as two equilateral triangles with heads facing one another with one point between them.

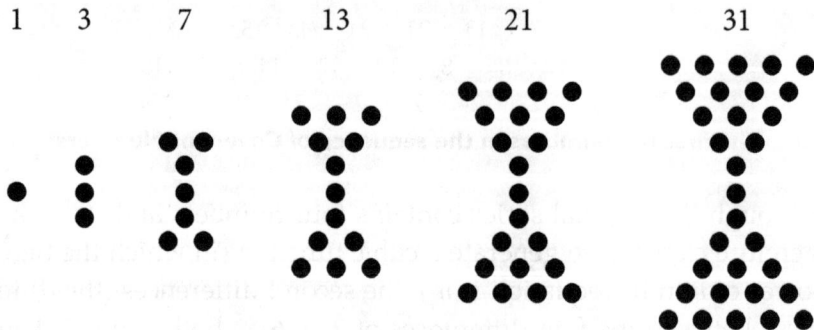

Qualitavely speaking, this geometric form symbolizes a masculine force descending and being integrated by an ascending feminine force. The point in the middle represents God entering between them to connect the two. This corresponds to the spelling of the letter *vav* (ואו) as we saw in the previous chapter, where each letter *vav* has a numerical value of 6 ($\triangle 3$) and the middle letter is the *alef*, with a numerical value of 1.

This also perfectly defines the marriage covenant, of which the sages state, "If they [the married couple] merit, then the Divine Presence is between them."[24] This symbol is therefore a symbol of the marriage covenant.

Before we continue, let's take note that although the number 137 does not appear explicitly in this series, the first three integers are 1, 3 and 7 in that particular order. This is of great significance when we consider that in physics, $\frac{1}{137}$ is the electromagnetic coupling constant, defined by the ratio between the speed of the electron and the speed of a photon (i.e., the speed of light), as mentioned in the Introduction. As such, it defines all interactions between the two. So, in practice, 137 is the number that defines the marriage, or the connection between the electron and the photon. It therefore defines the relationship between energy and matter, the spiritual and material aspects of the universe.

This provides us with a clue to the secret of 137 and the accelerating, expanding universe as it is limited and contained by an attractive force.

Reinforcing the Connective Force

All of the numbers in the above-mentioned series are defined by $2 \triangle n \perp 1$, as shown. This means that all of the numbers in the series are odd numbers.

Any odd number (n) has a mid-point that is defined by $[(n-1)/2] \perp 1$. In the case of the series of covenant numbers the mid-point is thus defined by $\triangle n \perp 1$.

The midpoints of the covenant numbers are represented below by one of the triangles together with the middle circle. The numbers can be counted either from the top down or from the bottom up; in either case, the results achieved are identical.

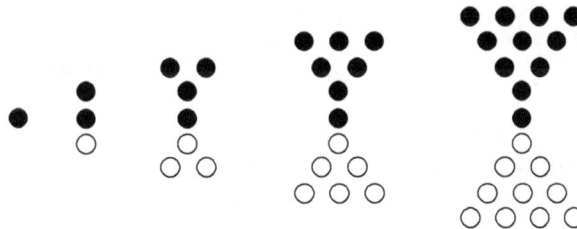

Covenant number	1	3			21
Mid-point	1	2	4	7	11

Let us now create a second series of numbers, defined qualitatively as the covenant number plus its mid-point, or algebraically by $3\triangle n \perp 2$. In this new series, 137 is explicitly present. It appears in the tenth position ($n = 9$), signifying that it is a holy number, "all tenths are holy to *Havayah*."[25]▼

	Covenant Number	Mid-Point	Covenant Plus Mid-Point ($3\triangle n \perp 2$)
1	1	1	2
2	3	2	5
3	7	4	11
4	13	7	20
5	21	11	32
6	31	16	47
7	43	22	65
8	57	28	85
9	73	37	110
10	91	46	137

▶ The fifth number in this series is 32, which is the number of times that "covenant" (בְּרִית) appears in the Torah without a prefix; the eighth number is 86, the numerical value of the Name *Elokim* (אֱ־לֹהִים), which appears 32 times in the account of creation; the sixth number in this series is 47, which is the sum of the numerical value of the two Divine Names *Havayah* (26) and *Ekyeh* (21), representing the father and mother principles, or higher fear and higher love.

The seventh number in the series is 65, which is the numerical value of the Name *Adni* (אֲדֹנָי). This corresponds to *malchut* (the *sefirah* of kingdom), the seventh *sefirah*, with which the Name *Adni* is associated. The sum of the first seven numbers until 65 is 182, which is the numerical value of Jacob (יַעֲקֹב). The number 182 equals 7 times 26, i.e., the average value of all seven numbers is 26, the numerical value of God's essential Name, *Havayah*. The sum of the numerical values of *Havayah* and *Adni* is 91, which appears here as the tenth number in the series of covenant numbers.

This series has feminine symmetry, which centers around 2-2, alluding

to the large letter *bet* with which the Torah begins. The rule for this type of symmetric series is that the average of the first 13 numbers must equal the eighth number in the series.

Indeed, the sum of the first 13 numbers is 1,118, which is the numerical value of the verse "Hear O' Israel, *Havayah* is our God, *Havayah* is one" (שְׁמַע יִשְׂרָאֵל הוי׳ אֱ־לֹהֵינוּ הוי׳ אֶחָד). 1,118 is the product of 13 [the numerical value of "one" (אֶחָד)] and 86 (the eighth number in the series, according to the rule), which is the numerical value of the Name *Elokim* (אֱ־לֹהִים). *Elokim* appears in the verse [in the form "our God" (אֱ־לֹהֵינוּ)].

Because this series is symmetrical, the number 1,118 appears as the thirteenth number on the left as well as the thirteenth number on the right. The sum of these 26 members of the series is thus 2 times 1,118, which equals 2,236, or 26 (the numerical value of *Havayah*) times 86 (the numerical value of *Elokim*). The double value of 1,118 alludes to proclaiming our faith in God with this verse twice every day, in the morning and in the evening.

Here we have found another, purely numerical way to derive 137 with slightly sophisticated mathematics.

The division of 137 into 46 and 91 is a significant one.▶ In Chapter 4, we mentioned that one of the four basic words with a numerical value of 137 is "origin" (מוֹצָא). The numerical values of the two syllables of this word are 46 (מוֹ) and 91 (צָא), respectively. "Origin" can relate to the origin of an individual soul, the origin of the Jewish People as a whole, or the single origin of a married couple.▶▶

Various Names of God and Kabbalistic intentions are alluded to by many other numbers in this new series. Of particular note to us here is that the sum of the first nine numbers is 378, the numerical value of *chashmal* (חַשְׁמַל). Since the next number is 137, the numerical value of Yofiel (יוֹפִיאֵל), the sum of all ten numbers equals the sum of *chashmal* plus Yofiel, which, as we saw in the previous chapter is equal to the numerical value of "prayer" (תְּפִלָּה).

This series is derived by adding the covenant numbers to their mid-points. This alludes to reinforcing the Divine connective force that bonds the two parties who make the covenant.

As the verse in Jeremiah states, the connection between God and the universe is also referred to as a covenant:[28]

So said Havayah, "If it were not for	כֹּה אָמַר הוי' אִם לֹא
My covenant, by day and by night,	בְרִיתִי יוֹמָם וָלָיְלָה
I would not have set the statutes of	חֻקּוֹת שָׁמַיִם וָאָרֶץ
heaven and earth."	לֹא שָׂמְתִּי:

The numerical value of this verse is 3,145, which equals 85 times 37, the companion to 137. However, when "statutes" (חֻקּוֹת) is written with an additional letter *vav* (חוּקוֹת), corresponding to the alternative spelling of "statute" (חוּק) in the Bible,[29] the numerical value is 3,151, which is exactly 23 times 137.▶▶▶ (See Chapter 12 for more about the relationship between 23, 37 and 137.)

The "statutes of heaven and earth," i.e., the laws of nature, depend

---

▶ The sum of the numerical values of the four Divine Names, *Kah* (15; יָ-הּ), *Havayah* (26; הוי'), *Kel* (31; אֵ-ל) and *Adni* (65; אֲ-דֹנָי) is 137. In Kabbalah, the order of these four Names is *Kah, Kel* and then *Havayah*, and *Adni*. The numerical values of these two pairs are 46 (*Kah* and *Kel*) and 91 (*Havayah* and *Adni*).

The sum of the numerical values of the four Divine Names, *Kah* (15 ; יָ-הּ), *Havayah* (26 ; הוי'), *Kel* The base number of the series formed from this particular sequence is 65, the numerical value of *Adni*. The seventh "All sevenths are beloved") number in the series is 1,096, which equals 8 times 137 (Chapter 8).

15	31	26	65	213	535	1096
	16	-5	39	148	322	561
		-21	44	109	174	239
			65	65	65	65

▶▶ In the Bible, there are two individuals with the name Motza (מוֹצָא, meaning "origin"), one is a descendant of Caleb[26] and one is from the tribe of Benjamin.[27] Motza of the tribe of Benjamin appears twice in Chronicles. His name ends the verse that begins with Achaz, and it begins the following verse that ends with name Atzel. The numerical value of Achaz (אָחָז) is 16 (4^2) and the numerical value of Atzel (אָצֵל) is 121 (11^2). The sum of their two names is thus 137, the numerical value of Motza (מוֹצָא).

▶▶▶ The Arizal sometimes calculates the full spelling of words, even when the word lacks a certain vowel letter.

upon God's "covenant by day and by night." This reinforces the Continuous Creation or constant Big Bang and Cosmic Hug theory proposed above. The male principle, the recurring Big Bang, corresponds to day, in which the sun rises at daybreak and expands until sundown. Nighttime corresponds to the hugging force. Because night is dark by nature, this force is not yet perceptible to science. This is suggestive of dark energy and dark matter, which remain an enigma.

In Jewish teachings, the sun, associated with daytime, is a masculine symbol. The moon, representing nighttime, is feminine. Just as the moon's gravity is responsible for the changing tides of the ocean, so too the feminine attractive force hugs the universe at night.

This theory thus takes into consideration both results of Einstein's equations, in that the universe is expanding (during the universal "day") and collapsing (at "night"). Taking our experience of day and night on earth as a parable for this universal day/night, we see that the heat of the sun is the expanding force of the universe, which contracts at night as the earth cools.

REFERENCE NOTES FOR CHAPTER 10

1. Deuteronomy 4:24.

2. See *Likutei Hagra* at the end of *Sefer Yetzirah* 2:25b. See also our book, *Living in Divine Space*, Supplementary Essay 10, p. 129.

3. *Sefer Yetzirah* 5:2.

4. The 72 Names that are derived from the three verses in Exodus 14:19-21. See *Chasdei David Hane'emanim* Part 7, p. 208.

5. Psalms 89:3.

6. *Tanya* Chapter 45.

7. *Tikunei Zohar, Patach Eliyahu.*

8. Song of Songs 2:6; 8:3.

9. Adapted from Brian Greene, *The Elegant Universe*, Chapter 10, Quantum Geometry.

10. Zachariah 2:8-9.

11. See *Shir Hashirim Rabah* 7.

12. *Psikta Rabati, Piska Deshabbat Verosh Chodesh*, 3.

13. See Rashi on Isaiah 58:14.

14. Genesis 28:14.

15. *The Elegant Universe*, Chapter 12.

16. *Megillah* 31a.

17. *Zohar Bamidbar* 168b; the *Zohar* relates this directly to Sarah's lifespan of 127 years.

18. *Avot* 1:3.

19. Genesis 15:18.

20. See Genesis 17:5

21. *Bereishit Rabah* 7:5.

22. *Berachot* 48b; 49a (according to Rashi's commentary).

23. Genesis 9:8-17.

24. *Sotah* 17a.

25. Leviticus 27:32.

26. I Chronicles 2:46.

27. Ibid 8:36, 9:42.

28. Jeremiah 33:25.

29. See for example Proverbs 8:27, 29.

MARRIAGE

Two Wives

One way of perceiving the current, conflicting state of connection between Torah and science is as two wives of the same husband in a bigamous marriage.▶ Each wife feels that the other wife is stealing her husband from her. Even if they were once as close as sisters, once they become wives to the same husband they are referred to in the Bible[1] as "trouble" (צָרָה) in relation to one another, because they compete for the same man.

The classic example of one man with two wives is Jacob, who married Leah and Rachel. As we have seen, Leah was philosophically inclined, while Rachel was pragmatic. Leah thus represents the hidden mysteries of the Torah and Rachel represents the revealed side that manifests in the practical application of it laws. Similarly, the two sisters represent Torah and science. Leah's and Rachel's relationship was refined by their sisterly love for one another, suggesting a rectified relationship between Torah and science.

Ultimately, rectification is achieved by completely eliminating the state of two wives. This is not accomplished by divorcing one of the wives, but by rectifying the bigamous attitude.▶▶ In our case in particular, this refers to rectifying our attitude towards Torah and science. As we have seen, the average numerical value of Leah (לֵאָה) and Rachel (רָחֵל) is 137. So, 137 plays an important role in the unification of Torah, corresponding to Leah, and science, corresponding to Rachel.

Maimonides was a Jewish medieval Torah sage and renowned physician to the Egyptian sultan. He obviously loved the Torah immensely. Yet, he expressed a fear that his love of the "maidservant," science, would outdo his love for his original "wife," Torah wisdom. Maimonides quotes a verse from Proverbs[4] that lists certain things

▶ Bigamy is permitted in the Torah. It was forbidden by the sages in later times.

▶▶ Adultery is an illegitimate state of bigamy that is prohibited by the Torah. Rebbe Nachman of Breslov[2] states that "adulterer" (נוֹאֵף) has the same numerical value as Kabbalah (קַבָּלָה)—137—to teach us that incorrect interpretation of ideas according to Kabbalah is tantamount to adultery. [3]

that are abominable to God, one of which is the phenomenon of a maidservant who overpowers and inherits a noblewoman's place. This was precisely what Maimonides feared would happen to him.[5]

Like Jacob and Maimonides, we too have fallen in love, as it were, with two wives. The ultimate solution to this conflict is to recognize how they are one and the same woman. Scientifically, the apparently impossible assignment of seeing two objects as one becomes feasible by abstracting some of the notions of M-Theory.

M-Theory unifies the five different versions of string theory. This unified theory hypothesizes that strings are 1-dimensional slices of a 2-dimensional membrane that vibrates in 11-dimensional space. One of the theoretical beauties of multi-dimensional space is that by taking a totally new perspective in a different dimension, we can perceive two objects as one. This property is referred to as super-symmetry.

We see this super-symmetry in the relationship of God, the Jewish People and the Torah. The Torah was one of seven things God created before He created the world.[6] In fact, it was the blueprint He used to create it.[7] Twenty-six generations later, God gave the Torah to the Jewish People.[8] Thus, at their super-symmetrical source, God, the Jewish People, and the Torah are one.[9]

The new perspective proposed by string theory and M-theory offers insight into our understanding of the apparent conflict between Torah and science. By elevating ourselves to perceive reality from the super-symmetrical origin that preceded the Torah and preceded creation, we will discover that Torah **is** science.

A Romantic Rendezvous

Just like two soul-mates who are attracted to one another until they finally meet and join in marital union, Torah and science are pulled towards one another by some mysterious force. Eventually, they will meet and unite. The point at which this meeting occurs is at the horizon.

In mathematics, the horizon is the infinite point of a coordinate system, where dimensions theoretically "kiss" one another. Most mathematicians believe that such infinite points exist. *Sefer Yetzirah* calls the infinite state of a coordinate system "depth"—for example, "depth of the south," "depth of the north," etc.[10]

Each of the ten *sefirot* is one extremity of a five-dimensional system (three space dimensions, one time dimension and one consciousness dimension) in which every coordinate has two infinite points.

The common soul-root of the couple in any marriage—or their super-symmetric origin—is reflected at the horizon, where heaven and earth kiss. Similarly, the Torah/science marriage is reflected at the furthest frontier, where the most cutting-edge scientific theories lie. The most conducive context for experiencing the kissing of Torah and science, heaven and earth, is in analyzing from a Torah perspective those scientific topics that lie on the horizon.

In order to meditate upon any concept, we must first relate to its Hebrew name. The two lettered "gate" of "horizon" (אֹפֶק) is *pei-kuf* (פ־ק). Other meanings of this gate can illuminate our understanding of the horizon concept. In particular, the root (פ־י־ק) is found in the verse from Proverbs[12] that is traditionally written at the head of the *ketubah*, the Jewish marriage contract:

> *"He who has found a wife has found goodness; he will produce good will from God."*
>
> מָצָא אִשָּׁה מָצָא
> טוֹב וַיָּפֶק רָצוֹן
> מֵהוי׳

The association between "he will produce" (וַיָּפֶק) and "horizon" (אֹפֶק), suggests that in order to produce good will from God—Infinity—a well-matched couple must always live at the horizon.

The origins of a spectral line doublet are in one single line (a phenomenon defined by 137, as mentioned in Chapter 1). This symbolizes the secret of marriage as the union of two souls with one common

▶ The numerical value of "Torah and science" (תּוֹרָה וּמַדָּע) is 731. The integers of 731 are 137 in the reverse order. Like a ray of light that is shone at a crystal-clear mirror, numbers with this mirror-image effect symbolize the best reflection of each other. Other examples of pairs that share this quality are "light" (אוֹר) and Shabbat (שַׁבָּת), whose numerical values are 207 and 702, respectively, and Rachel (רָחֵל) and "the Land of Israel" (אֶרֶץ יִשְׂרָאֵל), whose numerical values are 238 and 832, respectively.

▶▶ The same root is used in the word meaning "to exert self-control" (לְהִתְאַפֵּק), which teaches us that one needs self-control when approaching the horizon, otherwise one may go too far beyond. The Talmud[11] relates a story about Ben-Zoma, a sage who went insane. At one point his friends saw him in deep meditation and they asked what he was contemplating. He replied that he was contemplating the distance between the firmaments. Since he was calculating cosmological "super-symmetry" in outer space, they concluded that he was already "on the other side" and they were correct in their assumption!

Another verse expresses the outcome of an unfavorable marriage, "I find this woman more bitter than death."[13] In this verse, the root "to find" is in the present perfect tense (מוֹצֵא), with identical letters to "origin" (מוֹצָא). The numerical value of this word, as mentioned, is 137.

A deeper Chassidic interpretation of this verse is, "I find myself more bitter than death with the woman." Those who are looking out for their own self-interests ("finding themselves") when they come into marriage, will perceive it as a bitter experience.

On the basis of these two verses, it was the tradition to ask a newly married man, "Have you 'found' or do you 'find?'" (מָצָא אוֹ מוֹצֵא).[14] The question was intended to help the groom realize that the outcome of his marriage depends on his perspective of his wife and his marriage. If he puts his self-interests before his wife—"I find myself..."—his life will be "more bitter than death." If he is prepared to live self-lessly and to "find" the good points in his wife, as in the verse, "He who has found a wife has found goodness," his marriage will be a good one, "and he will produce good will from God."

Obviously, the same applies to a wife's attitude towards her husband.[15]

The total numerical value of the six alternating words beginning with the first word of the verse (וּשְׁתֵּים אֲבָנִים יְהוֹשֻׁעַ הַיַּרְדֵּן מַצַּב הַכֹּהֲנִים) is 1,781, which equals 13 times 137. The total numerical value of the five words interposed between these six words (עֶשְׂרֵה הֵקִים בְּתוֹךְ תַּחַת רַגְלֵי) is 2,209, which equals 47.

The numerical value of "and they have remained" (וַיִּהְיוּ) is 37 (the companion number to 37) and the numerical value of "until" (עַד) is 74, 2 times 37.

origin. Via their separate journeys to this world, each soul must rediscover its origin by connecting to their other half.

"Origin" (מוֹצָא) appears in Numbers in a verse which literally reads: "Moses wrote the origins of their journeys, according to God's command; these are the journeys to their origins."[16] This verse is followed in the Torah by a list of the 42 places from which the Children of Israel departed and the place of their arrival during their sojourn in the wilderness.

Rabbi Shneur Zalman of Liadi, the founder of the Chabad Movement, explains that this verse alludes to the origin of the soul and its journey through this world throughout life. The literal reading of the verse suggests that the ultimate purpose of the soul's journey is to return to its origin.[17]

The Jewish People finally entered the Land of Israel following their 42 journeys through the wilderness. As they stood at the east bank of the Jordan, the priests stood in the river valley as the Jordan split so that the Jewish People could pass through it. At the point in the river where the priests had stood, Joshua erected a monument of twelve stones.[18]

And twelve stones did Joshua erect within the Jordan beneath the position of the legs of the priests who carried the Ark of the Covenant; and they remain there until this day.	וּשְׁתֵּים עֶשְׂרֵה אֲבָנִים הֵקִים יְהוֹשֻׁעַ בְּתוֹךְ הַיַּרְדֵּן תַּחַת מַצַּב רַגְלֵי הַכֹּהֲנִים נֹשְׂאֵי אֲרוֹן הַבְּרִית וַיִּהְיוּ שָׁם עַד הַיּוֹם הַזֶּה

The numerical value of this entire verse is 5,754, which equals 42 times 137, which is also the numerical value of "origin" (מוֹצָא). This refers to the 42 origins of their journeys in the desert as described in the verse, "Moses wrote the origins of the journeys … and these are the journeys to the origins" (וַיִּכְתֹּב מֹשֶׁה אֶת מוֹצָאֵיהֶם לְמַסְעֵיהֶם... וְאֵלֶּה מַסְעֵיהֶם לְמוֹצָאֵיהֶם).[19]

"The position" (מַצָּב) is the masculine form of "monument." Thus, the verse is clearly referring to the erection of a monument. As we shall see in the next chapter, "monument" (מַצֵּבָה) is another word with a numerical value of 137.▸

God, the Matchmaker

The relationship between the massless photon, the smallest quantum of energy, and the electron, the smallest matter particle, is essentially a marriage. An electron can integrate a photon and become energized to a higher quantum state. The number 137 is the Divinely sent "matchmaker" between the two. In a vacuum, a single photon moves at the speed of light. The highest speed that an energized electron can reach is $\frac{1}{137}$th of the speed of light.[20] Thus, 137 essentially marks the quantum leap from energy to matter and the connection between the masculine photon and the feminine electron.

More than any other phenomenon in the world, marriage is an act of Divine Providence. The sages teach us that in each of the three parts of the Bible there is one verse that explicitly states that matches are made in heaven.[21]

In the Torah

The verse in the Torah that relates to this phenomenon is the statement made by Rebecca's family before she leaves to marry Isaac. Betuel and Laban, Rebecca's father and brother—far from being righteous individuals—stated in complete sincerity, "This matter has come from God" (מֵהוי' יָצָא הַדָּבָר),[22] thereby acknowledging that this was Divine Providence.▸▸

As explained in Kabbalah, Isaac and Rebecca were the only truly monogamistic couple. Abraham married Sarah, who later gave him her maidservant, Hagar, as a wife. After Sarah's passing, Abraham married Keturah.[23] Jacob married two sisters and their two maidservants. Thus, the most perfect monogamous match among the

▸ The total numerical value of the first verse in the Book of Joshua is 3,562, which equals 137 times 26 (the numerical value of *Havayah*, the essential Name of God). The numerical value of the final verse in Joshua is 4,111, which is equal to 30 times 137 [10 times "chaos" (תהו)] plus the additional unit (the *kolel*).

▸▸ The numerical value of the unique phrase, "This matter has come from God" (מֵהוי' יָצָא הַדָּבָר) is 378, which is the numerical value of *chashmal* (חַשְׁמַל).

265

▶ With reference to Isaac's love for Rebecca the verse states, "And she became his wife and he loved her" (וַתְּהִי לוֹ לְאִשָּׁה וַיֶּאֱהָבֶהָ). The first three letters of this phrase (ותה) are the letters of "chaos" (תהו). The numerical value of of the next 13 letters is 411, which is the numerical value of "chaos" (תהו). The numerical value of the entire phrase is thus 822, the numerical value of "coming into being" (הִתְהַווּת), which equals 6 times 137.

▶▶ The sum of the numerical values of their names, Isaac (יִצְחָק) and Rebecca (רִבְקָה) is 515, the numerical value of "prayer" (תְּפִלָה). As we have seen in Chapter 9, 515 is also the sum of 137 and 378 (chashmal, חַשְׁמַל).

Rebbe Schneur Zalman of Liadi, the Alter Rebbe[26] teaches that the marriage of Isaac and Rebecca exemplifies the secret of the rectification of the world. It alludes to the perfect unification of male and female. This refers to the union of the Divine Names, Mah (45) and Ban (52), two Names of God that correspond to the male and female principles of the universe. The total numerical value of their names, together with the two Names of God (יִצְחָק רִבְקָה מה בן) is 612, which is the numerical value of "covenant" (בְּרִית). This alludes to the perfect messianic marriage covenant, in which both partners are of equal spiritual stature.

Rebecca corresponds to the Name Ban (בן), which refers to her "son" (בֵּן), i.e., Jacob. The numerical value of "Jacob son of Rebecca" (יַעֲקֹב בֶּן רִבְקָה) is 541, the numerical value of Israel (יִשְׂרָאֵל), the name that God gave to Jacob. This emphasizes Jacob's role as Rebecca's son in particular.

Patriarchs is that of Isaac and Rebecca. Rebecca was born while Isaac was bound to the altar, when Abraham was 137 years old and Isaac was 37 years old. It was then that their match was made in heaven.

The first mention of love in the Torah is Abraham's love for Isaac. The next time love is mentioned is Isaac's love for Rebecca. This is the first time that love is mentioned in the Torah with reference to a man and woman. Thus, just like Isaac was passively involved in Jacob's love for Rachel (see Chapter 7), so too, Abraham's love for his son bred Isaac's love for Rebecca. ◀

Isaac and Rebecca waited ten years after their marriage until Rebecca was mature enough to bear children and then beseeched God in heartfelt prayer.[24] Isaac and Rebecca are the only couple in the Torah who explicitly prayed together, "And Isaac beseeched God, opposite his wife."[25] ◀◀

In the Prophets

The verse from the prophets that states that meeting one's match comes straight from God (in this case, even when the match is inappropriate) appears in the story of Samson. While still a bachelor living with his parents, Samson saw a Philistine maiden and wanted to marry her. Because he honored his parents, he asked them to arrange the match between them. His parents, after certain resistance, gave in to his request and approached the girl's family. At this stage of the proceedings in the Bible, the prophet states: "But his father and his mother did not know that from God was this."[27] The final word in this phrase, "this" (הִיא) literally means "she." It is not the word that we would expect to find in this context. As mentioned in Chapter 1, this is the only combination of three letters whose filled value equals 137.

In the Writings

The verse in the Writings that states that a match comes only from God is the verse in Proverbs discussed in Chapter 8, "House and wealth are the inheritance of fathers but from *Havayah* [comes] an intelligent woman."[28] As mentioned, the numerical value of "an intelligent woman" (אִשָּׁה מַשְׂכֶּלֶת) is 1,096, which equals 8 times 137. Since this phrase has 8 letters, the average value of each letter is 137.

Thus, in each of the three Divinely ordained marriages in the Torah, Prophets and Writings, we find an implicit reference to the number 137.

More About Marriage

Isaac and Rebecca's marriage, the first example from the Torah of a Divinely ordained match, is the epitome of a most kosher match. By contrast, Samson's marriage to the Philistine maiden was a most non-kosher marriage, which ended abruptly when the Philistines incinerated the woman and her father.[29] Nonetheless, their match was also Divinely ordained.

It is apparent from the context of the story in the Torah that Rebecca was a righteous woman who was clearly worthy of joining Abraham's household. However, there is no explicit term in the text that describes her righteousness. The verses referring to Samson's wife do not describe her; her name is never mentioned. Yet, we are taught that she converted when he married her.[30]

Ruth, who married Boaz, was also a righteous convert. She confirms the statement that an intelligent woman is not an inheritance of fathers. The sum of the numerical values of Boaz (בֹּעַז) and Ruth (רוּת) is 685, which equals 5 times 137.▸ The Zohar states that Boaz and Ruth, together with the other souls mentioned in the Book of Ruth, were the souls with whom God consulted before creating the world.[31]

Another couple in the Bible that relates to 137 is King Saul's daughter Michal (מִיכַל) and her second husband, Paltiel (פַּלְטִיאֵל). The total numerical value of their names in primordial numbering is 1,096, the numerical value of "an intelligent woman" (אִשָּׁה מַשְׂכֶּלֶת), which equals 8 times 137. Michal and Paltiel withstood the greatest challenge of

▸ The ordinal value of Boaz (בֹּעַז) is 25 (כה) and the ordinal value of Ruth (רוּת) is 48 (מח), the sum of which is 73, the numerical value of "wisdom" (חָכְמָה) which is a permutation of these letters (כה and מח). The numerical value of "a covenant of wisdom" (בְּרִית חָכְמָה) is equal to 685, which equals 5 times 137.

▸ In number theory, the midpoint is the middle of an odd number. If, for example, there are seven candles lined up, the midpoint is the fourth candle, which has three candles before it and three candles after it. An even number has no middle number, but it can be split into two equal halves. The Arizal notes that the number 7 is of particular significance in this context, since it is the midpoint of 13. The number 7 relates to Shabbat which has six days before it—during which we prepare for the coming Shabbat—and six days that follow that receive blessing from Shabbat, reaching a total of 13 days, with Shabbat, the seventh day, in the middle. As explained in Chapter 9 with reference to the spelling of the letter *vav* (ואו)..

In physics, vector forces of quantum numbers are added by multiplying each of the numbers by its pair in the second vector. Similarly, when two words have the same number of letters, the numerical value of each letter can be multiplied by the numerical value of the parallel letter in the second word, and the products added together.

In the case of Dinah (דינה) and Job (איוב), the dot product $(1 \cdot 4 + 10 \cdot 10 + 6 \cdot 50 + 2 \cdot 5)$ is 414, which is the numerical value of, "God's infinite light" (אור אין סוף) and other phrases. 414 is the product of 6 times 69, which is the numerical value of Dinah (דינה), as mentioned above. Thus, through her marriage to Job, Dinah is multiplied by 6. Dinah was the only daughter of Jacob and Leah. Her marriage to Job balanced the scales to make her equal to her six maternal brothers.

▸▸ The Zohar states, "If [the nation of] Israel will observe one Shabbat, they will immediately be redeemed."[41]

a celibate marriage. King Saul had promised to give Michal to the soldier who triumphed over Goliath.[32] When David succeeded in killing the giant, Michal told Saul that she loved David and they married.[33] Nonetheless, Saul erroneously annulled their marriage[34] and gave Michal to Paltiel instead.[35] After King Saul's demise, David sent Saul's son to bring his wife Michal back to him.[36] Knowing that she was married to David, Paltiel had never laid a hand on her. Although the sages praise Paltiel for this,[37] it was Michal, the intelligent woman, who infused him with the power to withstand this challenge.

Job's wife is not named explicitly in the text. Nonetheless, the Aramaic translation[38] states that her name was Dinah. The numerical values of Job (אִיוֹב) and Dinah (דִינָה) are 19 and 69, respectively. These two numbers are the midpoints of the two numbers 37 and 137. The midpoint of 37 is 19, the numerical value of Job (אִיוֹב), and the midpoint of 137 is 69, the numerical value of Dinah (דִינָה). ◂

Thus, in the Bible, 137 is a coupling constant that joins the male and female aspects of creation in marriage. The measure of connection between the husband and wife is the level of Divine Presence that resides between them, which depends on the extent to which they have refined themselves spiritually.[39]

Shabbat and Redemption

One of the optimal times for marital union in sanctity is on Shabbat. Every Shabbat has a female aspect (the eve of Shabbat) and a male aspect (the day of Shabbat). The act of intimacy on Friday night, when these two aspects unite, reinforces the union between husband and wife.

The Talmud states, "If they [the Jewish People] will observe two Shabbats, they will be redeemed."[40]◂◂ Every week begins with Shabbat and ends with Shabbat. The first of these corresponds to the woman and the second to the man. Thus, two Shabbats also represents the marital union between man and wife.

Two Shabbats further relate to a mathematical series called "the union of two Shabbats series."▸ The first Shabbat is represented by n and the second Shabbat is $n + 7$, i.e., the first Shabbat plus another seven days until the next Shabbat. Each expression is squared to indicate the inter-inclusion of each Shabbat within the other. Thus, the complete algebraic expression for the series is $f(n) = n^2 + (n + 7)^2$.

The first 13 integers produced by this series (beginning from $n = -3$) are:

-3	-2	-1	0	1	2	3	4	5	6	7	8	9
25	29	37	49	65	85	109	137	169	205	245	289	337

The third number of the series is 37 ($n = -1$) and the eighth number ($n = 4$) is 137. The sum of the first 13 numbers is 1,781, which equals 13 times 137.▸▸

▸ When 137 is added to the sum of the numerical values of "man" (311 ,אִישׁ) and "woman" (306 ,אִשָּׁה), the sum is 754. Similarly, when we add 137 to the numerical value of "groom" (458 ;חָתָן) and "bride" (55 ;כַּלָּה), the sum is 650. When added together, these two sums total 1,404, the numerical value of two times Shabbat (702 ;שַׁבָּת).

▸▸ This series contains the squared numbers of 5, 7, 13, 17 (25, 49, 169, 289). Other square numbers produced by this series are: $n = 48$ [the numerical value of "brain" (מֹחַ)] $f(n) = 73^2$ [the numerical value of "wisdom" (חָכְמָה) squared]; and $n = 65$ [the numerical value of the Name Adni (אדנ״י)] $f(n) = 97^2$ [the numerical value of "time" (זְמָן), or Meheitavel (מְהֵיטַבְאֵל) squared].

REFERENCE NOTES FOR CHAPTER 11

1. I Samuel 1:6; see also *Petichta D'eichah* 24.
2. *Chayei Moharan* 526.
3. *Likutei Moharan* II 18, citing *Zohar, Hakdamah 5a*.
4. Proverbs 30:23.
5. See *Igrot Rambam* (in Hebrew) Shilat edition 5755, Vol. 2 p. 502.
6. *Pesachim* 54a; *Nedarim* 39b.
7. *Zohar Shemot* 161a.
8. *Bereishit Rabah* 1:4.
9. *Zohar Vayikra* 73a.
10. *Sefer Yetzirah* 1:5.
11. *Chagigah* 15a.
12. Proverbs 18:22. Another, almost identical verse states, "He who has found Me has found life, he will produce good will from God" (ibid 8:35); see our book in Hebrew, *Shaarei Ahavah Veratzon*, p. 259 ff.
13. Ecclesiastes 7:26.
14. *Berachot* 8a.
15. See The Mystery of Marriage, Chapter 1.
16. Numbers 33:2.
17. *Likutei Torah, Parashat Masei* 96:1.
18. Joshua 4:9.
19. Numbers 33:2. See also, Appendix 1.
20. See our book in Hebrew *Shechinah Beineihem*, p.176, fn. 16.
21. *Moed Katan* 18b.
22. Genesis 24:50.
23. Ibid 25:1.
24. Rashi on Genesis 25:26.
25. Genesis 25:21.
26. *Likutei Torah Vezot Habrachah* 96:3-4.
27. Judges 14:4.
28. Proverbs 19:14.
29. Judges 15:6.
30. *Metzudat David* on Judges 14:2.
31. *Tikunei Zohar* 31 (75b).
32. I Samuel 25:17.
33. Ibid 18:20; 27.
34. *Sanhedrin* 19b.
35. I Samuel 25:44.
36. II Samuel 3:15.
37. See *Vayikra Rabah* 10.
38. Job 2:9.
39. *Sotah* 17a.
40. *Shabbat* 118b.
41. *Zohar Chadash Yitro*..

LOVE AND FEAR

The Fine-Structure Constant

Before we continue, let's recap the physical significance of 137.

Every chemical element absorbs certain frequencies of light and excludes others. This produces the spectral lines that appear when light is shone via various elements. Spectral lines occur in relation to the number and location of the electrons that orbit the nucleus in the atom. Since each chemical element has a specific number of electrons, each has its own unique set of spectral lines, or "atomic fingerprints," as they are sometimes called. The number that governs this phenomenon is a pure dimensionless number— 0.007297—that is very close to the reciprocal of 137 (i.e., $\frac{1}{137}$).

At the time when this was discovered, two leading physicists, Neils Bohr and Wolfgang Pauli, believed that an integral relationship exists between physics and psychology (see Chapter 6). Bohr introduced the principle of complementarity (see Chapter 2), which states that, in the quantum world, phenomena exist in complementary pairs. Each member of the pair apparently opposes or contradicts the other but, nonetheless, they are codependent. The most important case of complementarity was identified when the uncertainty principle proved that the position of a particle and its momentum cannot be accurately measured simultaneously. Thus, position and momentum are a complementary pair. The same is true about the very nature of light, which can be observed either as a wave or as a particle, but never as both at the same time.

Another complementary pair is matter and antimatter. Matter and antimatter particles are mutually exclusive. They are created under certain conditions when a single photon (or another elementary particle) collides with another particle. For example, an electron is created together with its anti-particle, a positron, when a photon collides with another particle. Essentially, this process creates matter (an electron) and antimatter (a positron) from energy (a photon). Like the super-symmetrical origin of the two souls of a married couple, the photon is the origin of the electron-positron pair. In the case of a spectral line, the origin is the single original line that splits in two.

All of these complementary phenomena—position/momentum, wave/particle, matter/antimatter—are governed by the fine-structure constant.

Kabbalah – the Spiritual Coupling Constant

When God created the world, He shone a single beam of light into the darkness. Yet, in the Torah's account of the creation of light, "light" appears twice, "And God said, 'let there be light,' and there was light."[1] This indicates that light has a dual quality. In Kabbalah, the two types of light are referred to as direct light and reflected light. Similarly, on the fourth day of creation, two luminaries were created—the sun and the moon. The sun represents direct light and the moon represents reflected light.◄ The duality of light alludes to the secret of the Divine Name *Elokim* (אֱ-לֹהִים), which is in the plural, indicating two Divine lights within one unified light source.▼▼

The Torah is also light, as the verse in Proverbs states, "For a *mitzvah* is a candle and the Torah is light."[6] At first, it appears that the Torah

▶ This also relates to the key phrase in the only passage in the Torah that contains exactly 137 words, "You shall surely appoint a king over you " (שֹׂום תָּשִׂים עָלֶיךָ מֶלֶךְ); see Chapter 4. There can be no king (direct light) unless there is a people (reflected light).

▶▶ "Morning" (בקר) is an abbreviation of "the splitting of light" [בְּקִיעַת אוֹר, referring in particular to "the morning star" (אַיֶּלֶת הַשַּׁחַר, the planet Venus when it appears in the east at dawn) which is, "Like two rays of light that rise in the east and illuminate"[2]]. Both rays of light emanate from the Higher *Elokim*, as in the verse in Psalms, "I will call to the Higher *Elokim*."[3] The numerical value of "I will call" (אֶקְרָא) is 302, which is the numerical value of "morning" (בקר). The two rays of light are "The light of *Havayah*"[4] and "The light of Israel,"[5] referring to "Let there be light" (direct light) "and there was light"(reflected light), as mentioned.

The sum of the numerical values of "the Higher *Elokim*" (אֱ-לֹהִים עֶלְיוֹן,

252), "the light of *Havayah*" (אוֹר הוי', 233), and "the light of Israel" (אוֹר יִשְׂרָאֵל, 748) is 1,233, which equals 9 times 137. The numerical value of "the Higher *Elokim*" (אֱ-לֹהִים עֶלְיוֹן) and two times "light" (אוֹר) is 666, which equals 18 times 37, the pair of 137, as we have seen throughout this book.

The sum of the numerical values of the two lights (אוֹר אוֹר, 414) that emanate from the Name *Elokim*, together with the numerical value of the Name *Elokim* (אֱ-לֹהִים, 86) is 500, which is the numerical value of the commandment to bear two children, "be fruitful and multiply" (פְּרוּ וּרְבוּ), (according to the ruling of the School of Hillel, the commandment is to have a son and a daughter, i.e., a complementary pair).

is a single beam of light, but, just as one spectral line splits in two, the Torah also contains two aspects. The first aspect is the Written Torah and the second is the Oral Torah, which includes all that Moses received through prophecy at Mt. Sinai, as well as the laws and interpretations that were passed down by word of mouth from generation to generation.▸ The Oral Torah has since been transcribed and comprises the Mishnah and the Talmud.

There is another, hidden aspect of the Torah that was revealed only after the destruction of the Second Temple, when direct prophecy was withdrawn. This is the prophetic wisdom of Kabbalah,[7] which unites wisdom (חָכְמָה) and prophecy (נְבוּאָה). As mentioned in the Introduction, the sum of the numerical values of these two words is 137, which is the numerical value of Kabbalah (קַבָּלָה), and the reciprocal of the fine-structure constant.

The revealed aspects (the Written and Oral Torah) and the hidden aspect of the Torah (Kabbalah) relate to fear and love, respectively. When we experience God's awe-inspiring and wondrous acts, we fear Him. This is the revealed aspect of the Torah. By contrast, love corresponds to the hidden aspect of the Torah. We are aroused to love something when we hear it praised, and we are inwardly attracted to experience it first hand. Fear is limited, while love is unlimited.[8]▸▸

The numerical value of "love" (אַהֲבָה) is 13, which is equal to the numerical value of "one" (אֶחָד). Love is an example of a phenomenon that reveals a fine-structure. The fine-structure is the inter-inclusion of love with fear. This includes the original unlimited attraction to one's beloved together with fear of detachment from the object of one's love.

A verse in Psalms that exemplifies the revelation of a fine-structure is "One spoke God [Elokim], I heard two, because God has might."[10] In particular, this verse relates to the fifth of the Ten Commandments, regarding the sanctity of Shabbat.[11] In the first account of the revelation at Mt. Sinai the verse commands: "Remember the Shabbat day to keep it holy."[12] In the second account it states, "Guard the Shabbat day to keep it holy."[13] Although God articulated one word, it simultaneously

▸ In Kabbalah, the division of the Torah into two aspects corresponds to the splitting of netzach (נֶצַח) and hod (הוֹד) into two separate sefirot from one original sefirah. The dot product of the two words (calculated by multiplying each letter of one word by the corresponding letter in the other word and adding their products) is 822 (50·5 ⊥ 90·6 ⊥ 8·4) or 6 times 137. Thus, each of the six letters of these two sefirot has an average value of 137.

▸▸ In general, the two primary Names of God, Havayah and Elokim correspond to love and fear, respectively. These two Names appear together in the prevalent phrase, "Havayah, your God." The numerical value of the phrase "Havayah your God" (הוי' אֱ-לֹהֶיךָ) is 92, which is also the numerical value of one synonym for "fear" (פַּחַד). 92 is the difference between 137 and 229, the sum of "love" (13; אַהֲבָה) and "fear" (216; יִרְאָה). The number that precedes these two numbers in the resulting linear series is 45.

$$45 \quad 137 \quad 229$$
$$92 \quad 92$$

These three numbers unite in the Torah portion that refers specifically to fear of God. It begins with the question, "What does Havayah your God ask of you other than to fear Him...," the verse continues, "...and to love Him."[9] The numerical value of the first word, "What" (מָה) is 45. The numerical value of the three words "What does Havayah your God ..." (מָה הוי' אֱ-לֹהֶיךָ) is 137. The sum of the numerical values of "love" (13; אַהֲבָה) and "fear" (216; יִרְאָה) is 229, as stated.

273

registered in the minds of the Jewish People as two distinct words, "Remember" and "Guard." This is an unfathomable feat that only God can accomplish. This quality of "spectral splitting" is referred to as "might" (עז). We shall see more of this word in Chapter 15.

Love and Fear

When considering complementarity in the realm of psychology, physicist Niels Bohr investigated the possibilities for four different complementary pairs: love-hate, life-death, light-darkness, male-female.

Let's examine the first of the pairs in the psyche that Bohr discussed: love and hate. Love is the power of attraction and hate is the power of deterrence. As such, the two are mutually exclusive polar qualities. In Kabbalah, we are taught that when two such polarities are in an unstable state of chaos, they are unable to look each at each other "face-to-face"[14] i.e., they are in a state of non-recognition that is characterized by each one fearing its opposite. Indeed, the Temple was destroyed (i.e., returned to a state of chaos) due to unwarranted hatred. It will be rebuilt (i.e., returned to a state of rectification, when equilibrium is resumed) as a result of unlimited love. The only way that love and hate can be reconciled is when "lovers of God hate evil."[15] This is the attitude of the consummately righteous person, whose unadulterated love of God is determined by his hatred of all evil. The more one's love of God is apparent, the more thoroughly one hates all that opposes and denies God.[16]

Rectified equilibrium in a complementary pair is a state of mutual recognition and appreciation. Through their acceptance of one another in love they are able to connect in a union of complements.

In Kabbalah, love (of good) is the all-inclusive emotion of the seven emotions of the heart. It is associated with the right, while fear (not hate) is associated with the left. When fear manifests as the fear of severance from good, it balances the attribute of love. Psychological stability is achieved when love and fear are incorporated in the psyche in this harmonious balance. If the complementary pair of love and fear is balanced, the other emotions follow naturally.

In general, love and fear refer to love and fear of God, which means directing both upwards, as we learn from the Torah, which states: "You shall fear your God,"[17] "And you shall love *Havayah*, your God with all your heart…"[18]▶

Although we should direct our love for God upwards, the most practical way for us to manifest it is by directing our love "downwards" towards all of God's creations. The downward vector force is the essential quality of love. As Rebbe Shneur Zalman of Liadi (founder of the Chabad Movement) taught, true love is to love whom one's beloved loves.[19] God loves all of His creation, or else He would not have created it, nor would He continually recreate it at every moment. Yet, God's special love is for His chosen people, Israel, as the Prophet Malachi states explicitly, "I love you, says *Havayah*."[20] The ultimate manifestation of loving God is to love the Jewish People.

By contrast, the vector force of fear, on the left, must only be directed upwards towards God. One of the foremost teachings of the Ba'al Shem Tov is that we must not fear anyone or anything besides God. This refers to awe, the natural experience of being mindful of the presence of God.

When either love or fear is experienced consciously, the other emotion is present in an unconscious state. Although they are a pair, only one of them can be in the forefront of a person's consciousness at any one time, while the other remains dormant in the unconscious. Translated into the language of modern physics, while one of the complementary qualities is being experienced, their polarity causes the other to collapse. As mentioned, a person cannot consciously love another while simultaneously experiencing fear in the other's presence. This is expressed in the Talmudic idiom, "The left [fear] deters, while the right [love] draws near."[21]

The sages state[22] that there is only one context in which we can consciously experience both love and fear simultaneously, without either

▶ The numerical value of the complete verse "And you shall love *Havayah* your God with all your heart and with all your soul and with all your might" (וְאָהַבְתָּ אֵת הוי' אֱ-לֹהֶיךָ בְּכָל לְבָבְךָ וּבְכָל נַפְשְׁךָ וּבְכָל מְאֹדֶךָ) is 1,644, which equals 12 times 137. The first five words have a numerical value of 959, which equals 7 times 137 and the last five words have a numerical value of 685, which equals 5 times 137.

In Chapter 1, we saw how 137 is associated in particular with *da'at* (the *sefirah* of knowledge). *Da'at* divides into two sets of five: the five *chasadim* of *da'at*, which are the origin of love (the *sefirah* of loving-kindness) and the five *gevurot* of *da'at*, which are the origin of fear (the *sefirah* of might). The division of the verse into two sets of five words reflects this. This teaches us that only by inter-including might within fear can one rise to the super-symmetric source of both in *da'at*. 12 times 137 (the sum of the entire verse) alludes to Moses' uniquely transparent prophecy, which he received with the word "This" (זֶה), which has a numerical value of 12. The numerical value of the initial and final letters of the first three words alone (ו ת א ת י ה) is 822, which equals 6 137. The numerical value of the first, middle and last letters of "And you shall love" (וְאָהַבְתָּ) spell "chaos" (תהו), the numerical value of which is 411, which equals 3 times 137.

function "collapsing." This is in the context of our service of God. Directing love towards the Jewish People and fear towards God is the initial stage necessary to create stability and equilibrium between these two qualities in the soul. Thus, love and fear are one example of a complementary pair that has a single root in God's service. This relates to the revelation of the fine-structure of love, which involves infinite attraction to God integrated with fear of detachment from His life-giving force.

This is illustrated in the "run and return,"[23] of the Divine spirits, which Ezekiel saw in his mystical vision. The spirits run towards the Infinite Light, but recoil from the danger of being utterly consumed in God's fire. This continuous "run and return" is the pulse of life. It is powered by the refined state of the fine-structure of love, which includes fear of completely losing one's identity in the identity of one's beloved.

The paradoxical revelation of the union of love and fear is reflected in the form of the letter *alef*, the first letter of the Hebrew *alef-bet*.[24]

אַ

As we see here, the form of the *alef* is an upright *yud* (י) on the right, an upside down *yud* on the left, and a letter *vav* (ו) between them. The upper *yud* represents the downward thrust of love (the letter *yud* is written from the top downwards), and the lower *yud* represents the tendency of fear to rise upwards. As we saw in Chapter 9, the *vav* is the letter of communication. When the letters of its name are spelled out in full (ואו אלף ואו), it has a numerical value of 137. Thus, the form of the letter *alef* alludes to 137 as the connecting force that unites love and fear as one entity. The numerical value of the letter *alef* (א) is 1, which further reflects its relation to union. Appropriately, the *alef* is the initial letter of "one" (אֶחָד) and also the initial letter of "love" (אַהֲבָה).

Cosmic Entanglement

In Kabbalah, every quality of the psyche is represented by an archetypal Jewish soul. Abraham, the first of the Patriarchs, is the archetypal soul who represents love. Abraham's son, Isaac, the second Patriarch is the archetypal soul who represents might, which implies awe or fear. God Himself refers to Abraham as, "Abraham My lover,"[25] while Jacob refers to God as "the fear of Isaac."[26]

Thus, the first two generations of Jews, father and son, represent the first two basic emotions of the psyche, love and fear. The sages state regarding God that we can simultaneously experience love and fear. The manifestation of this innovative statement was accomplished by these two souls during the last of Abraham's ten trials—the *Akeidah*, the Binding of Isaac. The literal significance of the *Akeidah* is Abraham's physical binding of Isaac to the altar in order to sacrifice him, as God commanded. However, the symbolic significance is the binding together of the two spiritual attributes of love and fear that were exemplified by these two souls.► Using modern science as a metaphor, the Binding of Isaac represents the cosmic entanglement of their two souls; one with the down-spin of love and one with the up-spin of fear. At the moment of the act, their love and fear of God were integrated and inter-included within each of their souls.

God tested Abraham, whose natural consciousness was one of love, in order to assess his fear of God. Indeed, when the angel told Abraham that he should not slay Isaac, the angel said, "Now I know that you fear God."[32]

Isaac, whose primary experience was fear, was willing to sacrifice his life to God. Only someone with the highest level of love and pure devotion can volunteer to such an act, as we are taught, "'You shall love *Havayah*, your God, with all your heart and with all your soul' – even if He takes your soul."[33] At that very moment, the souls of Abraham and Isaac became cosmically "entangled" as a complementary pair. From then on, it became possible for every descendant

► From a somewhat different perspective, Abraham and Isaac represent the complementary pair of momentum and position. Abraham represents momentum, as the verse states, "And Abraham travelled back and forth to the south."[27] Isaac represents position, as the verse states, "And Isaac sojourned in Gerar."[28] The root of the name of the place where he chose to settle, Gerar (גְּרָר) is related to a synonym for "fear" (מִגּוּר). Sitting/settling as a characteristic of fear (the quality associated with Isaac) is indicated in one of the permutations of the first word of the Torah, *Bereishit* (בְּרֵאשִׁית), which permutes to read, "sit in fear [of God]" (שַׁב יְרֵאת).

The balance between Abraham's sense of love, which catalyzed his constant movement, and Isaac's sense of fear, which fixed him in one position, is balanced by Jacob, of whom the verse states, "And Jacob dwelt in the Land of his father's sojournings, in the Land of Canaan" (וַיֵּשֶׁב יַעֲקֹב בְּאֶרֶץ מְגוּרֵי אָבִיו בְּאֶרֶץ כְּנָעַן).[29] The numerical value of the entire verse is 1554, which equals 42 times 37. The numerical value of the middle word in the verse, "[the] sojournings [of]" (מְגוּרֵי) is 259, which equals 7 times 37, which is 1/6 of the numerical value of the entire verse.

This is the secret of the verse describing the peace that will prevail at the end of days, "And the wolf shall sojourn with the lamb."[30] Then supernal fear will be revealed and there will no longer be any cause to fear even wild animals, "For the earth will be filled with knowledge of God, etc."[31]

of Abraham and Isaac to experience these two apparent opposites of love and fear of God simultaneously without the expected "collapse." As we shall see, the root "love" (א־ה־ב) appears 42 times in the Torah and the root "fear" (י־ר־א) appears 95 times, totaling 137. The Binding of Isaac thus alludes to the ability to experience these two basic emotions of love and fear simultaneously in our conscious service of God.[34] In the context of true Divine service, love and fear are in a state of superposition. This is the secret of true peace between opposites.[35]

As we saw in Chapter 3, at the time of the Binding of Isaac Abraham was 137 years old. This connects this pair of two spiritual attributes, love and fear, and the fine-structure constant of physics, which governs the splitting of spectral lines and the coupling of complementary pairs. Recall that Isaac was 37 at the time of the *Akeidah*.▾

Sarah and Abraham were physically unable to bear children.[36] However, God changed each of their names and they miraculously gave birth to Isaac. Indeed, we find that the "coupling constant," 137 comes into play here too. The numerical value of Abraham's full name, when calculated in primordial numbering is 959, which equals 7 times 137.◄◄ This relates to the first five words of the verse, "You shall love *Havayah*, your God, with all your heart and with all your soul and all your might" (וְאָהַבְתָּ אֵת הוי' אֱ-לֹהֶיךָ בְּכָל), as mentioned previously.

▶▶ By this method, each letter is calculated as the sum of the numerical values of all letters until that letter. Thus, אַבְרָהָם equals 1 (א) + 3 (ב) + 795 (ר) + 15 (ה) + 145 (מ) = 959.

Another phrase with a numerical value of 959 is "from this we will take to serve *Havayah* our God" (כִּי מִמֶּנּוּ נִקַּח לַעֲבֹד אֶת הוי' אֱ-לֹהֵינוּ).[37] This verse refers to taking cattle and sanctifying it in God's service. This also alludes to Abraham who was a holy soul born from Terach, an idolater.[38]

Abraham was the first of the Patriarchs and Naftali is the last of the tribes of Israel mentioned in various contexts in the Torah.[39] Jacob blessed Naftali, "Naftali is a swift gazelle"[40] "Gazelle" (אַיָּלָה) is a permutation of Elijah (אֵלִיָּה), alluding to the swift arrival of the Prophet Elijah to announce the final redemption. The numerical value of "Naftali is a swift gazelle" (נַפְתָּלִי אַיָּלָה שְׁלֻחָה) is 959. Thus, we see that the end (the final redemption with the prophet Elijah) is wedged in the beginning (Abraham, the first Patriarch).

▶ The sum of the numerical values of "the God of Abraham" (אֱ-לֹהֵי אַבְרָהָם, 294) and "the God of Isaac" (אֱ-לֹהֵי יִצְחָק, 254) is 548, which equals 4 times 137, the average numerical value of each word. 685 is an inspirational number, i.e., the sum of two consecutive square numbers (a phenomenon recognized as significant both in Kabbalah and in mathematics). The two consecutive numbers in this case are 19 and 18 (i.e., $19^2 + 18^2 = 685$).

The sum of 19 and 18 is 37, which was the age of Isaac at the time of the *Akeidah*.

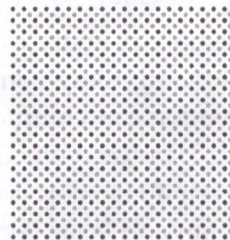

Illustration of $19^2 + 18^2$

In some places in the Prophets and Scriptures the letter *tzadik* (צ) of Isaac's name is replaced with the letter *sin* (ש) and pronounced "Yischak" (יִשְׂחָק).▸ When the numerical value of this alternative spelling is calculated by the same method of primordial numbering, it equals 1,781, which is equal to 13 [the numerical value of "love" (אַהֲבָה) and "one" (אֶחָד), as mentioned above] times 137.

Love, Fear and Communication

As mentioned in the previous chapter and in Chapter 4, one basic Hebrew word with a numerical value of 137 is "origin" (מוֹצָא). This word appears in various contexts in the Torah, particularly with reference to language and speech.

Each of the twenty-two letters of the Hebrew *alef-bet* are pronounced with one of the five vehicles of speech in the mouth: the throat, the palate, the tongue, the teeth and the lips. Each of these speech transmitters is an "origin." The ability to communicate through speech, the basic form of "coupling" between humans, depends on these five origins. Translated into a mathematical equation, the five origins of speech are 5 times 137 [the numerical value of "origin" (מוֹצָא)], which equals 685. This number is the sum of "Abraham," "Isaac," "love" and "fear" (אַבְרָהָם יִצְחָק אַהֲבָה יִרְאָה). ectified verbal communication demonstrates an integral harmony. Through it, one expresses love, and also fear of losing the love of ones beloved. The latter is achieved by remaining sensitive to their needs. This is the meaning of the Talmudic statement, "the left [fear] should always deter and the right [love] attract."

42 Stages of Love

As mentioned in the previous chapter, the Jewish People passed through 42 camps during their travels through the desert. We can incorporate this sense of constant advance towards a goal by considering every phase of our journey through life as a relative state of

▸ The numerical value of Isaac (יִשְׂחָק) in this spelling is 418. The numerical value of Abraham (אַבְרָהָם) is 248. The sum of these two numbers is 666, which is the triangle of 36, which itself is the triangle of 8. For more on the number 666 and its relationship to 137, see Appendix B.

Isaac's name literally means, "he shall laugh" (יִצְחָק) because his birth was such an absurd phenomenon. The alternate spelling implies playfulness, which is considered a higher level of Isaac's spiritual stature.

► The 42-lettered Divine Name is encoded in the *Ana Bekoach* prayer. The beginning of this prayer refers to the greatness of God's right hand, referring to His attribute of great love.

One significant word with a numerical value of 95 is "the King" (הַמֶּלֶךְ), in whose presence we experience awe.

redemption. Rebbe Nachman of Breslov exemplified this idea, saying, "Wherever I go, I am always on my way to the Land of Israel." The Tzemach Tzedek, the third Rebbe of Chabad, taught that we can create the atmosphere of the Land of Israel wherever we are. Every Jewish soul should sense an innate attraction to the Holy Land and yearn to be there. Advancing towards the Land of Israel, and all of our journeys through life, are expressions of love.◄

The first Jew to travel was Abraham, who "traveled back and forth to the south."[41] The south represents love (on the right-hand side), and all of the journeys of Abraham, the archetypal soul of love, were "back and forth." Abraham was like the Divine spirits, who run in love towards the Infinite Light, but recoil in fear of being utterly consumed in God's fire. This pulse symbolizes the manifestation of fear in refined love, as we saw above. Similarly, the Jewish People's sojourn in the desert illustrates the inter-inclusion of fear within refined love.

The 42 stages through which they passed while traveling through the desert to the Land of Israel indicate that there are 42 levels of love. Indeed, the root of "love" (א-ה-ב) appears in the Torah exactly 42 times. The 42 appearances of "love" in the Torah correspond to the 42 camps of the Jewish People during their travels from bondage to freedom. The Divine Providence working through nature was not apparent in the wilderness. There, God led the Jewish People with open miracles. By contrast, the Land of Israel is "a Land which *Havayah*, your God, always looks after, where the eyes of *Havayah*, your God, are upon it from the beginning of the year to the year's end."[42] Love and attraction are expressed in the desire to see one's beloved at all times. Therefore, this verse is an expression of God's love for the Land of Israel, which manifests in His special providence over the Land and its inhabitants. Similarly, we express our love for God by always seeking Him out and perceiving His providence in all that we do. Yet, as long as we are constrained in the confines of our own personal Egypt, God's providence is hidden from us. Since we love

God, we constantly desire to reach the state of revealed providence that is apparent in the Land of Israel.

"Fear" (ירא) is the complementary root to "love." It appears in the Torah 95 times, more than double the amount of times that love appears. The sum of 42 (appearances of "love") and 95 (appearances of "fear") is 137. Here, we see that the mysterious number of modern physics is associated with the entanglement of all the love in the Torah with all of the fear in the Torah.▼

A Monumental Task

In Hebrew, 42 and 95 are represented by the Hebrew letters *mem bet* (מב) and *tzadik hei* (צה), respectively. When interlaced with one another, they form the word "monument" (מַצֵּבָה), which appears a number of times in the Torah.▶▶

The first person to erect a monument was Jacob. On his way to his uncle Laban, Jacob slept one night with a stone beneath his head. That night, he dreamt his prophetic dream of a ladder that reached heavenwards. When he awoke, he realized that the place was holy ground. He then proclaimed the stone to be a monument, and vowed

▶▶ Moses constructed 12 monuments—one for each of the twelve tribes of Israel. These monuments are represented numerically by 12 times 137 (the numerical value of מַצֵּבָה), which is 1,644. This is the value of the entire verse that directs us to love God, "And you shall love *Havayah*, your God, with all of your heart and with all of your soul and with all of your might" (וְאָהַבְתָּ אֵת הוי' אֱ-לֹהֶיךָ בְּכָל לְבָבְךָ וּבְכָל נַפְשְׁךָ וּבְכָל מְאֹדֶךָ).[46] As mentioned, the phrase "*Havayah* your God" indicates both love and fear. The seemingly redundant preposition (אֵת) also indicates the inclusion of fear in love.[47] Each of the three ways of loving God alludes to a quality of fear of God that is integrated with love: "with all your heart—with both the good and the evil inclination (i.e., by overcoming your evil inclination)"; "with all your soul—even if He takes your soul [in death]"; "and with all your might—with all your resources."[48] All these acts require a great deal of courage, which is the outer manifestation of fear in reality. Together, the 12 monuments indicate the collective Jewish consciousness of loving and fearing God.

▶ Pregnancy is a primary example of merging love and fear. On the one hand, it is the fruit of a married couple's love. On the other hand, fear manifests in the need to protect the pregnancy. Indeed, these two continue once the child is born. Nurturing the child's growth is a manifestation of love, and protecting the child from harm is a manifestation of fear. In particular, pregnant women and children require nurturing and protection.[43]

There are three significant moments during pregnancy.[44] The first is the moment of conception (within 3 days after marital union). The next stage is when the zygote takes on human form (40 days after conception). The final moment is the moment of birth, after 273 days of pregnancy. These three moments correspond to the three lower Worlds: *Beriah* (the World of Creation), *Yetzirah* (the World of Formation) and *Asiyah* (the World of Action). The moment of marital union corresponds to *Atzilut* (the World of Emanation).[45]

First, let us note that the mid-point of 273 is 137.

Let us now analyze these numbers as a mathematical series (beginning with 0, the moment of marital union) and further expand it to seven members of the series, by calculating the finite differences:

0	3	43	273	846	1915	3633
	3	40	230	573	1069	1718
		37	190	343	496	649
			153	153	153	153

The sum of the first seven members in this series is 6,713, which is equal to 49 (7^2) times 137, meaning that the average of each of the seven numbers is 959, which equals 7 times 137. As we saw above, 959 is the numerical value of Abraham in primordial numbering, amongst others. This is an allusion to the significance of the number 137, the inverse of the electromagnetic coupling constant, in the formation of a child, the fruit of a couple's love.

▶ In the verses relating to Jacob's dream, the two words "set in [the earth]" (מֻצָּב) and "standing [over it]" (נִצָּב), share the same root as "monument" (מַצֵּבָה). The numerical value of "set in [the earth]" (מֻצָּב) is 132. The numerical value of "standing [over it]" (נִצָּב) is 142. Their sum is 274. The average value of these two key words in the passage relating to Jacob's dream is thus 137, the same as "monument" (מַצֵּבָה).

▶▶ The Prophet Chagai prophesied, "…before they placed a stone by a stone in the hall of *Havayah*."[50] His prophecy reflects the stage of placing the stones that preceded building the Temple. The phrase "stone by a stone" (אֶבֶן אֶל אָבֶן) equals 137. The filled value of "stone" (אלף בית נון) is 629, which equals 17 times 37, the companion to 137. The filled value of "by" (אלף למד) is 185, which equals 5 times 37. The numerical value of the complete phrase when spelled in full is thus 1,443, 39 times 37.

▶▶▶ The three-stage process from stone, to monument, to house, is a classic example in of the concept "point-line-area,"[51] mentioned above in Chapter 9 with reference to the three levels of spelling the letter *vav*. A monument is the intermediate stage between the initial point of origin and the full manifestation of a phenomenon.

that he would make it into a house for God.◄[49] Here we see a three-stage development from stone-to-monument-to-house. The ultimate objective is a house of God—a Temple—but for the stone to develop into a Temple, it requires an intermediary "monument" stage.◄◄

Love and fear are both dynamic emotions. Love represents the forward dynamic, while fear represents the backward dynamic. By coupling the two in the correct balance, we establish a firm and stable psychological state of being. Metaphorically speaking, this is the intermediate stage between a stone (often a symbol for an unemotional heart) and a home (in which the inhabitants feel free to express their love). In order to construct a house of God, Jacob's stone must first become a monument. This stage is the realization of all the love and fear of the collective Jewish soul. The monument (מַצֵּבָה), with a numerical value of 137, thus binds together the 42 appearances of love and the 95 appearances of fear in the Torah.

Unlike a stone (which is portable) and a building (within which people can find shelter and comfort), a monument is permanent and unchanging, but offers no shelter as a dwelling place.◄◄◄ The journeys of the Israelites reflect movement and advance. Their consciousness was one of constantly leaving Egypt (fear) and being attracted to the Holy Land (love). By contrast, the rigid stability of a monument is formidable and thus represents fear. When fear is experienced alone, it is liable to degenerate into foreign fears and becomes negative. Indeed, the Canaanites would worship idols on the monuments they built. Therefore, once the Temple was constructed, monuments became superfluous and even undesirable. In the times of the Patriarchs, monuments were beloved but, since the building of the Temple, the Torah forbids all worship – even of God – that involves a monument.[52] The transferred status of a monument from beloved to prohibited reflects a unique transition that is not found in other contexts.

The *Tanya* refers to love and fear as two wings, symbolizing the power to ascend in flight.[53] Jacob's monument symbolized the ladder he saw in his dream, "A ladder was set in the earth and its head reached the heavens, and behold, God's angels were climbing up and down it, and behold *Havayah* was standing over it."[54] Jacob's ladder represents the ladder of prayer, which must be firmly positioned on earth, i.e., in the physical world.[55] The "wings" of love and fear of God enable us to ascend the ladder, heavenwards.

Alternatively, a monument symbolizes temporal continuity. It marks the sturdy reality of the present that links the past to the future. In modern Hebrew, "monument" (מַצֵּבָה) also means "gravestone." The association with time is particularly noteworthy when we recall that in physics, *alpha* indirectly controls the lifespan of an elementary particle. Significantly, 137 is most explicit in the Torah as a lifespan (see Chapter 7).

137 and the Attributes of the Heart

As mentioned, love and fear are the two primary emotions of the heart. The other emotions are branches of those two roots. The emotion that follows love and fear is compassion (רַחֲמִים), the motivating power of *tiferet* (the *sefirah* of beauty). Compassion is the intermediary quality that unites love and fear. It corresponds to Jacob's soul root. The Hebrew root of "compassion" (ר־ח־ם) also means "womb" (רֶחֶם), where the fruit of a couple's love is nurtured (love) and protected (fear).[56] In Kabbalah it is explained that the two words share the same root because, "The womb of the mother extends to the heart of the son." *Tiferet* (the *sefirah* of beauty) receives its inspiration from the womb of the mother figure, *binah* (the *sefirah* of understanding). A third meaning is the name of a bird (רָחָם), varyingly identified as an Egyptian vulture, or a magpie etc., whose consumption the Torah prohibits. Altogether, this root appears 23 times in the Torah.▶

Just as 37 is the companion number to 137, in Kabbalah 23 is the companion to 37. This is particularly noticeable in the pair of words relating to the two highest levels of the soul, *chayah* (חַיָּה, 23) and *yechidah* (יְחִידָה, 37), which manifest in the two hollows of the heart as

▶ The sages explain that the bird (רָחָם) is so-called because its arrival coincides with the beginning of the rainy season. Rain is considered to be a primary manifestation of God's compassion.[57] Another opinion[58] is that the name is intended to humiliate the bird because it is overly compassionate towards its offspring (but cruel to others). Nonetheless, it announces God's compassion. If it sits on the ground (not habitual for this bird) and shrieks, it is a sign that the arrival of Mashiach is imminent.[59] Of the 23 times that the root ר־ח־ם appears in the Torah, it appears 9 times in the context of "compassion," 12 times in the context of a woman's womb and twice in the context of the impure bird with that name. The sum of the squares of these numbers ($9^2 + 12^2 + 2^2$) is 229, which is the sum of the numerical values of "love" (אַהֲבָה) and "fear" (יִרְאָה).

283

▶ The sum of 23 (the number of times the root ר־ח־ם appears in the Torah) and 137 (the number of times "love" and "fear" appear in the Torah) is 160, which is the numerical value of Cain (קַיִן). 37 is the numerical value of Abel (הֶבֶל).

▶▶ Netzach and hod correspond to the two "kidneys" (בָּטְחוֹת). The root of this word means "trust"(ב־ט־ח). The plural form implies two types of confidence, or trust.

▶▶▶ The 69 appearances of "sincerity" (ת־ם) are equal to 3 times 23, the number of times that "compassion" (ר־ח־ם) appears in the Torah. Altogether, "sincerity" and "compassion" appear a total of 92 times. There are 137 appearances of "love" and "fear." Thus, the total number of these four roots is 229, the sum of the numerical values of "love" (אַהֲבָה) and "fear" (יִרְאָה).

"joy" (23, חֶדְוָה) and "weeping" (37, בְּכִיָה), respectively. Mathematically, 23 and 37 are the golden ratio of 60. ◀

On the Kabbalistic Tree of Life, netzach (the sefirah of victory) appears to the right, below chesed. Its inner motivation is confidence (בִּטָחוֹן). There are two types of confidence: active and passive. ◀◀ Active confidence refers to proactively taking the initiative to realize our ambitions and trusting that God will give us the power to succeed, as the Torah states, "He is the one who gives you the power to succeed."[60] The root "trust," or "confidence" (ב־ט־ח) appears 8 times in the Torah.

The sefirah that follows netzach is hod (the sefirah of acknowledgment); it is situated to the left, beneath gevurah (the sefirah of might). Hod relates to passive trust, which means having absolute confidence that God will rectify reality in the way that we understand it should be rectified according to His Torah. The inner attribute of hod is sincerity (תְּמִימוּת). Throughout the Bible, sincerity often appears in the context of walking with God, or in the path of God.[61] This relates to always acting on God's behalf. The Hebrew root of "sincerity" (ת־ם) appears 69 times. ◀◀◀ This number is the mid-point of 137, meaning that the sincerity of hod is the midpoint of the union of love and fear, which together appear 137 times. This indicates that the only way to unify complementary opposites is with simple sincerity.

The attribute associated with the sixth sefirah, yesod (the sefirah of foundation) is truth. In the psyche, it manifests as a drive to fulfill our mission or dreams in life, and it motivates self-realization. As such, truth is associated with Joseph, the archetypal soul who interpreted his own dreams and those of others and saw them come true. "Truth" (אֱמֶת) appears in the Torah 11 times.

The attribute associated with the final sefirah—malchut (the sefirah of kingdom)—is lowliness (שִׁפְלוּת); this is the attribute associated with King David who is the archetypal soul of malchut. He said of himself, "and I am lowly in my eyes."[62] Lowliness refers to a sense of humility, which brings us to feel unworthy of any greatness bestowed upon us. As mentioned with reference to quantum geometry, the Zohar states,

"He who is small is big, and he who is big is small."[63] The more that lowliness features in our psyche, the greater we deserve to be. Indeed, a true king must have the greatest sense of lowliness. The root for "lowliness" (ש־פ־ל) appears only 5 times in the Torah.

To summarize:

gevurah (might)		*chesed* (loving-kindness)
fear (י־ר־א)		love (א־ה־ב)
95		42
	tiferet (beauty)	
	compassion (ר־ח־ם)	
	23	
hod (acknowledgment)		*netzach* (victory)
sincerity (ת־ם)		confidence (ב־ט־ח)
69		8
	yesod (foundation)	
	truth (א־מ־ת)	
	11	
	malchut (kingdom)	
	lowliness (ש־פ־ל)	
	5	

The total number of times that these attributes appear is 248, which is the numerical value of Abraham (אַבְרָהָם).▶ Abraham is the archetypal soul-root associated with love, the first and all-inclusive attribute of the seven emotions of the heart. In the realm of sanctity, love accompanies all the other emotions.▶▶ This is alluded to by the 248 times that the first six emotions, from love to truth, appear in the Torah.▶▶▶ They verify the initial state of love.

The sages teach us that there are 248 "limbs" in the human body, which is true of the male physique, but a woman has a number of extra limbs, i.e., skeletal bones. There are different opinions as to how many more limbs a woman has than a man. The maximal opinion is that a woman has five extra limbs, bringing the total to 253 limbs in the female.[66]▶▶▶▶ The root "lowliness" (ש־פ־ל), associated with the

▶ The numerical value of "in the image of God" (בְּצֶלֶם אֱ־לֹהִים) is also 248. The first man was created in the "image of God." However, the first human being who manifested that image after the primordial sin of Adam and Eve was Abraham.

▶▶ This we learn from the verse in Psalms,[64] "Daily does *Havayah* command His loving-kindness." The six emotions associated with the six days of the week are all accompanied by the first emotion, loving-kindness.[65]

▶▶▶ Note that Jacob's wife, Leah, had six sons and one daughter. Similarly, Rebbe Nachman of Breslov modeled the first of his stories on a king who had six sons and one daughter. The six sons represent *Zeir Anpin*, and the one daughter represents *malchut* (the *sefirah* of kingdom).

▶▶▶▶ In Leviticus, the impure *racham* (רָחָם)[67] is in the masculine form and has a numerical value of 248. However, in Deuteronomy,[68] the fifth book of the Torah (the fifth is relatively feminine), this bird is referred to in the feminine form (רָחֲמָה) with an additional *hei* (5), bringing the numerical value to 253, the numerical value of the phrase, "originates from the mouth of *Havayah*" (מוֹצָא פִי הוי׳) The letter that concludes the phrase is the final *hei* of God's Name, which has a numerical value of 5.

► The sum of the numerical values of "bread" (78 ,לֶחֶם) and "wheat" (22 ,חִטָּה) is 100, which equals 10^2. The sum of a perfect square alludes to a perfect association between the two concepts.

253 is the triangle of 22. There are 22 letters in the Hebrew alphabet. The numerical value of "wheat" (חִטָּה), the origin of bread, is 22. The sages express the relationship between wheat and speech in the statement that a child only begins to speak once they have tasted grain.[71] The prime factors of 253 are 23 [the number of times that the root of "compassion" (רי־חם) appears in the Torah] and 11 [the number of times that "truth" (אֱמֶת) appears in the Torah].. The sum of these two prime numbers is 34, the prime "root" of 137.

►► The numerical value of the idiom "Torah from heaven" (תּוֹרָה מִשָּׁמַיִם) is 1,096, which equals 8 times 137.

►►► Friday is the final day of masculine consciousness. The Shabbat Queen is the manifestation of *malchut* (the *sefirah* of kingdom). Masculine consciousness incorporates the 248 appearances of the six masculine attributes from love to truth. By awaiting the feminine consciousness of *malchut*, we include the five appearances of lowliness, bringing the total to 253, the number that reflects the rectified awareness that our sustenance is directly from the mouth of God (see Chapter 4).

inner attribute of *malchut*, appears in the Torah exactly 5 times, thus completing the five extra limbs of the feminine physique.

"Love" and "fear" appear 137 times in the Torah, corresponding to the numerical value of "origin" (מוֹצָא). The total of all the appearances of the seven roots in the Torah is 253, which is the numerical value of the phrase, "originates from the mouth of *Havayah*" (מוֹצָא פִּי הוי׳) in the verse, "Not by bread alone does man live but by all that originates from the mouth of *Havayah* does man live."[69] This refers to the source of Divine energy in food (bread).

This verse, stated in the context of the manna in the desert, is the most important verse in the Torah pertaining to bread. According to one opinion, the primordial sin involved eating wheat[70] before Shabbat.◄ This was due to a flaw in Adam's faith. He believed that bread sustained him. God commanded Adam and Eve to wait three hours before eating of the Tree of Knowledge.[72] Had they done so, their consciousness would have been elevated to a higher plane. They would have been aware that the bread they ate originated from the word of God. rectifying their sin is by becoming aware of the twenty-two letters—articulated by God's mouth—that generate the bread. By means of these letters, God constantly re-creates the bread that sustains us (and all of creation). In the desert, the Israelites were aware of this because the manna was clearly heaven-sent. Before eating bread we bless God, Who, "extracts bread from the earth." Before eating manna they blessed He "Who brings down bread from heaven."[73] The heightened spiritual consciousness provided by the manna allowed those who ate it the privilege of receiving the Torah directly from heaven.[74]◄◄ This is the rectified feminine consciousness that Adam and Eve would have achieved had they waited until Shabbat.

The first eleven of the 39 categories of work forbidden on Shabbat are deduced from the breadmaking process, beginning with plowing the wheat field, and culminating with baking the dough. Here, these eleven categories correspond to the eleven appearances of "truth" in the Torah. Truth is the final *sefirah* of *zeir anpin*. ◄◄◄

The 137 times that love and fear appear in the Torah, thus represent the beginning of a rectified awareness that our food emanates from the mouth of God. This consciousness must permeate all seven emotive powers of the soul, culminating in *malchut* (the *sefirah* of kingdom), which corresponds to the mouth and to the seventh day of the week; Shabbat.▼

▶ 253—the number that represents the complete array of the attributes of the soul as they appear in the Torah, including *malchut*—is the numerical value of Avner (אַבְנֵר). Avner was originally King Saul's chief-of-staff.[75] After Saul's death, Avner accepted King David as king. He brought King Saul's daughter and King David's wife, Michal, back to King David. Later, David's nephew Yoav ben Tzruyah killed Avner before King David could appoint him as his own chief-of-staff.[76] Avner was instrumental in the establishment of the Davidic line of rule that will ultimately culminate with *Mashiach*. Indeed, in Kabbalah we are taught that Avner is destined to return as the chief-of-staff of the *Mashiach*.

The sum of the numerical values of Avner (אַבְנֵר) and *Mashiach* (מָשִׁיחַ) is 611. This number is the numerical value of Torah (תּוֹרָה) and the phrase "run and return" (רָצוֹא וָשׁוֹב). As mentioned, this phrase relates to the inter-inclusion of love and fear in our service of God.

The dot product of the two names Avner (אַבְנֵר) and *Mashiach* (מָשִׁיחַ) is 2,740 (1·40 + 2·300 + 50·10 + 200·8), which equals 20 times 137.

The numerical value of Avner (אַבְנֵר) is 253, as mentioned. The numerical value of *Mashiach* (מָשִׁיחַ) is 358. These two numbers appear as a pair in the series of triangular numbers that are the midpoint of another triangular number. The generalized function for this equality is $2n(n \pm 1) - y(y \pm 1) = 2$. This equation can be solved using Pell's Equation.

The first five pairs in the series that results from this equation are 1-1, 7-10, 43-61, 253-358, 1477-2089. The diagram below is a figurative illustration of △10, also showing △7 (7-10 is the second pair in the series). The black dot completes the triangle of 7 and is the 28th point from both ends of the triangle, i.e., the midpoint of △10.

Similarly, △253 (i.e. the sum of all numbers from 1 to 253) is equal to the midpoint △358. This is the fourth pair of the series. The sum of the five smaller numbers leading to △253 (1, 7, 43, 253, 1,477) is 1,781, which equals 13 times 137.

The sum of the five larger numbers (1, 10, 61, 358, 2,089) is 2,519, which equals 11 [the number of times the root "compassion" (ר-ח-ם), appears in the Torah] times 229, which equals "love" (אַהֲבָה) plus "fear" (יִרְאָה).

REFERENCE NOTES FOR CHAPTER 12

1. Genesis 1:3.
2. *Yerushalmi Yoma* 3.
3. Psalms 57:3.
4. Isaiah 2:5.
5. Ibid 10:17.
6. Proverbs 6:23.
7. For a detailed outline of the evolution of Kabbalah, see the Introduction to our book, *What You Need to Know About Kabbalah*.
8. See *Noam Elimelech, Chukat* (second explanation).
9. Deuteronomy 10:12.
10. Psalms 62:12.
11. Rashi on Exodus 20:8.
12. Exodus 20:8
13. Deuteronomy 5:12.
14. See beginning of *Sifra Detzniuta*.
15. Psalms 97:10.
16. *Tanya* ch. 10.
17. Leviticus 19:14, 32 etc.
18. Deuteronomy 6:5.
19. Quoted in *Hayom Yom* for 28th Nissan.
20. Malachi 1:2.
21. *Sotah* 47a; *Sanhedrin* 107b.
22. *Sifri Va'etchanan* 32. See also *Keter Shem Tov* 1:36.
23. Ezekiel 1:14.
24. For more on the secrets of the Hebrew letters, see our book, *The Hebrew Letters*.
25. Isaiah 41:8.
26. Genesis 31:42.
27. Ibid 12:9
28. Ibid 26:6.
29. Ibid 37:1. See *Pelach Harimon, Bereishit*, p. 19.
30. Isaiah 11:6.
31. Ibid, v. 10.
32. Genesis 22:12.
33. Rashi on Deuteronomy, 6:5.
34. As quoted by the Ba'al Shem Tov in the name of Nachmanides.
35. See *Likutei Moharan* 80.
36. *Yevamot* 64b.
37. Exodus 10:26.
38. See *Bamidbar Rabah, Chukat* 19.
39. See for example, Numbers 2:29 and 7:78.
40. Genesis 49:21.
41. Genesis 12:9.
42. Deuteronomy 11:12.
43. *Sotah* 47b; *Sanhedrin* 107b.
44. Rabbi Abraham Abulafia, *Imrei Shefer* Part 3.
45. See our book, *The Mystery of Marriage* p. 328.
46. Deuteronomy 6:5.
47. See *Tosfot Rosh, Kidushin* 57a.
48. *Mishnah Berachot* 9:5.
49. Genesis 28:22.
50. Chagai 2:15.
51. See *Etz Chayim, Sha'ar* 30, chapter 7; see also Rabbi Dov Ber Schneersohn, *Torat Chayim*

65d, notes 52 and 53, and the introduction to our book, *Living In Divine Space*, pp. 4-6.

52. *Rashi*, Deuteronomy 16:22.

53. *Tanya* chs. 40, 41 etc.

54. Genesis 28:12, 13.

55. *Zohar Bereishit* 261b; *Zohar Vayikra* 306b etc.

56. *Etz Chaim* 31:3; 32:1.

57. *Rashi* on *Chulin* 63a.

58. *Radak, Sefer Hashorashim*.

59. *Yalkut Shimoni*, Zachariah 10.

60. Deuteronomy 8:18.

61. See Genesis 17:1; Psalms 15:2; 84:12; 101:2, 6 etc.

62. II Samuel 6:22.

63. *Zohar Bereishit* 122b; *Zohar Bamidbar* 168b.

64. Psalms 42:9.

65. See *Zohar Emor* 103a.

66. See *Bechorot* 45a and Rashi ad loc. According to Rabbi Akiva, the extra limbs are two "hinges," two "doors," and a "key."

67. Leviticus 11:18

68. Deuteronomy 14:17.

69. Deuteronomy 8:3.

70. *Berachot* 40a.

71. Ibid.

72. *Shemot Rabah* 32:1.

73. *Bnei Yissachar, Ma'amarei Hashabbatot 3, Birkat Hashabbat*.

74. *Mechilta, Beshalach* 4.

75. I Samuel 17:55.

76. II Samuel ch. 3.

THE TIME DIMENSION

Time and Space

Until about a century ago, time was thought to be infinite. The real timeline stretched from infinity in the past to an infinite future. Like the waves and ripples on a river, events on this timeline appear as disruptions and upheavals on the surface of time. They have no effect on its passage.

One of the greatest innovations of general relativity is that time is not an absolute constant. In order for time to be constant it must be inherently linked to the three dimensions of space. The four dimensions of space-time are experienced in our human perception in a unified way. Once Einstein had formulated the theory of general relativity, it became clear that time and space had a finite beginning.

The Big Bang theory is a specific cosmological model within the framework of general relativity. It upholds the theological view of creation ex-nihilo. This theory leaves the atheist scientific community with the uncomfortable feeling that the world may not be as deterministic as it should be if there was no God to create it. They prefer to seek a different type of world, in which time is smooth and unchanging.

The device that physicists have found to iron out this quandary is imaginary time. Imaginary time is not imaginary in the sense that it is unreal. It simply runs in the same direction as space, perpendicular to real time. In essence, imaginary time is a way of looking at the time dimension as if it were a dimension of space: one can move forward and backward along imaginary time, just as one can move right and left in space. This solves the dilemmas of black holes, and other irregularities in the universe, producing a smooth universe without singularities. Incorporating imaginary time in the equations of quantum mechanics avoids many technical difficulties, and essentially annuls the distinction between time and space. The Big Bang, for example, appears as a singularity in "regular time." But, when visualized with imaginary time, the singularity is removed and the Big Bang functions like any other point in spacetime.

In such a universe, the need for God is eliminated. In the words of Stephen Hawking, who popularized the concept of imaginary time:

> So long as the universe had a beginning, we could suppose it had a creator. But if the universe is really completely self-contained, having no boundary or edge, it would have neither beginning nor end: it would simply be. What place, then for a creator?[1]

The Kabbalistic Perspective on Time

Scientists imagine that imaginary time precludes the presence of a Divine Creator. But, in fact, it reinforces His existence as the Creator of both real and imaginary time.

Kabbalah teaches us that the concrete dimensions of space contain time. Time is the inner dimension of reality that motivates it to expand. Thus, time penetrates space as an inner dimension within an outer dimension. From this perspective, time is relatively male and space is relatively female.

Joseph is the master of both dimensions of time. He was a dreamer, and one of his special talents was dream interpretation.[2] He correctly interpreted the dreams of the butler and the baker while he was imprisoned in Egypt. Later, he interpreted Pharaoh's dream, which earned him the privilege of becoming second-in-command to Pharaoh himself. Joseph's wisdom in solving the butler's, the baker's and Pharaoh's dreams is related to time.

The visions we see in our dreams are usually symbols. By correctly interpreting the symbols in a dream, we can discover their profound psychological meaning.[3] There are two methods by which we can interpret the meaning of a symbol. The first and most common method is by deciphering the linguistic phenomena in the dream. This method is abundant in the Talmud in its discussions on dream interpretation. It is used today in psychoanalysis, psychodrama, art therapy and other forms of psychotherapy. Using this method, the objects in a dream are interpreted via etymologocial associations. Joseph was proficient in seventy-one languages,[4] which made him an expert in this realm.

The second method of dream analysis is to abstract the physical objects in the dream and interpret them as temporal symbols. This is true in particular when there is a numerical symbol in the dream. When a number has no other explanation, it must relate to an abstract concept such as time. Joseph used this method to decipher the butler's and baker's dreams. He interpreted the butler's three vine twigs and the baker's three baskets as three days. He then proceeded to solve their dreams according to symbols and their literal and linguistic meanings. Joseph used the same method to interpret Pharaoh's dream. After initially perceiving the equivalence between Pharaoh's two dreams, Joseph interpreted the seven cows and the seven ears of grain as seven years.

Joseph's unique method of dream interpretation in the time dimension proved correct in all cases. He was sensitive to time as the inner dimension of reality.

Days and Years

Here we must ask, why did Joseph interpret one set of dreams as days and the other as years? What is the difference between a day and a year?

The numbers here tell the story. The principle division of the ten *sefirot* is into three intellectual soul powers and seven emotive attributes.[5] The intellectual powers illuminate the soul with the light of day, as in, "and God called the light day."[6]▶ Light is constant; it reflects our subjective sense of infinity. The concepts of light and day symbolize the unchanging common denominator, the eternal aspect of time that pervades all of reality. In contrast, the changing seasons are reflected in the year.▶▶

▶ The numerical value of "light" (אוֹר) is 207, which is equal to the numerical value of "infinity" (אֵין סוֹף). Both terms are often used with reference to God, who never changes, as the verse in Malachi states, "I am God, I have not changed."[7]

▶▶ In Hebrew, the two words "year" (שָׁנָה) and "change" (שִׁנּוּי) share the same root.

293

Thus, day is the inner, unlimited, eternal dimension of time. Years represent the external, limited, changing aspect of time that corresponds to the seven emotive attributes. They constantly fluctuate between revelation and darkness.

Each of the two versions of Pharaoh's dream contains a double seven (seven fat cows and seven thin cows; seven fat ears of grain and seven thin ears of grain). The dream was also repeated, indicating that the seven years would begin very soon.[8] This double quality is one of the unique characteristics of the seven attributes, which correspond to seven letters of the *alef-bet* associated with the six emotive *sefirot* and *malchut* (the *sefirah* of kingdom).◀ They are *bet, gimel, dalet, kaf, pei, reish* and *tav*. Each of these seven letters have two distinct forms, with or without a center point: *bet* (ב-בּ), *gimel* (ג-גּ), *dalet* (ד-דּ), *kaf* (כ-כּ), *pei* (פ-פּ), *reish* (ר-רּ)◀◀ and *tav* (ת-תּ). The double form represents the good and the bad inclinations. The "hard" letter, emphasized with the *dagesh* (center point) represents the untamed chaotic powers of the soul. The "soft" letter, without the *dagesh*, represents the powers of the soul when rectified. The two inclinations battle in the psyche to actualize the emotive attributes. But, sometimes they cancel each other out; sometimes the emotion is diverted away from actualization; and sometimes, it manifests in reality.

Thus, Joseph interpreted the ministers' dreams as days and Pharaoh's dream as years.

Two Dimensions in the River of Time

There are two accounts of Pharaoh's dream in the Torah. The first account states, "And behold, Pharaoh stood on the river."[10] When Pharaoh recalled his dream to Joseph, he added the word "bank" (שְׂפַת): "And behold, I was standing on *the bank of* the river."[11] Joseph recognized that the river is the river of time. He realized that Pharaoh had added this word to make his dream sound more plausible.◀◀◀ Only someone in control of time can stand "on" the river.

Pharaoh is the king, relating to the feminine *malchut* (*sefirah* of

▶ The three days in the ministers' two dreams that Joseph interpreted were the same three days for both of them. The years were a double set of seven, one of plenty and the second of famine. This indicates that in the three upper *sefirot* time is more condensed than in the seven lower *sefirot*.

The ministers' dreams were subjective, referring to their personal destinies. Regarding individuals, each can have a different fate during a single time period. In contrast, Pharaoh is the king of the Egyptian people. As such, his dream relates to the collective. One nation cannot undergo two circumstances simultaneously. Therefore, there was one fate for the first period of seven years, then the fate of the collective reversed from good to bad. The first seven years were years of plenty and the seven years that followed were years of famine.

▶▶ Although the letter *reish* does not usually receive a *dagesh*, there are a number of incidences in the Bible where it appears with a *dagesh*, e.g., "A gentle reply turns away wrath" (מַעֲנֶה רַּךְ יָשִׁיב חֵמָה).[9]

▶▶▶ This is stated in a figurative interpretations of the verse in Psalms, "A '[river] bank' that I did not know did I hear" (שְׂפַת לֹא יָדַעְתִּי אֶשְׁמָע).[12]

kingdom), corresponding to space. From his static positition on the river bank, he identified with the changing aspects of the seven cows and the seven ears of grain. These correspond to the seven lower *sefirot*, the emotive attributes. They also relate to the seven Kings of Chaos (see Chapter 1), who reflect the trials and tribulations of life, "and he ruled … and he died."[13] Pharaoh so completely identified with the changing aspect of time that he readily recognized Joseph's interpretation of the seven cows and seven ears of grain as years.▶

Time includes an outer dimension and an inner dimension. The external dimension of time that Pharaoh was in tune with is dynamic and changing. It corresponds to real time. When Pharaoh recalled his dreams, he told Joseph that he had seen himself standing on the river bank, perpendicular to the river current. He was in the space dimension, watching as the superficial, turbulent aspect of time flowed by. He could not visualize himself standing on the river of time itself, nor sense the inner current of time that flows unperturbed within. Thus, he could not tap into it, nor could he control or direct it.

The inner dimension of time is a more profound undercurrent that is smooth and direct. It corresponds to imaginary time. In order to tap into this uniform aspect of time, one must stand upon the river of time and control it as Joseph did. He represents the rectified masculine aspect of time. Because he stood "on" the river of time, he was able to recognize the true significance of Pharaoh's dream.

By successfully controlling his sexual desires, Joseph rectified his masculine condition and was thus able to control real-time and synchronize it with the smooth imaginary time aspect. He was thus worthy of becoming a true leader, essentially taking over Pharaoh's place as king. Such a righteous individual is able to master both aspects of time and to realize their unity.

Tapping into imaginary time is the rectification of chaos. As mentioned in Chapter 2, "chaos" (תֹהוּ) is the onlooker's subjective sense of astonishment as he observes the changing upheavals of time.[14] The turbulence of unrefined emotions resonates with these changes

▶ In Pharaoh's dream, the time units are related to his essence as king of Egypt. The numerical value of "year" (שָׁנָה), which is 355. This is also the numerical value of "Pharaoh" (פַּרְעֹה). It is also the number of days in a lunar year. The numerical value of the words "And behold he stood upon the river" (וְהִנֵּה עֹמֵד עַל הַיְאֹר) is 496, the numerical value of *malchut* (מַלְכוּת).

▶ Chanukah is the Festival of Lights. The thirteen synonyms for light correspond to the thirteen words in the blessing made before lighting the Chanukah candles. The Arizal writes that, on each of the first seven days of Chanukah, we should have one of God's Thirteen Attributes of Compassion in mind. On the eighth day, we complete all thirteen by reading the passage of the twelve princes to the end before reading the passage that begins: "When you elevate the candles…." In this way, on the eighth day of Chanukah, we have in mind all thirteen attributes.

Chanukah is also the festival of "beauty" (יֹפִי). The eight Hebrew synonyms for beauty correspond to the eight days of Chanukah.[15]

▶▶ When the numerical value of each of the letters of the essential Name of God, *Havayah* (10, 5, 6, 5) is squared, their sum equals 186, which is the numerical value of the Name *Makom* (מָקוֹם), which means "space."

and generates our astonishment, preventing us from penetrating the underlying order. A righteous individual, such as Joseph, has rectified the emotional powers of his psyche. In a calm state of balance and equilibrium, he withdraws his astonishment and watches the proceedings without involvement. Once he is confident that he can intervene in a way that is in tune with the smooth order that underlies the chaotic superficial appearance of the river of time, he may dive beneath the surface to control the spontaneous order as it emerges. Appropriately, the righteous are referred to as "fish," which ride upstream or downstream in the undercurrent of the river, reflecting the ability to travel backward or forward in time. The evil eye has no power over the righteous, just as it has no power over the fish.

The first and last letters of "space," *makom* (מָקוֹם) are two letters *mem* (מ..מ). The letters *kuf-vav* (קו) that lie in between them spell "line" (קו). Thus, figuratively, "space" (מָקוֹם) is like two banks of a river with a line flowing between them. The "line" that flows through the two banks is the ray of light that God infused into the vacuum after the initial contraction of His Infinite Light. This is illustrated by two of the 13 Hebrew synonyms for "light."◀ The foremost synonym is "light" (אוֹר), which relates linguistically to the River Nile (יְאֹר) as it appears in Pharaoh's dream. The fourth synonym is "current" (נְהָרָה), which has the same root as "river" (נָהָר). This synonym is the most generic Aramaic translation of "light" (נְהוֹרָא).

Makom (מָקוֹם), which means "space," is one of the Names of God. Indeed, "He [God] is the space of the world, but the world cannot contain Him."[16]◀◀ In Kabbalah, absolute Divine space is depicted as a square. The created universe is a circle inside the square, and time is a line that extends from the top of the circle.

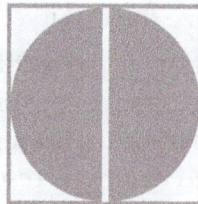

In this Kabbalistic depiction of space and time, the square relates to the union of space-time: the area of the circle within the square represents the vacuum, i.e., physical space created by the contraction. The circumference relates to the annual cycles that rise and fall with the seasons. These are the years of "real" time.

The line that penetrates the circle is the ray of light that God shone into the world to illuminate it. This ray of light corresponds to the "imaginary" time line, which is direct and unchanging. Here there are no upheavals, and no great changes or irregularities. This aspect of time corresponds to days; the inner current, the eternal, infinite aspect of time.▼

The three intellectual powers of the soul, representing days, illuminate the seven emotive attributes. They infuse their light into the cycle of the seven sabbatical years and similarly, the seven days of the week. Unlike years and months, which are marked by changes in the sun and moon, there is no astrophysical indication that a week has passed. Nonetheless, the seven-day week is universally accepted. This is a profound mystery in Kabbalah.▶▶

The seven days of the week illustrate how the three intellectual powers of the soul penetrate and illuminate the seven emotive attributes, which represent years. The word "days" (יָמִים) is often used in

▶▶ Shabbat—like the Jewish People—is above astronomy. The universality of the seven-day week is only by merit of the Jewish People who sanctify the Shabbat. They are the super-astronomical sign that a week has seven days. Without the Jewish People, the seven-day week would collapse.

▶ The synonym for "light" that means "current" (נְהָרָה) is the fourth of the thirteen synonyms. It corresponds to the river that exuded from Eden to water the garden and then split into four river-heads.[17] This relates to the special Kabbalistic ratio of 1:4, which is indicated by the numerical values of the letters *alef-dalet* in "vapor" (אד), as in the verse "and a vapor rose from the earth."[18] These two letters are the initial letters of "man" (אָדָם). The "vapor" that rose from the earth is the arousal of the earth from below that preceded the creation of man. The 1:4 ratio can manifest in either direction, reflecting either evaporation or condensation. Thus, we find a similar ratio in evaporation/condensation processes described in the Torah, when a physical entity becomes spiritual and vice versa.

One significant example in our discussion of the two dimensions of time is the ratio between the numerical value of "days" (יָמִים), which is 100 (10^2), and "years" (שָׁנִים), which is 400 (20^2).

There are many illustrations of this ratio within creation. One of the most beautiful examples is the numerical ratio between the higher, spiritual "tree of life" (עֵץ הַחַיִּים), which represents the Torah, and the "the tree of knowledge of good and evil"(עֵץ הַדַּעַת טוֹב וָרָע), the lower tree that represents human intellect and scientific logic.

The numerical value of "the tree of knowledge of good and evil" (עֵץ הַדַּעַת טוֹב וָרָע), is 932, which equals 4 times 233, which is the numerical value of "the tree of life" (עֵץ הַחַיִּים). 233 is also equal to the numerical value of the phrase "annulling the time dimension" (בִּטוּל מֵמַד הַזְּמָן). The annulment of the time dimension occurs in human consciousness, represented by the tree of knowledge.

This is also the ratio between *Adam Kadmon* and the four spiritual Worlds. The inital unity divides into two and then four.

the Bible to emphasize "complete" years. A year full of days represents the changing aspect of time that is full of the infinite aspect. This can be achieved by those who master time and do not allow time to master them. One of the clearest examples of this is with reference to Joseph himself. After Joseph successfully interpreted the butler's and the baker's dreams, the Torah states, "And it came to pass after two years of days [i.e. two complete years]."[19] The completeness of the "two years" (שְׁנָתַיִם) is emphasized by the additional word "days" (יָמִים).[20]

On Rosh Hashanah, the day the Jewish year begins, we catch a glimpse of how the eternity of day is integrated into the changing years. ◄ Rebbe Shneur Zalman of Liadi[21] stresses that, on the eve of Rosh Hashanah, the previous year comes to a complete end. The light that was projected into the world returns to its source, having completed its mission. The new year begins at the moment of the *shofar* blast on the morning of Rosh Hashanah, bringing with it a new light that has never before been experienced. If the same projection of light from the previous year would continue into the following year, time would be eternal. But, the year comes to an end, and is therefore temporal and given to change.

In the Torah, the first day of the year is the first day of the Hebrew month of Tishrei. The sages set Rosh Hashanah as two days, the first and second of Tishrei. These two days are referred to as "one long day."[22] The year comes to a definite end. It represents the temporality of real time. In contrast, the two days of Rosh Hashanah roll into one another. They represent eternity. This is how imaginary time infuses our natural experience of time. Thus, the two aspects of eternal days and temporal years unite in the Hebrew calendar to carry us along the river of time. ◄◄

The two days of Rosh Hashanah become one. The opposite is also true. A complete single day is two days. This is apparent in the unique Biblical expression, "a day or two" (אִם יוֹם אוֹ יוֹמַיִם).[23] This does not mean, "either one day or two days," but one day that is like two days, (i.e. a complete day, from one sunrise to the next).[24] The numerical value of

▶ At the climax of the *Musaf* prayer of Rosh Hashanah, the word "Today" is repeated in a unique prayer, "Today, strengthen us, etc." The numerical value of "today" (הַיּוֹם) when filled (הי יוד ויו מם) is 137. The final appearance of this word in the Pentateuch relates to Moses' burial place, "And no person knows the place of his burial, unto this day" (וְלֹא יָדַע אִישׁ אֶת קְבֻרָתוֹ עַד הַיּוֹם הַזֶּה). The numerical value of "unto this day" (עַד הַיּוֹם) is 135, alluding to the 135 times "today" appears in the Pentateuch. It appears twice more with the prefix "Like today" (כְּהַיּוֹם), bringing the total to 137.

▶▶ The numerical value of the phrase that relates to the "dual" aspect of light, "'Let there be light' and there was light" (יְהִי אוֹר וַיְהִי אוֹר), is 470. This is also the numerical value of "a time" (עֵת).

"or two days" (או יומים) is 113. This is also the numerical value of "one day" (יום), when its letters *yud-vav-mem* are spelled in full (יוד ואו מם). The numerical value of "year" (שָׁנָה) is 355.▶ The ratio of the numerical values that relate to "year" (355) and "day" (113) is $^{355}/_{113}$, which equals exactly 3.141593. This lends us an amazing insight into the meaning of the value of *pi* (π)—the numerical constant that indicates the ratio of a circle's circumference to its diameter (which is 3.1415927 correct to 7 decimal places).

This fraction $^{355}/_{113}$—known to the ancient Chinese (480 BCE) but only discovered in the West a few centuries ago—is the most accurate fractional approximation for *pi* using whole numbers less than 1,000, correct to six decimal places.

This relates to the figure—described above—that illustrates time penetrating space. The line represents the light of day as it penetrates the cyclic representation of the year (the circle), that is encompassed by the square of space. Thus, *pi* represents the ratio between temporal time (a year) and eternal time (a full day) within space.▼▼

In this scheme of days and years, we see a clear picture of day as a transmitter (*mashpia*) and a year as a receptor (*mekabel*). The three intellectual powers of the soul transmit light and the seven emotive attributes are the receptors. This echoes the male-female relationship mentioned above between *yesod* and *malchut*. Indeed, in Hebrew,

▶ The number 113 is an inspirational number i.e., the sum of two consecutive squares. It equals 7^2 plus 8^2. The number 355 is the sum of six consecutive squares (5, 6, 7, 8, 9, 10) i.e., the sum of three consecutive inspirational numbers. The intermediate inspirational number in this triad is 113. The sums of the square roots are 15 (for 113) and 45 (for 355). 15 is the numerical value of the Name *Kah* (יה). 45 is the numerical value of the Name [*Havayah* when its letters are filled with *alef* (יוד־הא־ואו־הא)discussed here more extensively with reference to time.

▶▶ The numerical value of "year" (שָׁנָה) is 355. The numerical value of "day" (יום) is 56. The sum of these two numbers is 411, the numerical value of "chaos" (תהו), "something from nothing" (יש מאין), etc., and equals 3 times 137. Thus, the union of year and day—temporal and eternal time—relates both to *pi* and to *alpha*.

The circumference of a circle with a diameter of 113 units is 355 units. Such a circle represents a year-full of days. Continuing this idea, the area of the square that surrounds the circle in the above illustration is 113^2, which equals 12,769 units. The area of the circle is π ($113\frac{1}{2}$)^2, which (rounded to a whole number) equals 10,029. The difference between the two areas, i.e., the area of the square that surrounds the circle, is 2,740, which equals 20 times 137.

In Kabbalah, the four corners of the square that surround the circle represent the "garment" (מלבוש) that enclothes the *tehiru*. Each corner represents one of the four Divine Names, *Ab*, *Sag*, *Mah* and *Ban*. The numerical value of "garment" is 378, the numerical value of *chashmal*.[25]

The numerical value of the verse in the Book of Samuel that pertains to God as eternal time, "The Eternal of Israel shall not lie and shall not regret" (וְגַם נֵצַח יִשְׂרָאֵל לֹא יְשַׁקֵּר וְלֹא יִנָּחֵם כִּי לֹא אָדָם הוּא לְהִנָּחֵם)[26] is 1,775, or 5 times 355, which equals the numerical value of "year" (שָׁנָה), relating to the temporal aspect of time. This equality is paradoxical. The numerical value of the entire verse in reduced numbering is 137. Two other verses that have a reduced value of 137 are the first of the Ten Commandments and the first verse in Exodus. 1,775 appears explicitly in the Torah with reference to the silver hooks that were attached to the pillars of the Tabernacle.[27] Kabbalah associates this number with the essential redemption of the Jewish People.[28]

"day" (יוֹם) is a masculine word, whereas "year" (שָׁנָה) is feminine. The objects that the ministers saw in their dreams, "twigs" (שָׂרִגִים) and "baskets" (סַלִּים) are also masculine, while "cows" (פָּרוֹת) and "ears of grain" (שִׁבֳּלִים) that Pharaoh saw in his dreams are both feminine words .

Order of Time

In Chapter 2, we discussed chaos and order as a complementary pair. We saw that there are those who believe that an initially chaotic state can result in spontaneous order.

In human physiology, *yesod* (the *sefirah* of foundation), corresponds to the procreative organ. *Yesod* (יְסוֹד) shares its two-letter root-gate (ס־ד) with "order" (סֵדֶר). Like a narrow tube that contains a number of marbles, *yesod* emits its contents in the order in which they were inserted. When rectified, *yesod* demonstrates the correct priorities of knowing what to do first. It creates a plan that resolves the trials and tribulations of temporal time. The order in which one carries out one's deeds bears vital significance.

Yesod receives its energy from all the *sefirot* above it. Its motivating force is the compulsion for self-actualization. When *yesod* receives its energy from the rectified emotive attributes, it projects them onto reality in a perfectly ordered holographic image. It appears organized at the external level. Thus, those who are oriented towards emotive experience will often acquire a rectified sense of real time. They will be systematic and habitual, with inherent order in their priorities.

In contrast, when it receives its energy from the intellectual powers of the soul, *yesod* appears chaotic. Its effect on reality is often revolutionary. The thought process of intellectually gifted individuals will probably be in a logical sequence (although others may not be able to access their logic or comprehend it). But, as in the common portrayal of the absent-minded professor, whose mind is preoccupied with cryptic riddles, they may be acutely chaotic in their day-to-day routines. It may take decades, or even centuries, before the insights of brilliant people are integrated as social norms. Rarely do we encounter individuals who are both ingenious and orderly.

We have considered the three intellectual *sefirot* and the seven emotive *sefirot* with reference to imaginary-eternal time and real-temporal time, respectively. Above these ten is the super-conscious *keter* (the *sefirah* of crown), which is enumerated when *da'at* (the *sefirah* of knowledge) is absent. This *sefirah* corresponds to the super-conscious source of the soul. It is above human perception and precedes creation. Kabbalah teaches us that although there is no concept of created time in *keter*, there is an inherent state of order, referred to as "order of times" (סֵדֶר זְמַנִּים).[29]

To be organized at both levels of time, it is necessary to reach up to the root of all order that is above time. At the level of "order of times," the concepts of real and imaginary time unite. The rectified *yesod* of the *tzadik,* who has achieved this level, reflects this. Such a righteous individual is in tune with and in control of time. He is master over time and time is no longer master over him.

Natural Units of Time

Measurements in the Torah are not arbitrary. Each Torah measurement is based on a natural phenomenon. Time measurements are included in that rule. A complete day is divided into twenty-four hours►► and each hour is divided into 1,080 parts of an hour. A part, according to one opinion, is the average time it takes to breathe one complete respiratory cycle (inhale and exhale).

Breath relates to life and the connection between man and God. The initial breath with which Adam was created was blown into his nostrils by God Himself, as the verse in Genesis states, "and He blew into his nostrils the breath of life."[30]

Breathing is the classic phenomenon via which we experience the presence of the Creator within us. The words for "breath" (נְשִׁימָה) and

► "Order of times" (סֵדֶר זְמַנִּים) is another phrase with a numerical value of 411, which equals 3 times 137.

►► According to the Talmud, there are two opinions regarding the division of night into "vigils." Nighttime is divided into either three or four divisions, all of which are defined by natural phenomena. By this standard, we can distinguish between the third and fourth hours of the night and the sixth and seventh. Following this division, we can further divide the night into twelve such hourly divisions. Since night is defined by twelve hours, so too daytime is also divided into twelve hours.

"[Divine] soul" (נְשָׁמָה) share the same Hebrew root. For this reason, the sages equate "breath" with "soul" in the verse from Psalms, "Every soul shall praise *Kah, hallelukah*."[31] With every breath that we take, we should praise the Creator. Thus, a part of an hour represents the fusion of the soul/breath and the body.

▸ Like the sum of the numerical values of "day" (יוֹם) and "year" (שָׁנָה), the sum of the numerical values of "part" (חֵלֶק) and "instant" (רֶגַע) is 411.

Each part is further divided into 76 instants.◂ Each instant is approximately $\frac{1}{23}^{rd}$ of a second. A synonym for instant is "a wink of an eye" (הֶרֶף עַיִן). Indeed, the human brain registers the projection of frames in a movie as a moving continuum only at a rate faster than 23 frames per second. From a phenomenological perspective, an instant is the smallest quantum of time. Just as a part relates to breathing, an instant is related to our ability to visualize a moment in the time-continuum. The Torah measure of distance is a Talmudic mile, which is equivalent to 2,000 *ama* (the length of the arm from the elbow to the tip of the middle finger)—approximately 1 kilometer. The speed of light is almost 300,000 km per second. Translated into natural units, the speed of light is approximately 1,000,000 Talmudic miles per part of an hour.

With every breath we take, light travels 1 million Talmudic miles. The speed of light not only incorporates the dimensions of space and time. It also alludes to a third dimension that relates to the human soul.

The connection between breath/soul and light manifests in praise. "Praise" (הַלֵּל) is the penultimate of the thirteen Hebrew synonyms for light in Hebrew. The root "praise" (הַלֵּל) appears thirteen times in this final chapter of Psalms.

REFERENCE NOTES FOR CHAPTER 13

1. Stephen Hawking, *A Brief History of Time* p. 107.

2. Genesis chapters 40-41.

3. For a detailed discussion on dream interpretation, see our book in English, *A Sense of the Supernatural.*

4. *Sotah* 36a.

5. *Sefer Yetzirah* 1:2.

6. Genesis 1:5.

7. Malachi 3:6.

8. Genesis 41:32.

9. Proverbs 15:1

10. Genesis 41:1.

11. Ibid 41:17.

12. Psalms 81:6.

13. Genesis 36:31-39.

14. Rashi on Genesis 1:2.

15. See our Hebrew volume, *Mivchar Shiurei Hitbonenut* 15 p. 151..

16. Rashi on Exodus 33:21.

17. Genesis 2:10.

18. Ibid 2:6.

19. Ibid 41:1.

20. Ibid, *Siftei Chachamim* ad loc.

21. *Tanya, Igeret Hakodesh* 14.

22. *Bartenura, Mishnah Eiruvin* ch. 6 m. 3.

23. Exodus 21:21.

24. Ibid, Rashi ad loc.

25. See first chapter of *Sefer Emek Hamelech.*

26. I Samuel 15:29.

27. Exodus 38:28.

28. See our book, *The Twinkle in Your Eye,* p. 102.

29. *Bereishit Rabah* 3:7.

30. Genesis 2:7.

31. Psalms 150:6.

CPT SYMMETRY

In 1963, the Swedish parliament decided that the direction of Swedish traffic would be switched from left to right hand drive. At 4:50 am on Sept. 3, 1967, after all the preparations had been made and an extensive campaign publicized, traffic in Sweden was stopped and redirected into the opposite lanes. At 5:00 am movement was resumed — with all traffic traveling on the righthand side of the road.

Reorientation

Reorienting the traffic in Sweden was accomplished without a hitch. But, physicists ask, what would happen if we reoriented the particles in physics. If all positive charges became negative, all compass directions reversed, or the direction of time transposed, what would happen? The symmetrical quality of three characteristics — charge (C-symmetry), spatial parity (P-symmetry), and time (T-symmetry) — has been investigated by physicists.

C-Symmetry

Every known particle has an associated antiparticle with the same mass and opposite electric charge. For example, an electron has negative electric charge, while its antiparticle is the positively charged positron. From our perspective of the laws of nature, there is no difference between matter and its symmetrical counterpart, antimatter. However, although most charges are symmetrical, the weak force has been proven to be asymmetrical. It thus violates charge symmetry.

P-Symmetry

Like the Swedish traffic reversal, parity symmetry relates to spatial direction in any orientation (up-down, left-right, backwards-forwards). Parity transformation produces a mirror-image reflection. Similar to charge symmetry, parity is conserved in electromagnetism, strong interactions and gravity, but is violated in weak interactions.

T-Symmetry

The present, as we perceive it, is a point on a one-directional continuum. Inversing time in this context is counter-intuitive because the effects of real time are clearly discernible. Consider a time-lapse movie of a plant growing. If we ran the movie in reverse, it would be clear that it does not depict a real process, i.e., time is inherently asymmetric. But, at the subatomic level, the laws of physics remain the same when time is reversed (this can be demonstrated experimentally in the laboratory). An electron orbiting an atom, or making a quantum leap as it integrates a photon, is a valid physical process in either time direction. In fact, one possible interpretation of the positron (the "anti-electron") is that it is an electron moving backward in time.

Physicist Stephen Hawking stated in his book, *A Brief History of Time*:

> *The laws of science do not distinguish between the forward and backward directions of time. However, there are at least three arrows of time that do distinguish the past from the future. They are the thermodynamic arrow, the direction of time in which disorder increases; the psychological arrow, the direction of time in which we remember the past and not the future; and the cosmological arrow, the direction of time in which the universe expands rather than contracts ... The psychological arrow is essentially the same as the thermodynamic arrow, so that the two would always point in the same direction ... The reason we observe this thermodynamic arrow to agree with the cosmological arrow is that intelligent beings can exist only in the expanding phase.* [1]

As mentioned in the previous chapter, "real" time is generally depicted as a horizontal line running between "past" in one direction and "future" in the other. For reasons of mathematical convenience, scientists have developed an "imaginary" time line that runs perpendicular to the normal time line, as mentioned. In essence, imaginary time allows us to consider time as a fourth dimension of space. Just as it is possible to walk south and then reverse one's direction to the north, so it is possible to travel backwards in imaginary time.

Simultaneous Symmetry

Initially, scientists believed that although charge and parity are not individually symmetric, if both were reversed in the entire universe, symmetry would be conserved. Under such circumstances, the universe would remain identical to its present state, and it would be impossible to differentiate between the old universe and the new universe. Later, it was calculated mathematically that this is not so. The resulting universe would only be identical to the universe we experience if all three symmetries of charge, parity and time were violated simultaneously. To date, no violations of CPT symmetry have been observed in the laboratory.

The symmetry of these three phenomena as a single unit is one of the most profound concepts of modern science. It connects the theory of relativity, the exclusion principle and its offshoot, particle spin.

Alpha is defined by an equation that includes the electric charge (*e*) and the speed of light (*c*), which incorporates both space (parity) and time. Thus, the mystery of 137 is defined by these three symmetries.

Three-Dimensional Letters

Sefer Yetzirah is the oldest known Kabbalistic text, attributed to Abraham, the first Jew.▶ The topic of this concise treatise is a profound analysis of the twenty-two letters of the Hebrew *alef-bet*. *Sefer Yetzirah* divides the twenty-two letters into three groups: one group of three letters, one of seven letters, and one of twelve letters.

Each of these three groups contains a spatial dimension (*olam*, lit. "world"), a time dimension (*shanah*, lit. "year") and a soul dimension (*nefesh*).[2] These three dimensions correspond to the three elements of CPT symmetry.

▶ The final version was edited by Rabbi Akiva, who was the rabbi and mentor of Rabbi Shimon bar Yochai, author of the Zohar.

The spatial dimension clearly corresponds to parity. The time dimension obviously corresponds to the time element. It remains that charge corresponds to the soul element. Indeed, charge is the energy force of a particle. Translated into the realm of living beings, this relates charge to life-force, the soul.

Souls, like charges, come in pairs. Positive and negative charges correspond to female and male energies, respectively. In an atom, the positively charged protons in the nucleus are much bigger than the negatively charged electrons that surround it. In this respect, they are similar to the ovum in a female, which is much larger than the male sperm (although proton-electron size ratio is massively larger than ovum-sperm size ratio).

Another allusion to male energy follows the Talmudic principle that "it is the way of a man to chase after a woman."[3] In this expression of the sages, "chase" (לַחֲזֹר) shares the same root as "cycle" (מַחֲזֹור). This corresponds to the constant motion of the electron around the nucleus. The same root is closely related to "lack" (חֶסֵר) referring to the electron's negative charge. An unmarried male senses that he is missing his other half and attempts to find her.

Like a woman, who is more inclined to stay at home, the positive charge in the atom is a static, attractive force. In the atom, the electron is in danger of being lured away by external sources of energy. The proton's positive charge draws the electron to it, preventing the electron from straying from its path as it whizzes around the atomic periphery. Without that positive charge, the electron would be influenced by forces outside the atom. Similarly, if the electron became detached from the atomic core, the atom would become unstable.

Kabbalah teaches us that in the future, the male and female aspects of reality will be inversed, as Jeremiah prophesied: "For God has created something new in the land, a female who circuits the male."[4] The physical result of such a reversal regarding the atom is antiparticles—positrons (bearing a positive charge) revolving around antiprotons (bearing a negative charge). This corresponds to the discovery of antimatter. "Antihydrogen" atoms, with a positron orbiting an antiproton, have been produced experimentally.

Annulling Space, Time, and Consciousness

From a Chassidic perspective, the necessity to combine parity, time and charge in order to create complete symmetry reveals that they all originate from one source. This origin is a point where there is no distinction between left and right (parity/space), past and present (time), and male and female (charge/soul). In Kabbalah, this point of singularity is called *omek reishit* (lit: "the depth of the beginning"). This is the point at which the inner power of selflessness in *chochmah* (the *sefirah* of wisdom) is annulled at its source in *keter* (the *sefirah* of crown). Beyond this level, all is one, and space, time and soul are indistinguishable from one another.

Space, time and soul correspond to three miraculous events witnessed by the Jewish People. They are: the Exodus from Egypt, the Splitting of the Red Sea, and the Giving of the Torah, respectively.

The Exodus from Egypt symbolizes freedom from the space dimension. Egypt (מִצְרַיִם) means "confines." Space restricts us within its dimensions. Until the moment of the Exodus, no slave had ever escaped the borders of Egypt. Thus, the Exodus of the entire Jewish People from the confines and boundaries of the Egyptian exile represents physical freedom and the annulment of the space dimension.

Rebbe Nachman of Breslov wrote that the Splitting of the Red Sea symbolizes the annulment of the time dimension.[5] The Pillar of Fire, which guided the fleeing Israelites in darkness, represents night. The Pillar of Cloud, which guided them during light hours, represents daytime. At the Splitting of the Red Sea these two became fused into one, thus anulling the time dimension.

The Giving of the Torah freed the Jewish People from the enslavement they experienced during the Egyptian exile. Up until the moment of the Giving of the Torah, they remained mentally subjugated to the Egyptian culture. But from that moment on, they received the resources necessary to become free-thinking human beings, subject only to God.[6] Then they became charged with a life-force that infused them with Divinely ordained purpose.

Time as Intermediary

As mentioned in the previous chapter, time has a changing aspect and an eternal aspect. Thus, time includes within it an aspect of each of the other two powers. The changing aspect of time connects it to space, and the eternal aspect of time connects it to the soul. Indeed, in Kabbalah, time is the intermediary factor that connects between space/parity and soul/charge.

For this reason, of the three general dimensions of reality— space, time and soul—we will concentrate on analyzing the time dimension. In doing so, we will contemplate the *mazalot* (signs of the Zodiac).

In essence, the *mazalot* are a purely astronomical phenomenon. However, they also form the basis of astrology. The Zodiac signs appear in all ancient traditions. Abraham was the first to comprehend their true significance. Through his knowledge of astrology, Abraham foresaw that his marriage to Sarah would not be fertile. But, God told him to step outside of astrological calculations and to relate to the stars on purely astronomical grounds.[7] God changed his name from Abram to Abraham and his wife's name from Sarai to Sarah. He then informed Abraham that he and Sarah would bear a son, thus teaching him that God channels His Divine influence, not through the stars, but through the letters of one's Hebrew name. This is the meaning of the Talmudic idiom, "There is no preordained fortune (*mazal*) for Israel" (אֵין מַזָּל לְיִשְׂרָאֵל).[8] Jewish individuals, descendants of Abraham, have the power to transcend all astrological influences.

The Ba'al Shem Tov offered a more profound interpretation of this prhase. He taught that it means, "The *mazal* of Israel is Divine noth-ingness" (אֵין מַזָּל לְיִשְׂרָאֵל)—that is, the *mazal* (soul-root, defining one's purpose in life and manifesting in one's fortune) of the Jewish People derives from the level of Divinity referred to as "Divine nothingness," that is beyond the fate dictated by astrological signs. To this very day, the Jewish People have an intrinsic nature that transcends the omens of the Zodiac, which have influence only over those souls who are not in touch with the Divine.

Nonetheless, the *mazalot* (the signs of the Zodiac) retain their significance in Jewish philosophy, Kabbalah and Chassidut. They describe the nature of the months in which they appear.

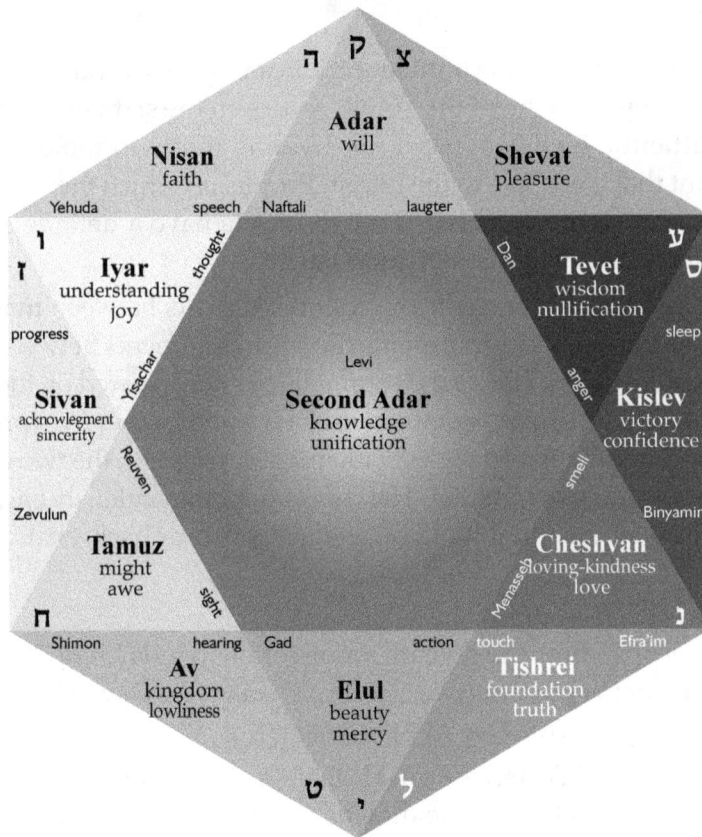

The Jewish Calendar
(showing the names of the months and associated letters, *mazal* and tribe)

The Significance of the *Mazalot*

The Hebrew names of the twelve *mazalot* (Zodiac signs) all refer to single entities, except for three that appear in pairs. The three *mazalot* in pairs are Gemini (תְּאוֹמִים), Libra (מאֹזְנַיִם) and Pisces (דָּגִים). Literally, these three names mean "twins," "scales" and "fish," and they correspond to parity, time and charge, respectively.

Gemini—Sivan

Gemini is the *mazal* of the third month, Sivan.◄ The festival of Shavuot, the Time of the Giving of the Torah, occurs on the sixth day of Sivan. The authentic Jewish symbol for Sivan is the twin tablets of the covenant that God gave to the Jewish People, on which the Ten Commandments were inscribed. The two tablets have a definite spatial orientation that relates to left-right parity.◄◄

The right tablet represents the commandments between man and God, and the left tablet represents the commandments between man and his fellow man. Five of the commandments appeared on the right tablet (mainly obligatory commandments) and the other five (prohibitive commandments) on the left tablet. Thus, together, the two tablets represent a union of right/left. The tablets were miraculously engraved in such a way that they could be read forwards or backwards.[10]◄◄◄

Libra—Tishrei

The *mazal* of Tishrei, the seventh month of the year is Libra. Tishrei is the month of the Days of Awe: Rosh Hashanah (the Day of Judgment) and Yom Kippur (the Day of Atonement).

The first day of Tishrei is Rosh Hashanah, the beginning of the solar year. This relates to the time dimension mentioned in *Sefer Yetzirah*, which is explicitly called "year" (שָׁנָה).◄◄◄◄ The scales, the symbol of Tishrei, symbolize that this is the time when God measures our merits of the past year and judges us according to our deeds. In this month, our merits and transgressions stand in the balance. God "balances the books" as it were, between good and evil.

► The Hebrew year begins in Tishrei, the seventh month. But, the Hebrew months are counted from Nisan.[9] This is part of the secret of the Hebrew calendar, which unifies the solar year with the lunar months.

►► Following the festival of Shavuot it was permitted to bring the festival sacrifices to the Temple for another six days. The penultimate day for actualizing this opportunity was the 11th of Sivan. The numerical value of 11 Sivan (יא סִיוָן) is 137. There is only one other day in the year that has a numerical value of 137 and that is 21 Kislev (כא כִּסְלֵו).

►►► The reduced numerical value of the first of the Ten Commandments "I am *Havayah* your God who brought you out of the Land of Egypt out of the house of bondage" (אָנֹכִי הוי' אֱ-לֹהֶיךָ אֲשֶׁר הוֹצֵאתִיךָ מֵאֶרֶץ מִצְרַיִם מִבֵּית עֲבָדִים) is 137.

►►►► One allusion to the connection between scales and time is that the numerical value of "scales" (מאֹזְנַיִם) is 148, which is the numerical value of "eternity" (נֵצַח). Someone who lives in a state of eternity, where "there is no time," has the ability to move freely backwards and forwards in time. 148 equals 4 times 37, which complements 137, as mentioned.

More profoundly, Tishrei is the optimal time of the year to tip the scales and reverse time symmetry by returning to God. The outcome of the judgment depends on our behavior during the first ten days of the year, the Ten Days of Repentance, which culminate on the holiest day of the year, Yom Kippur. These first ten days of the month of Tishrei are our opportunity to rectify our deeds and reverse the judgment by turning our sins into merits.

Rosh Hashanah and Yom Kippur are "Days of Awe." The *teshuvah* accomplished during these days is primarily through fear. Such repentance corrects the impression our past transgressions have stamped on the present (and future). At this level, any transgressions that we committed intentionally are transformed into unintentional transgressions.

The week of Sukot, the festival that follows Yom Kippur, is a week of rejoicing.▶ By rejoicing, we reach an even higher level of *teshuvah* (repentance and return to God) through true love. On Sukot, our *teshuvah* through love has the power to take us back in time. When *teshuvah* is accomplished through love and joy it corrects our sins at their source in the past. In this case, our intentional transgressions transform into good deeds.[11]

Thus, during Tishrei, the first month of the solar year, we receive a special boost of power that takes us backwards in time to rectify all our past deeds.

The name "Tishrei" (תִּשְׁרֵי) begins with the last three letters of the *alef-bet* (*reish-shin-tav*) in reverse order. This alludes to the ability to return to the past to rectify all misdeeds at their source. This property connects the future to the past via a two-way channel. A righteous individual who has never sinned travels along the arrow of time in one direction. He moves up and away from the initial point of creation when all was united, towards the ultimate redemption when all will once again be one. In contrast, the *ba'al teshuvah* (one who has sinned but has repented) opens up a new track in the opposite direction. Such an individual returns to the past to correct his misdeeds. By

▶ There are four consecutive verses in the Torah (Lev. 23:40-43) that relate to the commandment of rejoicing on Sukot. They include the commandment to take the four species and the commandment to dwell in a *sukah* for seven days.

The numerical value of the first and third verses is 10,001, which equals 73 (the numerical value of "wisdom" (חָכְמָה) times 137. This relates to the division of 137 the numerical value of "Kabbalah" (קַבָּלָה) into "wisdom" (חָכְמָה) and "prophecy" (נְבוּאָה), i.e., the numerical value of these two verses is thus "wisdom" times "wisdom-prophecy."

The numerical value of the second and fourth verses is 9,727, which equals 71 times 137.

The sum of the numerical values of all four verses is 19,728, which equals 144 (122) times 137.

doing so, a *ba'al teshuvah* breaches the chasm that his sins opened between the past and the future, fusing the two together in a single moment of repentance. This idea is expressed in the Talmudic dictum, "There is no precedence or subsequence in the Torah."[12] Kabbalah and Chassidut explain that the source of the Torah is at a point above time. Therefore, it connects the past and the present with the future, without any necessity for ordered continuity. Someone who has purified himself by returning to God and His Torah can achieve the level at which his prayers are answered even before he articulates them, as the Prophet Isaiah states in the Name of God, "Before they call Me [in prayer], I have already answered."[13]

When explained in this way, the concept of *teshuvah* throws light on the many-worlds theory (see Chapter 6), which claims that our consciousness is simultaneously present in any one of a number of different universes depending on the choices we have made. Whereas science offers no way to relay our consciousness between each reality state, *teshuvah* allows us to exchange our present reality for that of a parallel universe.

Pisces—Adar

The third month of this group is the twelfth month of the year, Adar, in which the holiday of Purim occurs. The symbol of Adar is Pisces or fish (דָּגִים), which is in the plural. We assume that this refers to the minimum plural of two fish.[14] The two fish are essentially unified, as in marital union. The marriage of Adam and Eve is described in the Torah as "and they became one flesh,"[15] which also refers to the child born of their union.[16] Indeed, God blessed the fish with the power of extreme proliferation, as the Torah states, "God blessed them, saying, "Be fruitful and multiply, and fill the waters of the seas."[17] Accordingly, the Talmud states that a wedding on Thursday, the day fish were created, is an auspicious sign for a fruitful marriage.[18] Thus, Adar is a month of fertile procreation. The root of Purim (פּוּרִים), the festival that occurs in the month of Adar, is related to "procreation" (פְּרוּ וּרְבוּ), which is the first commandment that God gave to mankind.

The month of Adar often gives birth to itself in a "pregnant" (leap) year, when a second month of Adar follows the first.▼

Many species of fish produce bio-electricity, either in small voltages for navigation, or for communication purposes. Some male fish also use their electrical powers to court females and deter rivals. In larger voltages (up to 500 volts!) some fish defend themselves with bio-electricity when under attack, or use it to stun their prey.

Purim is packed with more charges and anti-charges than any other holiday in the Jewish calendar—Mordechai and Haman, Esther and Vashti, etc. When Haman was hanged on the tree fifty cubits high, his charge reversed and became contained within Mordechai's positive charge.▶▶ The Talmud relates that Haman's descendants eventually converted to Judaism and studied Torah in Bnei Brak.[24]

The souls of the righteous are incarnated in the form of fish.[25] Indeed, the righteous are the most highly charged souls of all.

Thus, like the three different types of symmetry in CPT symmetry, these three months and their Zodiac signs have a common theme of duality or complementarity and correspond to charge (Adar), parity (Sivan) and time (Tishrei).

▶▶ Haman's charge-reversal is indicated by the numerical value of "cursed is Haman" (אָרוּר הָמָן), which equals 502, but is equal to the numerical value of "blessed is Mordechai" (בָּרוּךְ מָרְדְּכַי). Similarly, the numerical value of the letters of Haman's name (הָמָן) when its filling letters are written in full (הא מם מם מם נון ואו נון אלף) is also 502.

The numerical value of Mordechai (מָרְדְּכַי) is 274, which equals 2 times 137.

▶ The literal meaning of "Purim" (פוּרִים), is "lots." This word is in the plural, although it refers to one lot that Haman drew.

The numerical value of the singular "lot" (פוּר) is 286, which equals 13 times 22. The numerical value of the plural "lots" (פוּרִים) is 336, which equals 7 times 48. The numbers 13 and 7 are relatively male and female [one simple explanation of this idea is that 7 is the mid-point of 13].[19] The sum of the other two integers, 22 and 48, is 70, the numerical value of "wine" (יַיִן) and also "secret" (סוֹד), referring to the statement of the sages, "when wine enters, a secret is revealed."[20] This phrase is related to Purim, when the custom is to drink wine. One interpretation of the Hebrew name of the Book of Esther—Megillat Esther (מְגִילַת אֶסְתֵר)—which is read on Purim, is "the revelation of the hidden" (גילוי נִסְתָּרוֹת).

The plural form of Purim also alludes to the two days of Purim, as mentioned explicitly in the Book of Esther, "therefore the Jews called these days 'Purim.'"

Purim is associated with Yom Kippur, which is one day that is considered as two, "anyone who eats and drinks on the ninth [of Tishrei] and fasts on the tenth, the Torah considers it as if he fasted two days."[21] Furthermore, Ezekiel refers to Yom Kippur as "the head of the year" (רֹאש הַשָּׁנָה, Rosh Hashanah).[22] Although the commandment referring to Rosh Hashanah is only for one day, in practice, we celebrate it for two consecutive days. Another verse upholds this, stating: "He will give us life from two days."[23] Here we see that Purim is connected not only to Yom Kippur, but to Rosh Hashanah as well. Appropriately, it is the custom to eat the head of a fish (or a ram's head) on Rosh Hashanah symbolizing that we should be "a head and not a tail." Others have the custom to eat fish on Rosh Hashanah, saying the prayer that, "we should be fruitful and multiply as fish, and the evil eye should have no influence upon us, as it has no influence upon fish."

The total numerical value of Rosh Hashanah (רֹאש הַשָּׁנָה) and Purim (פוּרִים) is 1,197, which equals 21 [the numerical value of the Divine Name Ekyeh (אהיה)] times 57 [the numerical value of "fish" (דָגִים)].

Three, Seven, Twelve

The ordinal numbers of Sivan, Tishrei and Adar in the calendar are 3, 7 and 12, respectively. They reflect the division of the twenty-two letters into three "mother" letters (*alef*, *mem* and *shin*), seven "double" letters (*bet*, *gimel*, *dalet*, *kaf*, *pei*, *reish* and *tav*) and twelve "simple" letters (*hei*, *vav*, *zayin*, *chet*, *tet*, *yud*, *lamed*, *nun*, *samech*, *ayin*, *tzadik* and *kuf*).[26] These three groups of letters also correspond to parity (3, Sivan), time (7, Tishrei) and charge (12, Adar). Thus, these three months reflect and include everything in the three states of world, year and soul.

The principle mathematical series in which the three numbers, 3, 7 and 12 appear as consecutive numbers can be deduced by taking the finite differences between them.

$$
\begin{array}{ccccc}
0 & \mathbf{3} & \mathbf{7} & \mathbf{12} \\
 & 3 & \mathbf{4} & \mathbf{5} \\
 & & 1 & \mathbf{1}
\end{array}
$$

This series is defined by the equation:

$$a = [n(n \perp 1)/2] - 3, \text{ which is equal to } \triangle n - 3$$

▶ The series of triangular numbers ($\triangle n$) represents the natural, or evolutionary, cause of events. This is the effect of Divine government through the union of the cyclic and the direct Divine emanations (*igulim* and *yosher*). The series $\triangle n - 3$ alludes to Divine government that removes the harshness of direct Divine emanation and is more in tune with the cyclic Divine emanation.

Expressed in geometrical form, this is the series of triangular numbers without their vertices:◀

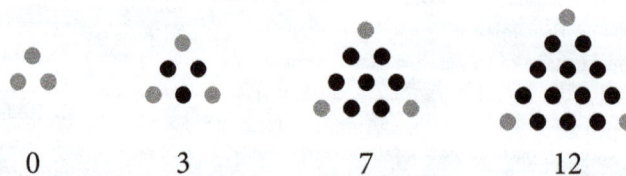

| 0 | 3 | 7 | 12 |

The appearance of 0 before the 3 (relating to the third month, Sivan) in this series, illustrates the rule of *Sefer Yetzirah* that "the end is wedged in the beginning."[27] By counting back from the first month (Nisan, represented by the number 1), we return to the month of Adar. This relates to the timeless question, "Before one, what do you count?"[28]

Zero refers to the three highest realms of the soul which are sometimes named "zero" (אֶפֶס) "nothing" (אַיִן) and "chaos" (תֹהוּ). Indeed, Purim is the most propitious time of the year to reach the level of "knowing nothing," which corresponds to the level of "zero." ▶

The Three Months and 137

Further analysis of these three months reveals a second phenomenon that considerably reinforces their connection to CPT symmetry. As mentioned, God created each of the twelve months of the year with one of the twelve simple letters. The letter of the third month, Sivan, is *zayin* (ז), which has a numerical value of 7. The letter of the seventh month, Tishrei, is *lamed* (ל) which has a numerical value of 30. The letter of the twelfth month is *kuf* (ק), which has a numerical value of 100. The sum of the numerical values of these three letters is 137. Thus, we find the mysterious number of modern physics related to the most profound subject of CPT symmetry, in the context of the *mazalot* (the signs of the Zodiac) the basis of the time cycle.

By dividing 137 into units (7), tens (30) and hundreds (100) in this way, we can consider these three integers as a quadratic series. We can then predict the next number in the series by taking the finite differences between them. ▶▶

```
31   7   30   100   217
  -24   23   70   117
     47   47   47
```

The sum of the first four members of this series is 354, the average number of days in a lunar year. The number in the series that appears to the left of the number 7 is also a positive integer (31). Adding this number to 354 raises the total to 385, the maximum number of days in the year. ▶▶▶ Thus, this series, generated by dividing 137 into hundreds, tens and units, is essentially connected to our measurement of time.

Capricorn, the kid goat, is the *mazal* of the tenth month of the year, Tevet. The numerical value of Tevet (טֵבֵת) is 411, which is the

▶ All three words, "zero" (אֶפֶס), "nothing" (אַיִן) and "chaos" (תֹהוּ), appear together in one verse, "All the nations like nothing against him and from zero and chaos he considered them" (כָּל הַגּוֹיִם כְּאַיִן נֶגְדּוֹ מֵאֶפֶס וָתֹהוּ נֶחְשְׁבוּ לוֹ).[29] The total numerical value of the three words (אֶפֶס, אַיִן, תֹהוּ) is 613, the number of commandments in the Torah.

▶▶ The fourth number in the series is 217, which equals 7 times 31 (the number that precedes 7 in the series). The ratio between 31 and 217 is thus 1:7, which is the secret of "then" (אָז), in which the numerical value of each letter is 1 and 7, respectively. This ratio reflects the ratio between one Shabbat and the next; the days from the birth of a baby boy to the day of circumcision, and the relationship between the first seven days of Chanukah and the eighth day.[30] It also reflects the ratio between the Jubilee year and the seven other time factors: sabbatical years, years, months, weeks, days, hours, and parts of an hour (see previous chapter).

▶▶▶ 385 is the numerical value of "Divine Presence" (שְׁכִינָה). The number 385 is the pyramid number of 10 (i.e., the sum of all squares from 1 to 10) and equals 5 times 77. Thus, the average value of the 5 numbers is 77, which is the numerical value of mazal (מַזָּל), referring to the astrological signs, the original basis for this meditation.

numerical value of the tenth word of the Torah, "chaos" (תֹהוּ). The average value of each of its three letters is 137. As discussed in depth in Chapters 1 and 2, chaos is particularly associated with the number 10. This is upheld by the Midrash that states that "a ten-year old boy jumps like a kid goat."[31]▼ The kid goat expresses the uncontained potential of a ten-year old boy. Chaotic energies are rectified when they are inter-included in the ten *sefirot*, the vessels that contain them without shattering. When rectified, the tenth is sacred.[32]

The sum of the ordinal numbers of the three months discussed thus far is 22 (3 plus 7 plus 12), the number of letters in the *alef-bet*. When we add 10, the ordinal number of Tevet, the tenth month, the total is 32. This reflects the division at the beginning of *Sefer Yetzirah* of the 32 pathways of wisdom into 10 *sefirot* and 22 letters.[33]

One relationship between the two months of Tevet and Adar is that the tribes associated with them were the two sons of Jacob from Rachel's maidservant, Bilhah. More significantly, Tevet is the first month of the winter season, and Adar is the third and last month of the winter season. The *mazalot* (Zodiac signs) of these two months are "kid goat" (גְּדִי) and "fish" (דָּגִים). All the letters of the *mazal* of Tevet are included in the *mazal* of Adar. This is a unique phenomenon that

---

▶ The numerical value of "a kid goat" (גְּדִי) is 17. 17 is also the numerical value of "good" (טוֹב), which shares its root with Tevet (טֵבֵת). The traditional Jewish blessing of "*mazal tov*" (מַזָּל טוֹב) is therefore particularly relevant to the month of Tevet.

The dot-product of the three letters of "kid goat" (גְּדִי) is 120 (3 · 4 · 10). Adding 17, the sum of the three letters, to 120, their product, yields 137. In this way, both the name of the month (טֵבֵת) and the name of its *mazal* (גְּדִי) relate to 137.

Sometimes the letter *dalet* (ד) is filled without the *tav* (ת) at the end of the name (דל). The numerical value of the filled *dalet* is then 34, the prime root of 137, which equals two times 17, the numerical value of "a kid goat" (גְּדִי). The numerical value of "kid goat" (גְּדִי) when its letters are filled in this way (גימל דל יוד) is 137. The sum of the numerical values of the filling letters alone is 120, the product of all three letters.

The three times this phrase appears in the Torah thus total 6 times 411, relating to the six possible permutations of "chaos" (תֹהוּ) or Tevet (טֵבֵת), which have three letters each.

In the Torah, "kid goat" (גְּדִי), appears three times in the prohibition against eating meat and dairy produce together, "Do not cook a kid goat in its mother's milk."[34] The three times allude to the three different prohibitions against eating, cooking and benefitting from meat and dairy together. The kid alludes to a fetus in its mother's womb, relating to the secret of Jonah in the stomach of the female fish.

The numerical value of the first four words of this phrase, "Do not cook a kid goat in milk" (לֹא תְבַשֵּׁל גְּדִי בַּחֲלֵב) is 822—twice the value of Tevet (טֵבֵת), or 6 times 137. The numerical value of the fifth word, "its mother's" (אִמּוֹ) is 47, the base of the quadratic series that begins 7, 30 and 100.

beautifully reflects the above-mentioned Kabbalistic statement, "The end is wedged in the beginning."[35] There is no other month or *mazal* whose name is included in another.

The letter associated with the month of Tevet is *ayin* (ע), which has a numerical value of 70. This represents the 70 years of man, reflecting the changing aspect of time that connects time with the soul when years are filled with days. Tevet thus introduces a fourth, asymmetrical aspect that relates to one-directional time, an element that infuses the other three with inherent symmetry.

This idea becomes apparent in the cubic series created from the numerical values of the names of these four letters, in the order the months appear on the Hebrew calendar: first comes the *zayin* (ז) associated with Sivan, then the *lamed* (ל) associated with Tishrei, and the *ayin* (ע) associated with Tevet, and finally, the letter *kuf* (ק) associated with Adar.

	זין	למד	עין	קוף		
158	**67**	**74**	**130**	**186**	193	102
-91	**7**	**56**	**56**	7	91	
	98	**49**	**0**	-49	-98	
	-49	**-49**	-49	-49		

The sum of the numerical values of *kuf* and *lamed*, the two numbers on either side of 130 (74 ⊥ 186) is 260 (2 times 130, or 10 times the numerical value of God's Name). The sum of the next two edges is also 260 (67 ⊥ 193 = 260). Any number added to the right of the series is always identical to the number subtracted from the left and vice-versa, thus this symmetry continues ad infinitum—i.e., the average of every pair of numbers equidistant from 130 will always be 130. Thus, the average number of the entire series is also 130, the numerical value of the letters that spell *ayin* (עֵיִן), meaning "eye."▸

The number 130 is the axis of symmetry around which this series revolves. The symmetry of the three months that come in pairs pivots around a concealed fourth month that appears in the singular!

▸ The sum of the seven numbers in the series with 130 in the center is 910., the numerical value of Tishrei (תִּשְׁרֵי), the first month of the year, which equals 7 (the numerical value of the letter *zayin*) times 130.

The sum of the five numbers from 74 to 102 is 685, which equals 5 times 137.

319

This is a fascinating point to consider with reference to CPT symmetry. None of the properties—neither charge, nor parity, nor time—is symmetric in every case. Yet, when all are considered together, they are perfectly symmetrical. From our Kabbalistic perspective, this means that there is a fourth, assymetric property that has yet to be discovered.◄

In Kabbalah, the three dimensions of spatial parity (*olam*), time (*shanah*) and charge (*nefesh*) correspond to the *sefirot* of *binah* (understanding), *tiferet* (beauty) and *malchut* (kingdom), respectively. The letter *ayin* (ע) of Tevet corresponds to a fourth aspect that includes them all. This is *chochmah* (the *sefirah* of wisdom), which corresponds to the quality of self-annulment, or selflessness. Indeed, self-annulment is the inner property associated with the month of Tevet. This suggests that the assymetric property that science must acknowledge is the point at which creation is annulled before the Creator. This manifests in the unknown state of particles before their measurement.

Cycles of Days, Months and Years

Let's take a deeper look at the three basic letters that produce the number 137, *kuf* (ק), *lamed* (ל) and *zayin* (ז), and how they relate to the division of time.◄◄

Each letter of the Hebrew alphabet has an ordinal number from 1 to 22 that refers to its place in the order of the twenty-two letters of the *alef-bet*.

The ordinal number of the letter *zayin* is 7, the same as its regular numerical value. The ordinal value of the letter *lamed* is 12, and the ordinal number of the letter *kuf*, the last of the twelve simple letters, is 19, which is the sum of the ordinal numbers of the *zayin* (7) and the *lamed* (12). The total ordinal value of the three letters is 38 (2 times 19). Two of these three letters (ק and ל) appear in the word *Kabbalah* (קַבָּלָה), which has a numerical value of 137. The ordinal numbers of the other two letters of *Kabbalah* (קַבָּלָה) are 2 (ב) and 5 (ה). Their sum is 7, which is equal to the ordinal value of the letter *zayin* (ז). Hence,

---

► As mentioned, the sum of the numerical values of the three letters associated with the three months discussed here (*zayin*, *lamed* and *kuf*) is 137. Adding 70, the numerical value of the letter *ayin* (ע) associated with Tevet brings the total to 207, the numerical value of "light" (אוֹר).

►► In Kabbalah, each of the twenty-two letters of the *alef-bet* has a mate with a similar visual form. Each "couple" contains a "male" letter and a "female" letter. The three letters *zayin*, *lamed* and *kuf* are all "female." The male pair of the letter *zayin* is *vav*, the male pair of the letter *lamed* is the *yud*, and the male pair of the letter *kuf* is the *hei*. The male counterparts of these three letters are the letters of the Divine Name *Havayah*.

the ordinal value of Kabbalah is also 38. The ordinal values of these three letters (38) thus self-refer to their regular numerical value (137, vis à vis Kabbalah.▶

The three numbers 7, 12 and 19 all have great significance with reference to time. There are seven days in a week and twelve months in a year. The number 19 is also very significant in the Hebrew calendar. To ensure that the month of Nisan, "the Month of Spring,"[36] appears in spring, the lunar year is adjusted to match the solar year in a 19-year cycle.▶▶ Time thus revolves in three concentric cycles of 7 days, 12 months and 19 years. Each 19-year cycle includes an inner division into 12 years with 12 months and 7 "pregnant" (leap) years that incorporate an additional month of Adar.▶▶▶ These are the ordinal numbers that correspond to the Hebrew letters with numerical values of 7, 30 and 100, which total 137.

In every 19-year cycle, the leap years are the third, sixth, eighth, eleventh, fourteenth, seventeenth and nineteenth years of the cycle. The first three leap years are in the first eight years of the cycle, followed by three years, followed once again by eight years (8, 3, 8). The sum of the squares of these three numbers (64, 9, 64) is 137.▶▶▶▶

The Centrality of Time, Space, and Soul

The sum of the numerical values "twins" (תְּאוֹמִים, 497), "scales" (מאֹזְנַיִם, 148), and "fish" (דָּגִים, 57) is 702, the numerical value of *Shabbat* (שַׁבָּת). This relates these three *mazalot* to the day of the week that is the ultimate consummation of time. As mentioned, there is no explicit astronomical indication that a week should be seven days long, nonetheless, the seven-day week is universally accepted. It is a constant reminder that God created the universe in six days and rested on the seventh. Thus, Shabbat is the axis around which time pivots. It stands like a monument of time in the week. The numerical value of "monument" (מַצֵּבָה) is 137, yet, when written in full (מם צדיק בית הא) its numerical value is 702, the numerical value of Shabbat (שַׁבָּת). This

---

▶ This is true for any four-lettered combination of letters that yields a numerical value of 137, as long as two of the letters have numerical values greater than 10 and two less than 10. Thus, all of the four-lettered words with a numerical value of 137 mentioned in this book (קַבָּלָה, אוֹפָן, מַצֵּבָה, מוֹצָא) have an ordinal value of 38. However, the ordinal value of other (meaningless) four-lettered combinations (e.g., כלפז), with a numerical value of 137, will have ordinal values greater than 38.

▶▶ In contrast, the Muslim calendar relies entirely on the lunar cycle and its months rotate throughout the seasons.

▶▶▶ 7 times 56 [the numerical value of "day" (יוֹם)] plus 12 times 312 [the numerical value of "month" (חֹדֶשׁ)] plus 19 times 355 [the numerical value of "year" (שָׁנָה)] equals 10,881, which equals 31 [the numerical value of the Name *Kel* (אֵיל)] times 351 [the triangle of 26, which is the numerical value of the essential Name of God, *Havayah*].

▶▶▶▶ Here again, we see a division of 137 into 73 (64 + 9), the numerical value of "wisdom" (חָכְמָה) and 64, which is the numerical value of "prophecy" (נְבוּאָה). 64 times 2 equals 2^7 and 9 equals 3^2. This shows that 137 is also the sum of $2^7 + 3^2$ (128 + 9). The general function for this is $2^n + 3^m$. There are exactly 37 (the companion number of 137) results for the equation $2^n + 3^m \leq 137$.

▶ The numerical value of the initial letters of the three *mazalot* under discussion (תְּאוֹמִים, מֹאזְנַיִם, דָּגִים) is 444, the numerical value of Temple (מִקְדָּשׁ). The number 444 equals 3 times "scales" (מֹאזְנַיִם), which is the center of the three *mazalot*.

▶▶ 861 is △41 and also 7 (the number of days) times 123 (the numerical value of "pleasure" (עֹנֶג).

▶▶▶ The numerical value of "the Holy Temple" (בֵּית הַמִקְדָּשׁ), 861, is the dot product (400·2 + 5·5 + 6·6) of "chaos" (תֹּהוּ) and "void" (בֹּהוּ). The sum of the numerical values of "chaos" (411, תֹּהוּ) and "void" (13, בֹּהוּ)—without the connecting *vav*—is 424, which is the numerical value of "Mashiach, son of David" (מָשִׁיחַ בֶּן דָּוִד).

▶▶▶▶ The total numerical value of the three months, Sivan (סִיוָן), Tishrei (תִּשְׁרֵי), Adar (אֲדָר) is 1,241, which is the product of 17 times 73. These two numbers are significant in creation. God created the world with wisdom (חָכְמָה), which has a numerical value of 73. The Goodly Name of Creation (אהריה, which has a numerical value of 17) appears as the initial letters of the words "[In the beginning God created] the heavens and the earth" (אֵת הַשָּׁמַיִם וְאֵת הָאָרֶץ), in the first verse of the Torah. The first creation was light which God saw to be "good" (טוֹב), which also has a numerical value of 17.

▶▶▶▶▶ The numerical value of *Yom Hakipurim* (יוֹם הַכִּפֻּרִים) is 411, the numerical value of "chaos" (תֹּהוּ), etc., which equals 3 times 137.

implies an inherent connection between the number 137 and the time factor of CPT symmetry with relation to Shabbat.

Just as Shabbat is the center of time, so the Temple, the holiest location on earth, represents the center of the space dimension. The Temple is the central point of Jerusalem, which is the center of the Holy Land.◀ It is the focal point via which we direct all of our prayers.

When we add 22 (3, 7, 12, the ordinal values of the letters associated with the three months) and 137 (100, 30, 7, the numerical values of the letters associated with them) to 702 (the sum of the numerical values of the three zodiac signs related to them) the total is 861,◀◀ the numerical value of "the Holy Temple" (בֵּית הַמִקְדָּשׁ).◀◀◀

We have here an indication of central time and central space. Is there something in these *mazalot* that could indicate a connection to a central soul?

The numerical value of the filled names of the three *mazalot* (דלת גימל יוד מם, מם אלף זין נון יוד מם, תו אלף וו מם יוד מם) is 1,790, which equals 5 times 358 [*Mashiach* (מָשִׁיחַ)], the all-inclusive soul of the Jewish People.

Each of the three months under discussion, Adar, Tishrei and Sivan, is related to symmetry. We can expect to find a relationship between them that relates them to one another as a single unit. Indeed, when we contemplate the festivals that appear in each of the three months we discover an inherent connection between them.◀◀◀◀

The Connection between Adar and Tishrei

The two days of Purim occur on the 14th and the 15th of Adar. When 14 and 15 are added to 336 (the numerical value of פּוּרִים) the total is 365, the number of days in a solar year. The two days of Purim therefore allude to the entire year, which begins on Rosh Hashanah, when it is the custom to eat fish, the *mazal* of Adar. In the Torah, Yom Kippur, which occurs in Tishrei, is called *Yom Hakipurim* (יוֹם הַכִּפֻּרִים), which can mean, "the day like Purim."◀◀◀◀◀ Indeed, despite the intrinsic differences between them, we are taught in Kabbalah and Chassidut that the spiritual properties of these two days are closely related.

On Yom Kippur, we read the Book of Jonah. The turning point of Jonah's experience came when he was swallowed first by a male fish and then by a female fish,[37] alluding to the symbol of the month of Adar.

The Connection between Sivan and Tishrei

The month of Sivan is also related to the month of Tishrei. The Revelation at Mt. Sinai was accompanied by the sound of the *shofar*. "It came to pass on the third day when it was morning, that there were thunder claps and lightning flashes, and a very powerful blast of a shofar ... And the sound of the shofar grew increasingly stronger..."[38] This description of the revelation at Sinai is an integral part of the Rosh Hashanah prayers. Rosh Hashanah is called, "the day of sounding [the shofar]" (יוֹם תְּרוּעָה) and the essential *mitzvah* of the day is to hear the shofar.

The creation of the world on Rosh Hashanah is also interlinked with the revelation at Mt. Sinai. The days of the week in creation are all stated in a general way, "second day," "fourth day," etc. However, the sixth day of creation (Rosh Hashanah), when man was created, is referred to with the definite article, "*the* sixth day."[39] Rashi interprets this deviation by stating that the world was created in Tishrei on condition that the Jewish People accept the Torah on the sixth day of Sivan—the day when God gave the Torah at Mt. Sinai.

The three *mazalot* associated with the months of Sivan, Tishrei and Adar are "twins" (תְּאוֹמִים), "scales" (מֹאזְנַיִם), "fish" (דָּגִים), as mentioned. Since they contain 16 letters, they can be written in a four-by-four square:

▶ Sincerity (תְּמִימוּת) is the inner power of *hod* (the *sefirah* of acknowledgment), associated with the left leg. In *Sefer Yetzirah*, each of the twelve months of the year is associated with a corresponding spiritual talent. The talent of Sivan is the sense of walking, or traversing space. We begin walking by stepping out with our left leg (especially a soldier in the army).

When rectified, *hod* brings a sense of inner peace and serenity in the space dimension. As the verse in Isaiah states, "Peace, peace to the distant and to the near."[40] This verse refers to inner peace in terms of location (distant and near), relating to the property of parity symmetry associated with the month of Sivan. Whether one is praised (from the right) or degraded (from the left), one remains at peace. The sages also state, "Anyone who studies the Torah altruistically makes peace in the entourage above [which was created with God's right hand] and in the entourage below [which was created with God's left hand],"[41] as another verse in Isaiah states, "If he shall uphold My stronghold he will make peace with Me; peace shall he make with Me."[42]

▶▶ The four verses that begin with each of the remaining four letters of *teshuvah* are, "I place God before me, always" (שִׁוִּיתִי הוי'),[47] "Love your fellowman as you love yourself" (וְאָהַבְתָּ לְרֵעֲךָ כָּמוֹךָ),[48] "In all your ways, know Him" (בְּכָל דְּרָכֶיךָ דָעֵהוּ)[49] and "Walk modestly with your God" (הַצְנֵעַ לֶכֶת עִם אֱ-לֹהֶיךָ).[50]

The numerical value of "Be sincere" (תָּמִים תִּהְיֶה) is 910, which is the numerical value of Tishrei (תִּשְׁרֵי).

מ	ו	א	ת
א	מ	מ	י
ם	י	נ	ז
ם	י	ג	ד

The right diagonal (the most significant diagonal of the square) spells out "sincere" (תָּמִים). This relates in particular to the *mazal* of Sivan, twins (תְּאוֹמִים). ◀ God calls the Nation of Israel, "My sincere one" (תַּמָּתִי).[43] The sages interpret this to mean, "They were sincere with Me at Sinai" (in Sivan). Appropriately, the Zohar states, "Do not read, 'My sincere one' (תַּמָּתִי), but "My twin" (תְּאוֹמָתִי)."[44] This indicates the vital importance of the trait of sincerity in our service of God, particularly in the time-reference. In particular, sincerity means acknowledging that God is in control and refraining from using methods of divining to tune into the future.[45]

This also associates Sivan to Tishrei (relating to the time dimension) and the service of *teshuvah* (repentance) which begins with the trait of sincerity. Rebbe Zusha of Anipoli taught that the five letters of *teshuvah* (תְּשׁוּבָה) are the initial letters of five phrases in the Torah, the first of which is "Be sincere with *Havayah*, your God" (תָּמִים תִּהְיֶה עִם הוי' אֱ-לֹהֶיךָ).[46] ◀◀

The Connection between Adar and Sivan

The Talmud[51] teaches us that when the Jewish People received the Torah in Sivan, they were so over-awed by the encounter with God that they were, in effect, coerced to accept it. However, on Purim, in the month of Adar, they accepted the Torah voluntarily.

This link between Adar and Sivan completes the vital connection between these three months as a unit. God created the world in Tishrei on condition that we accept the Torah in Sivan. The culmination of our acceptance of the Torah was on Purim, in the month of Adar. Thus, these three months, which relate to CPT symmetry, are intrinsically related to the creation of the world.

Gifts to the Priests

During the Temple times, when an animal was slaughtered for secular purposes, the owners were required to give certain parts of the animal as gifts to the *kohanim* (priests). The three parts were the foreleg (זְרֹעַ), the stomach (קֵבָה) and the cheeks (לְחָיַיִם).[52]

The initial letters of these three are *zayin* (ז), *lamed* (ל), and *kuf* (ק), the same letters that refer to the months of Sivan, Tishrei and Adar. The sum of the numerical values of these three letters is 137.

The Foreleg

The initial letter of "foreleg" (זְרֹעַ) is *zayin* (ז), associated with the month of Sivan. The symbol of Sivan is the "twins," or the two tablets that Moses brought down from Mt. Sinai in his arms. The forelegs of the animal, like arms, represent an extension either to the right or left. They correspond to parity. The arm is used explicitly to represent space in the Torah, as we see from the phrase, "and beneath the arms of the world,"[53] which refers to the arms of God that "hug" the universe (see Chapter 10).

The Cheeks

The initial letter of "cheeks" (לְחָיַיִם) is *lamed* (ל), associated with the time dimension and the month of Tishrei. They are the only gift that appears in the form of a pair. In Kabbalah, the cheeks correspond to the seventh attribute of compassion, "and truth" (וֶאֱמֶת), which also corresponds to Tishrei, the seventh month of the year.

The cheeks represent the energy of the scales, time and *teshuvah*, which all refer to the *mazal* of Tishrei. A *ba'al teshuvah* (repentee) undergoes a complete metamorphosis and becomes a new person. The light of their renewed personality is reflected in their cheeks when they smile. In Kabbalah, the cheeks are associated with two rosy red apples,[54] especially the cheeks of the person who blows the *shofar* on Rosh Hashanah, the first day of the month of Tishrei. Appropriately, it is the custom to eat apples (dipped in honey) on Rosh Hashanah.

"Cheeks" (לְחַיִם) is written with the same letters as, "to life!" (לְחַיִּים), the traditional wish that Jews make before drinking together. During the first ten days of Tishrei, special supplications are added to the *Amidah* prayer, asking that God remember to give us life.◄ This relates once again to the blessing of life that is renewed every year in the month of Tishrei. Then, we are judged whether or not we are worthy to complete the lifespan that was bestowed upon us at birth.

The Stomach

The initial letter of "stomach" (קֵבָה), is *kuf* (ק), the letter related to charge, which corresponds to male-female energies and to the symbol of the months of Adar, which is two fish. The stomach is the primary body organ associated with digestion. The letter *kuf* is the twelfth and last of the simple letters. Similarly, in the analysis of the twelve physiological systems of the body, the digestive system is the twelfth.[55] The stomach is the principle organ involved in ingesting new life force.◄◄ The digestive process assimilates useful elements from the environment into the body and expels waste products.

The sifting processes that take place in the stomach affect the potency to bear offspring. Thus, this function is closely related to the reproductive system. "Stomach" (קֵבָה) also translates as "womb."[56] In this way, the stomach closely relates to charge.

Another synonym for "womb" is "grave" (קֶבֶר).[57] This is an acronym for "stomach-uterus" (קֵבָה־רֶחֶם). The remaining letters (הח"ם), are an acronym for "life" (הַחַיִּים). These two words thus relate to the growth of new life in the womb, and also to the revival of the dead from the grave.[58]

The relationship between fish and the stomach is particularly

► Including the additional supplications of the Ten Days of Repentance, "life" (חַיִּים) appears a total of thirteen times in the *Amidah*.

The sum of the numerical values of "prayer" (515, תְּפִלָּה) and "for life" (לְחַיִּים, 98) is 613, the number of commandments in the Torah. Indeed, we are taught that our prayers during the Days of Repentance have the power to rectify our commitment to all 613 commandments.

►► The numerical value of "stomach" (קֵבָה) is 107, which relates to charge and parity [the letters *kuf* (ק) and *zayin* (ז)]. The *lamed* (ל), relating to time, is omitted. Charge and parity are asymmetrical (as individuals and as a pair). Only when time-reversal is considered in the equations does symmetry return, meaning that time is the controlling factor of CPT symmetry.

apparent in the Book of Jonah, read on Yom Kippur. Jonah was first swallowed, by a male fish (דָּג).[59] The following verse states that he was in the intestines of a female fish (דָּגָה).[60]▼ The Midrash[61] explains that Jonah did not pray while he was in the stomach of the male fish because he had plenty of room, and did not feel restricted. But, God made the male fish vomit Jonah out into the mouth of a female fish. Conditions were cramped inside the female fish because of all the eggs inside her. It was there that Jonah prayed to be released.►►

Here we see two fish, one male and one female, which relates to the symbol of the marital pair of the fish. The stomach therefore, is particularly related to the *mazal* of fish, the symbol associated with the month of Adar and to charge symmetry.

To summarize this chapter:

►► When Jonah prayed from the stomach of the fish, he said, "and [God] answered me from the belly of the grave."[62] In this case, the words "belly/stomach" (בֶּטֶן) and "grave" (שְׁאוֹל) are synonyms for the words mentioned (קֵבָה־קֶבֶר).

CPT symmetry	Soul, Space, Time	Month	Zodiac Symbol	Priestly Gifts	Letter	Numerical value	Ordinal value
Parity	Space עוֹלָם	Sivan סִיוָן	Twins תְּאוֹמִים	Forearm זְרֹעַ	ז	7	7
Time	Time שָׁנָה	Tishrei תִּשְׁרֵי	Scales מאֹזְנַיִם	Cheeks לְחָיַיִם	ל	30	12
Charge	Soul נֶפֶשׁ	Adar אֲדָר	Fish דָּגִים	Stomach קֵבָה	ק	100	19

► The numerical value of "[male] fish" (דָּג) is 7, the ordinal number of the letter *zayin* (ז). It is associated with the month of Sivan, the *mazal* of twins, and the forearm of the sacrifice. The numerical value of a "[female] fish" (דָּגָה) is 12, which is the ordinal number of the letter *lamed* (ל), associated with the month of Tishrei, the *mazal* of scales, and the cheeks of the sacrifice. As we have already seen, the sum of 7 and 12 is 19, which is the ordinal value of the letter *kuf*, the letter that corresponds to the month of Adar and number of years in the cycle that unites the lunar year with the solar year.

The product of the letters of "[male] fish" (דָּג) is 12 (4 times 3), the numerical value of "[female] fish" (דָּגָה). The product (4 · 3 · 5) of the letters of "[female] fish" (דָּגָה) is 60. The sum of these two numbers is 72, which is the highest filled value of God's Name, and the numerical value of "loving-kindness" (חֶסֶד).

The number 12 associated with "fish" alludes to the twelve stones that were arranged on the High Priest's breastplate in three columns of four stones each. The numerical value of Aaron the Priest (אַהֲרֹן הַכֹּהֵן), who was the first to wear the breastplate, is 336, which is the numerical value of Purim (פּוּרִים).

The *mitzvah* of giving the gifts to the priests appears in a short paragraph that contains only three verses:[63]

And this is to be the stipend of the priests from the people, from the offerers of sacrifices—whether it is an ox or sheep, he shall give the priest the arm, the cheeks and the stomach.	וְזֶה יִהְיֶה מִשְׁפַּט הַכֹּהֲנִים מֵאֵת הָעָם מֵאֵת זֹבְחֵי הַזֶּבַח אִם שׁוֹר אִם שֶׂה וְנָתַן לַכֹּהֵן הַזְּרֹעַ וְהַלְּחָיַיִם וְהַקֵּבָה.
The first of your grain, your wine, and your oil, and the first of the fleece of your sheep, you shall give him.	רֵאשִׁית דְּגָנְךָ תִּירֹשְׁךָ וְיִצְהָרֶךָ וְרֵאשִׁית גֵּז צֹאנְךָ תִּתֶּן לוֹ.
For Havayah, your God has chosen him [the priest] from among all your tribes to stand and serve in the Name of Havayah, him and his sons, all the days.	כִּי בוֹ בָּחַר הוי' אֱלֹהֶיךָ מִכָּל שְׁבָטֶיךָ לַעֲמֹד לְשָׁרֵת בְּשֵׁם הוי' הוּא וּבָנָיו כָּל הַיָּמִים:

The total numerical value of the first words of each verse (וְזֶה, רֵאשִׁית, כִּי) is 959, which equals 7 times 137. The total numerical value of the last words of each verse (וְהַקֵּבָה, לוֹ, הַיָּמִים) is 259, which is 7 times 37. As mentioned, the two numbers 137 and 37 are a numerical pair. In particular, they refer to Abraham's age at the time of the Binding of Isaac (137) and Isaac's age at that time (37), as discussed in Chapter 7. In the Kabbalistic correspondence between the Patriarchs and the *sefirot*, Abraham represents the right (the first words, on the right, equal 7 times 137), and Isaac represents the left (the last words, on the left, equal 7 times 37).

So, we see that these three verses that command us to give gifts to the priests are of great interest to our quest to discover the connection between Torah and science in general, and to CPT symmetry in particular.

REFERENCE NOTES FOR CHAPTER 14

1. Stephen Hawking, *A Brief History of Time*, p. 115.

2. See at length in our book *The Hebrew Letters*.

3. *Kiddushin* 2b.

4. Jeremiah 31:21.

5. *Likutei Moharan* II 79.

6. *Avot Derabi Natan* 2:3.

7. *Shabbat* 156a; *Nedarim* 32a.

8. Ibid; ibid.

9. Exodus 12:2.

10. *Shabbat* 104a.

11. *Yoma* 86b.

12. This idea, with specific reference to *teshuvah*, repentance, is discussed at length in our Hebrew book, *Teshuvat Hashanah,* pp. 117-132

13. Isaiah 65:24.

14. See *Midrash Aggadah Bereishit* 28; accoording to the rule that the minimum plural is two.

15. Genesis 2:24.

16. Ibid, Rashi ad loc.

17. Genesis 1:22.

18. Rashi on *Ketubot* 5a.

19. See our book (in Hebrew) on mathematics, *Einayich Breichot Becheshbon,* Chapter 3.

20. *Eiruvin* 65a.

21. *Rosh Hashanah* 9a.

22. Ezekiel 40:1.

23. Hosea 6:2.

24. *Gitin* 57b.

25. *Kanfei Yonah* Part 2, 104.

26. *Sefer Yetzirah* 1:2.

27. Ibid 1:7.

28. Ibid.

29. Isaiah 40:17.

30. See also, *Magid Mereishit Acharit* pp. 109-111.

31. *Kohelet Rabah* 1:3.

32. Leviticus 27:32.

33. *Sefer Yetzirah* 1:1-2.

34. Exodus 23:19, 34:26; Deuteronomy 14:21.

35. *Sefer Yetzirah* 1:7.

36. Deuteronomy 16:1. See also, Exodus 13:4, 23:15, 34:18.

37. Jonah 2:1-2; Rashi on verse 1.

38. Exodus 19:16, 19.

39. Rashi on Genesis 1:31.

40. Isaiah 57:19.

41. *Sanhedrin* 99b.

42. Isaiah 27:5.

43. Song of Songs 5:2; 6:9.

44. *Idra Zuta Devarim* 296a.

45. Rashi on Deuteronomy 18:13.

46. Deuteronomy 18:13.

47. Psalms 16:8.

48. Leviticus 19:18.

49. Proverbs 3:6.

50. Micah 6:8.

51. *Shabbat* 88a.

52. Deuteronomy 18:3.

53. Ibid 33:27; see Ibn Ezra ad loc.
54. *Zohar Chadash Shemot* 177a.
55. See our book, *Body, Mind and Soul*, p. 84-85.
56. See Numbers 25:8.
57. *Mishnah Ohalot* 7:4, etc.
58. See *Berachot* 15b; *Sanhedrin* 92a.
59. Jonah 2:1.
60. Ibid 2:2.
61. *Yalkut Shimoni, Yonah, remez* 550.
62. Jonah 2:3.
63. Deuteronomy 18:3.

THE GRAND FINALE

"The theory of relativity occurred to me by intuition, and music was the driving force behind that intuition."
(Albert Einstein)

The Enigma of 137

Since Sommerfeld's discovery of the fine-structure constant, it has remained one of the greatest enigmas of modern science. Eddington and Pauli spent years of study trying to fathom its source. Born said that 137 is the dominating factor for all natural phenomena.[1] And Feynman said, "All good theoretical physicists put this number up on their wall and worry about it."[2] As physicist James Gilson[3] writes,

> *The fine-structure constant, has been for many years a source of scientific and philosophical questions regarding its value and significance ... The mysteries of α are very much also in the domain of the philosophy of science and, while the technical progress that has occurred is a first step, there is still a long way to go with those more tantalising philosophical aspects of what the fine-structure really represents and how understanding it can help with the big questions of space-time...*

The sophisticated mathematics of string theory is unable to explain its origin. Yet, there is one important place where this number appears naturally in pure number theory. It is the numerator of the fifth harmonic number—$^{137}/_{60}$.

The Harmonic Series

The harmonic series is a mathematical series beginning with the number 1 and decreasing progressively to $\frac{1}{2}$, $\frac{1}{3}$, $\frac{1}{4}$, etc.

$$\sum_{n=1}^{\infty} \frac{1}{n} = 1 + \frac{1}{2} + \frac{1}{3} + \frac{1}{4} + \frac{1}{5} + \dots$$

A harmonic number, H_n, is the sum of all the numbers in the series until the nth number. Thus, H_2 equals $1 + \frac{1}{2} = \frac{3}{2}$ and H_3 equals $1 + \frac{1}{2} + \frac{1}{3} = \frac{11}{6}$:

$$1 \qquad \frac{3}{2} \quad \frac{11}{6} \quad \frac{25}{12} \quad \frac{137}{60}$$

As we can see from the sequence above, the fifth harmonic number (H_5) in its most compact form is $\frac{137}{60}$. ◄

▶ Adding 1/6, the minor third, gives 147/60. 147 is Jacob's lifespan. The sum of 147 and 60 is 207, the numerical value of "light" (אור) etc.

One particular significance of the harmonic series in mathematics is its counter-intuitive quality of divergence. Since the fractions in the harmonic series become smaller as the series progresses, we might expect it to converge to a particular limit. This is the case with similar series, for example:

$$S = \sum_{n=1}^{\infty} 10^{-n} = 0.1 + 0.01 + 0.001 + 0.0001 + \dots$$

This series clearly converges to 0.1111111111… which is equal to ⅑. Nonetheless, there is a most elegant proof in mathematics that the series of harmonic numbers diverges to infinity. ◄◄ This was proven in the Middle Ages but the proof was later forgotten. It was re-established a few centuries later.

▶▶ One proof for the divergence of this series is by the comparison test. The terms in the series are grouped together in the following way: sum = (1) + (1/2) + (1/3 + 1/4) + (1/5 + 1/6 + 1/7 + 1/8). Since the sum of each grouping of terms is ≥ 1/2, continuing this action by taking twice as many terms in each successive group produces an infinite series in which the sum of each of the groups ≥ 1/2. The series 1 + 1/2 + 1/2… clearly diverges to infinity, therefore the harmonic series also diverges. From a Kabbalistic perspective, the harmonic series represents the secret of a whole and infinite halves.

The divergence of the harmonic series gives rise to many apparent paradoxes. Unlike the pseudo-paradoxes that sometimes result from modern calculus, these are true paradoxes. One such paradox is illustrated in the following example:

A worm advances 1cm per minute along a rubber band a meter long. At every minute, the band is stretched by another meter. Although we would imagine that the worm will never be able to return to the end of the rubber band, this can be proven to be counter-intuitive. The time it would take for it to reach its *origin—although very great—is computable. Because the rubber band is stretched proportionally at all points, the 1 cm that the worm has advanced after one minute stretches to 2 cm as the length of the rubber band stretches to 2 meters. As the worm completes the second minute, it has advanced 2 cm (the length now behind the worm) $\perp$ 1 cm (the distance the worm has advanced in the second minute), totaling 3 cm. As the band stretches to 3 meters, the 3 cm of rope behind the worm stretches proportionately to 4½ cm, meaning that after 2 minutes, although it has moved only 2 cm forward, the worm has advanced 4½ cm from the origin. This is a function of the harmonic series, because, after n minutes, the worm has advanced n cm (the distance that the worm actually moved) $1 \perp \frac{1}{2} \perp \frac{1}{3} \perp \frac{1}{4} \perp \frac{1}{5} \ldots \perp \frac{1}{n}$ cm (the amount by which the rope behind the worm has stretched). Since the worm is progressing, and its advance shares the same proportional growth as the growth of the rope, the worm's advance will increase each time by smaller and smaller increments. At a certain stage, the worm will eventually reach the end of the rubber band!*▸

Halachah and Kabbalah

From a Torah perspective, a converging infinite series relates to the concrete limitations of the revealed aspect of the Torah (*halachah*, i.e., Jewish law). In comparison, a diverging series corresponds to the infinite scope of the concealed aspect of the Torah (Kabbalah).

As mentioned in previous chapters, reality is created in complementary pairs, symbolized by the large *bet* of *Bereishit* ("In the beginning"), the Torah's account of creation. Every phenomenon has two facets.

Usually, only one aspect of a phenomenon can be perceived at any one moment. However, a wise person is able to envisage the two sides as a single unit. In the case of *halachah* and Kabbalah, the two terms are a perfect complementary pair; they coexist simultaneously. Nonetheless, in general, a person can either contemplate the mystical side, or study the legal side, but usually never both at once.

▸ On a smaller scale: If the rubber band was 2 cm long, growing by 1 cm at every minute, then after 5 minutes the rubber band would be 7 cm long, but in the same time, the worm would have already advanced by 5 $\perp$ 137/60 cm > 7cm.

We already know that the numerical value of Kabbalah (קַבָּלָה) is 137. A lesser known fact is that the numerical value of *halachah* (הֲלָכָה)—the revealed aspect of Torah—is 60.

The subjective convergence and objective—but counter-intuitive— divergence of the harmonic series, immediately indicates a merging of these two aspects of the Torah. The fifth harmonic number, $137/60$, pinpoints this correspondence.

As mentioned in Chapter 2, the passage commanding the donation of olive oil for the menorah in the Temple is the only passage in the Torah that contains exactly 137 letters. This relates to abundant light—spiritual energy. Similarly, there is only one passage in the Torah that contains exactly 60 letters: the Priestly Blessing, relating to maximal physical abundance (see Chapter 3). These two passages thus perfectly reflect the fifth harmonic number, $137/60$, as the connecting link that allows us to perceive "both sides of the coin" simultaneously.◄ Aaron, the High Priest, is the soul who connects these two.

The Arizal points our attention to the fact that on weekdays, the *Pesukei D'Zimrah* ("Verses of Song") section of the morning prayers begins with the psalm that opens with the phrase, "Call out to *Havayah*, all the earth" (הָרִיעוּ לַהוי' כָּל הָאָרֶץ).[4] The initial letters of this phrase spell *halachah* (הֲלָכָה). *Halachah* proclaims earthliness to God. However, on Shabbat, this recitation begins with the phrase, "Bow to *Havayah* in holy splendor" (הִשְׁתַּחֲווּ לַהוי' בְּהַדְרַת קֹדֶשׁ).[5] The initial letters of this phrase (read backwards) spell *Kabbalah* (קַבָּלָה). Shabbat and the six week days are one of the complementary aspects of creation. Kabbalah reflects the Shabbat dimension of the Torah while *halachah* represents the week-day dimension. For this reason, the Arizal would study six facets of the revealed aspect (*halachah*) of every topic that he learnt in Torah and one corresponding revelation of Kabbalah. A true sage, who harmoniously integrates these two aspects of the Torah, experiences a sense of Shabbat throughout the week.[6]

Studying *halachah* requires a talent for direct rational reasoning.

▸ From a purely allegorical point of view, Einstein's famous equation, $E=mc^2$ means that the Torah (the essence of light, c^2) "equals" Kabbalah/*halachah* (E/m). This is the secret of the fifth harmonic number 137/60.

Studying Kabbalah requires a perception of infinity and unity, without which the student of Kabbalah is in danger of materializing the Divine.

The Ba'al Shem Tov revealed a new dimension of Kabbalah, called Chassidut. The talent required to study Chassidut is the talent to bear paradoxes, i.e., the ability to contain the infinite energies of Kabbalah within maximum finite vessels, without falling into the trap of materialization.[7] This concept is illustrated perfectly by the fifth harmonic number, $^{137}/_{60}$.

The sum of the numerical values of Chassidut (488 ,חֲסִידוּת) and *halachah* (60 ,הֲלָכָה) is 548, which equals 4 times 137. Adding Kabbalah (137 ,קַבָּלָה) brings the total to 685, which equals 5 times 137. This is equal to the numerical value of Kabbalah (137 ,קַבָּלָה) when written "back" and "face" (see Chapter 1 for an explanation of this form of calculation):

ק קב קבל קבלה קבלה בלה לה ה

This relates to the secret of the five origins of speech in the mouth. Chassidic teachings facilitate the safe verbalization of Kabbalistic mysticism.

Thus, the harmonic series, and the fifth▶ harmonic number in particular, teach us that, through the prism of Chassidut, both sides of the coin can be perceived simultaneously. The intellectual teachings of Chassidut connect the mystical super-conscious aspect of Kabbalah with the practical application of *halachah*. By incorporating the correct Kabbalistic intentions in our halachic observance of the commandments, and in our every day lives, we achieve a harmonious union between the two.

The first appearance of a permutation of the letters of *halachah* (הֲלָכָה) in the Torah is "which goes" (הַהֹלֵךְ),[8] referring to the third river that flowed out from the Garden of Eden. Because their letters are identical, these two words have the same numerical value of 60. This is the 137th word of the second Genesis account of creation. It is the

▶ In Hebrew, "fifth" (חֲמִישִׁי), has the same letters as "messianic" (מְשִׁיחִי) indicating that there is something messianic about the fifth harmonic number.

first reference to the union of Kabbalah and *halachah*, and to the fifth harmonic number, $^{137}/_{60}$.

Similarly, the first permutation of Kabbalah (קַבָּלָה) in the Torah is "her voice" (בְּקלה).[9] This relates to sound. There are exactly 60 letters that precede this word in that Torah verse.◄

▶ The numerical value of the 60 letters that precede "her voice" (וַיֹּאמֶר אֱ־לֹהִים אֶל אַבְרָהָם אַל יֵרַע בְּעֵינֶיךָ עַל הַנַּעַר וְעַל אֲמָתֶךָ כֹּל אֲשֶׁר תֹּאמַר אֵלֶיךָ שָׂרָה שְׁמַע) is 115 times 37, the companion number of 137.

Music to Your Ears

The fine-structure constant relates directly to the speed of light. In contrast, the harmonic series, as its name suggests, bears an important relationship to music. This turns our attention to the significance of 137 in the context of sound.

Isaac Newton and subsequent 18th century mathematicians suggested that the seven-color light spectrum[10] is governed by the same ratios that underlie the diatonic music scale. This theory was seized upon by musicians and scientists, who sought evidence of universal laws common to music and light. However, research in this field was later abandoned.

Music is produced by vibrating objects, such as the strings and pipes of various instruments. When an object such as a string is set into motion, it vibrates not only at its full length (which produces the fundamental pitch), but in half, thirds, fourths, and so on. The simultaneous frequencies produced by these vibrations are called harmonics. The full-length vibration, the fundamental, is the first harmonic and is usually the easiest to identify. However, it can usually only be isolated from all overtones by electronic means.

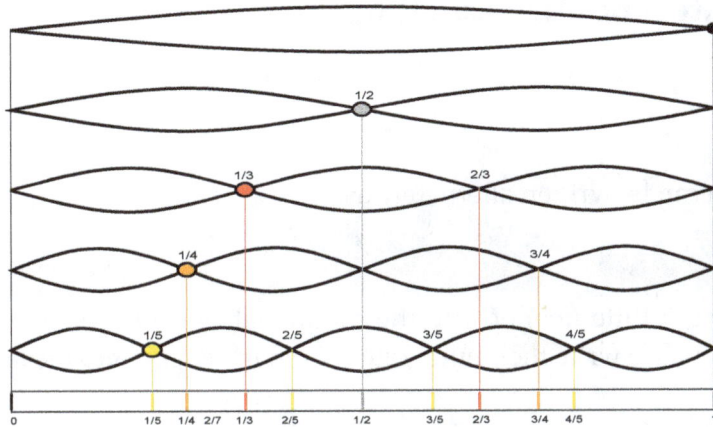

The first five harmonics

For example, playing an open violin string is defined by 1, the fundamental pitch. However, when the open string is played, it also vibrates at other points, producing overtones together with the fundamental. Each progressive overtone has the same tone as the fundamental, but the ½ tone adds one octave, the ⅓ adds a major fifth, the ¼ adds a major fourth, the ⅕ adds a major third, after which the ⅙ adds a minor third. The ability to identify the different timbres of various instruments playing the same note is possible because each instrument emphasizes different harmonic overtones. Thus, the quality of sound depends more on the overtones than on the pitch of the music.

Since the harmonics depend on the vibrations of the string at different points according to the series of 1, ½, etc., as discussed above, the harmonic series is an important link between mathematics and music.

The relationship between 137 and music implies a strong connection to superstring theory (see Chapter 11). This potential GUT claims that all the elementary particles in the universe are formed of infinitesimally tiny strings that vibrate at varying frequencies.

Counter-intuitive Counting

The first five integers of the harmonic series are:

$$1 + \tfrac{1}{2} + \tfrac{1}{3} + \tfrac{1}{4} + \tfrac{1}{5}$$

Which can be written alternately as:

$$\tfrac{60}{60} + \tfrac{30}{60} + \tfrac{20}{60} + \tfrac{15}{60} + \tfrac{12}{60}$$

Delving a little deeper into the mathematics of this sequence of numbers, we write their numerators from the smallest to the largest.

$$12 \quad 15 \quad 20 \quad 30 \quad 60$$

We then take the finite differences between them to construct a four-dimensional series from which we can derive the next number in the series.

12	15	20	30	60	137	300
	3	5	10	30	77	163
		2	5	20	47	86
			3	15	27	39
				12	12	12

Here, we see that the next number in the series is 137, the sum of the first 5 numbers. In a counter-intuitive way, by calculating the differences between the numbers, we arrive at their sum! ◀

In the entire array of sigmas for this sequence, the original sum yields 137; the third sigma (the integral) totals 300 (the number following 137 in the four-dimensional series constructed from the finite differences); the fourth sigma totals 600 and the fifth sigma totals 1,100 and so on, as the array below illustrates. ◀◀ A quick analysis of the array confirms that all of the numbers in the array are present in the above series together with the finite differences.

▶ The sum of the first 13 numbers in the series is 18,551, which equals 13 times 1,427. This is also the sum of the numerical values of "repentance" (תְּשׁוּבָה), "prayer" (תְּפִלָה) and charity (צְדָקָה).

▶▶ In the table of sigmas (see opposite page), 137 lies between 60 and 300. These two numbers are the regular numerical value of the letter *samech* (ס), which equals 60 (the denominator of the harmonic number), and the numerical value of the filling spelling of the same letter (סמך מם כף), which equals 300. This suggests that 137 is related to this letter, which is the only letter of the *alef-bet* that does not appear explicitly in creation. The first appearance of the letter *samech* (ס) in the Torah is in "that surrounds" (הַסֹּבֵב), which is appropriate for the circular shape of the *samech*.

In the fourth sigma, the numbers 137 and 300 appear again, for no apparent reason, and the total is 600, which is equal to 10 times 60, the numerical value of the letter *samech*, and 2 times 300, the numerical value of the filling of the filling of the letter (as mentioned above).

The sum of the five final numbers: 60 137 300 600 1100 is 2,197, which is 13³—a perfect cube. The sum of the remaining 20 numbers in the array is 1,815, which is the number of letters in the 34 verses of the seven days of creation.

Numerators of harmonic series	12	15	20	30	60
Second sigma	12	27	47	77	137
Third sigma	12	39	86	163	300
Fourth sigma	12	51	137	300	600
Fifth sigma	12	63	200	500	1,100
Total	**60**	**195**	**490**	**1,070**	**2,197**

The Echo of Creation

We also see the fifth harmonic in the echo of creation. As explained in Chapter 2, in the Torah there are large letters and small letters. The first account of creation begins with a large letter *bet* in the first word of the Torah, *Bereishit* (בְּרֵאשִׁית). The next unusual letter appears a chapter later, in the phrase, "when they were created" (בְּהִבָּרְאָם).[11]▶ In a Torah scroll, the letter *hei* (ה) in this word is smaller than the other letters. The numerical value of the letter *hei* is 5. A novel interpretation for this small letter is that the world was created with the fifth harmonic number, which reflects the harmonious combination of the three aspects of the Torah: Kabbalah, *halachah* and Chassidut, as mentioned previously.▶▶

An "echo" (הֵד), is another form of overtone. This word appears once in the Bible. It refers to the reverberation of sounds that is heard between mountains.[16] An allusion to the fifth harmonic number can be seen in the numbers 5 and 4, the numerical values of the two letters *hei* and *dalet*, which form this word. These represent the five tones of the fifth harmonic number, of which one is the pitch and four are overtones.

The numbers 4 and 5 relate to two Names of God, *Elokim* (אֱ-לֹהִים) and the essential Name, *Havayah* (הוי'). The Name *Elokim* has five letters and the Name *Havayah* has four.

▶ One permutation of the letters of "When they were created" (בְּהִבָּרְאָם) means "with Abraham" (בְּאַבְרָהָם). The additional letter *bet* raises the regular numerical value of "Abraham" (אַבְרָהָם) from 248 to 250. This is equal to the numerical value of another phrase that refers to Abraham in Isaiah, "great light" (אוֹר גָּדוֹל).[12] Although the numerical value of this phrase is different from Abraham, its primordial value is equal to the primordial value of "Abraham"—959. Indeed, the sages teach us that God's first saying in creation, "Let there be light,"[13] refers to the revelation of the soul-root of Abraham.[14]

Abraham is also called, "the greatest of giants" (הָאָדָם הַגָּדוֹל בָּעֲנָקִים).[15] The numerical value of this phrase is 370, which equals 10 times 37, the numerical companion of 137, as we have seen.

▶▶ This relationship between the fifth harmonic number and creation is echoed in the first five verses of Genesis, which include the creation of light. The total number of letters in these five verses is 197 (137 ± 60).

Brass Instruments

The harmonic series is of particular significance for brass instruments. A pianist or xylophone player can obtain only one note from each key. To obtain a different note, a string player holds the string tightly in a different place. This creates a vibrating string of a new length, with a new fundamental. But, without changing the length of the instrument, a brass player achieves the different overtones by playing the harmonics of the instrument with his own breath. The notes are produced by depressing the valves, but the pitch and its overtones are produced by the way the musician introduces air into the instrument through his lips. The brass player is the most conscious of the overtones.

During the era of the First Temple, idolatry was instituted as the religion of the land. The doors of the Temple were locked for a prolonged period. King Hezekiah reinstated the Temple services at the beginning of his reign and the Jews were once again able to celebrate the festivals. ◄

The first festival they observed was Passover, as related in Chronicles, "And the Children of Israel who were present in Jerusalem observed the Festival of Matzot for seven days with great joy; the Levites and the priests praised God day by day, with mighty instruments to God."[18]

This verse describes the tremendous joy with which the people celebrated together with the priests and the Levites, who played music on, "mighty instruments" (כְּלֵי עֹז) i.e., trumpets. The numerical value of this phrase is 137.

"Might" (עֹז) is one of many synonyms for "strength," each of which has its particular "overtones" with reference to the general concept. "Might" implies the power to simultaneously bear two opposite states, particularly with reference to the sense of hearing. This is illustrated in Psalms by the verse, "One spoke God [*Elokim*], I heard two, because God has might."[19] In our context, "a mighty instrument" relates to the trumpet blast as a sound that includes other sounds within it—the fundamental pitch and secondary overtones.

▶ The numerical value of Hezekiah (חִזְקִיָּהוּ) is 136, the numerical value of "sound" (קוֹל). King Hezekiah's name appears in the Hebrew Bible in four different variants (חִזְקִיָּה חִזְקִיָּהוּ יְחִזְקִיָּה יְחִזְקִיָּהוּ). The average numerical value of all four names is 138, the value of the name Menachem (מְנַחֵם). One of the proposed names of *Mashiach* is Menachem ben Hezekiah.[17]

The artisan of the Tabernacle in the wilderness was Betzalel. The artisan who helped King Solmon build the first Temple was Chiram. The sum of the numerical values of Betzalel (153 ,בְּצַלְאֵל) and Chiram (חִירָם, 258) is 411, 3 times 137.

The numerical value of "[musical] instrument" (כְּלִי) is 60. The sum of the numerators preceding 60 in the harmonic series (12 15 20 30) is 77, the numerical value of "mighty" (עֹז). Thus, the phrase "mighty instruments" clearly alludes to the harmonic number, $^{137}/_{60}$. ▶

Two Halves of One Whole

The Maggid of Mezeritch, successor to the Ba'al Shem Tov, interprets another, more common name for "trumpet" (חֲצֹצְרָה), as an acronym of the two words, "half a form" (חֲצִי צוּרָה). Indeed, in the Temple, the trumpets were always blown in pairs.

Another mitzvah that necessitates uniting two halves is the commandment that every individual donates a half-shekel (מַחֲצִית הַשֶּׁקֶל) coin to the Temple every year. Rabbi Moshe Alshich explains in the name of Rabbi Shlomo Alkavetz, that each contributor should remember that without their fellow Jew, they are merely one half of a person. The numerical value of "half" (מַחֲצִית) is 548, which is 4 times 137, indicating once more the power of 137 as a spiritual coupling constant.

Tone and Image

In Hebrew, the two-letter gate "shadow" (צֵל) also means "tone" (צְלִיל). The only time this word appears in the Bible, it relates to a loaf of barley bread burnt on coals until it produces a hollow sound.[22] This suggests that sound is related to the darkness that precedes light. The common gate of these two words implies that every tone has a "shadow," or an overtone.

As explained, every musical note contains a conscious experience of the pitch, together with the unconscious experience of the tone. The tone is the entirety of all the overtones, the "shadows" of the original tone. The subtle implications of the overtones in a person's voice in speech, for example, transmit information to the listener in a way that his words cannot convey.

▶ "A mighty vessel" (כְּלִי עֹז) refers to the statement, "God found no better vessel (כְּלִי) to hold blessing for Israel other than peace, as the verse states, 'Havayah gives might (עֹז) to His people, Havayah blesses His people with peace.'[20]"[21]

Another important word that shares this gate is "image" (צֶלֶם). The first time the word appears in the Torah is in the story of the creation of man, "And God created man in His image, in the image of God, did He create him."[23] The second time it appears is eight chapters later in the verse prohibiting manslaughter, "Because in the image of God did He make man."[24] The third time this word appears is in Psalms, "But, in the image walks man" (אַךְ בְּצֶלֶם יִתְהַלֶּךְ אִישׁ).[25] ◄

Both of the first two appearances of this word relate to man as he was created in the image of God. Accordingly, one interpretation of the third appearance is that the soul of man reflects an image of the Divine. Rashi cites this interpretation in his commentary. but dismisses it. Instead, he explains that "image" (צֶלֶם) is an acronym for "the shadow of death" (צֵל מָוֶת). This implies that man lives out his entire life in darkness, never knowing when his end will be.

These two authentic interpretations of the same phrase complement one another perfectly.

According to Rashi's interpretation, "image" (צֶלֶם) is derived from the word "shadow" (צֵל), with an additional final *mem* at the end of the word. The numerical values of the words "tone" (צְלִיל) and "image" (צֶלֶם) are equal—both are 160. Thus, we can expand on Rashi's interpretation by suggesting that every individual is accompanied by the aura of his unconscious "overtones." According to the interpretation that "the image" is the image of God, the numerical equality of the two words "tone" (צְלִיל) and "image" (צֶלֶם) and their common gate imply that the image of God—a purely "visual" effect—can be compared to a musical tone, or even an entire spectrum of musical tones, complete with an inherent fine-structure of overtones. Like one's face or fingerprint, the voice of every individual is unique.

The ultimate expression of man as a reflection of the Divine manifests in the righteous individual, the *tzadik*. There are two types of *tzadik*, the higher *tzadik* (represented by Joseph) and the lower *tzadik* (represented by Benjamin).[28] This fine-structure is expressed in the phrase, "Justice, justice shall you pursue" (צֶדֶק צֶדֶק תִּרְדֹּף).[29] "Justice"

► The numerical value of the idiom "in the image of God" (בְּצֶלֶם אֱלֹהִים) is 248, the numerical value of Abraham (אַבְרָהָם). The numerical value of "But, in the image walks man" (אַךְ בְּצֶלֶם יִתְהַלֶּךְ אִישׁ) is 959, which equals 7 times 137, the primordial value of Abraham (אַבְרָהָם). The numerical value of "image" (צֶלֶם) is 160. The sum of these two numbers is 1,119. The addition of a second "image" (צֶלֶם) alludes to the two appearances of "image" in the verse referring to the creation of man "And God created man in His image, in the image of God did He make him."[26]

All this is indicated in the verse, "If I saw light shining and the moon becoming brighter" (אִם אֶרְאֶה אוֹר כִּי יָהֵל וְיָרֵחַ יָקָר הֹלֵךְ).[27] The sum of the numerical values of "If I saw" (אִם אֶרְאֶה) is 248, which is the numerical value of Abraham (אַבְרָהָם). The numerical value of the entire verse is 1,119, which is also the numerical value of Abraham (אלף בית ריש הא מם) when the letters are written in full.

(צֶדֶק) is the root of *tzadik* (צַדִיק).▼ Its repetition corresponds to each of the two righteous individuals.

Atzilut (אֲצִילוּת) is also conjugate to "shadow" (צֵל) and to "tone" (צְלִיל). *Atzilut* (the World of Emanation) is the highest of the four spiritual worlds that correspond to the four unconscious levels of overtones. In music, *Atzilut* corresponds to the super-conscious harmony of all the tones at their source in the fundamental pitch.▶▶

Ascent to God through Prayer and Song

In Aramaic, "prayer" (צְלוֹתָא) also derives from the root meaning "shadow" (צֵל). The two letters, *samech* (ס) and *tzadik* (צ) are both pronounced with the teeth and are therefore easily interchangeable. When pronounced in this manner, "image" (צֶלֶם) becomes "ladder" (סֻלָם), the word used in Hebrew to refer to the musical scale.

The Zohar teaches us that the ladder Jacob saw in his dream is the ladder of prayer.[36] Prayer is the means by which we ascend to the spiritual worlds, or higher levels of consciousness. The four stages of prayer begin in mundane reality and ascend to heaven. During prayer, we access the hidden recesses of the unconscious mind by meditating upon the absolute unity of God. Following the introductory levels, we affirm our faith by declaring that "God is one!" when reciting the *Shema*. This stage allows us to gain access to the highest levels of the mind, and touch upon the super-conscious.

▶▶ Realizing the spiritual unity of *Atzilut* in mundane reality is called, "ennobling" (לְהַאֲצִיל). One conjunction of this root has a numerical value of 137 (וַיַאְצֶל).[30] This word divides into 17 (וַיַא) and 120 (צֶל), the same division mentioned above with reference to the sum and product of "a kid goat" (גְדִי).

Those individuals who have touched the level of *Atzilut* are called "the nobles of the children of Israel" (אֲצִילֵי בְנֵי יִשְׂרָאֵל).[31] The numerical value of this phrase is 744, equal to the numerical value of the phrase in Genesis, "in Our image, resembling Our likeness" (בְּצַלְמֵנוּ כִּדְמוּתֵנוּ).[32] Each of the three words in the phrase has an average value of 248, the numerical value of Abraham (אַבְרָהָם), which is also the numerical value of "in the image of God" (בְּצֶלֶם אֱלֹהִים).

▶ "Justice" (צֶדֶק) is an abbreviation for the unique Biblical phrase, "scrawny and thin" (צַנְמוֹת דַקוֹת),[33] referring to the seven thin stalks of grain that swallowed up the seven plump stalks of grain in Pharaoh's dream. The seven stalks correspond to the seven colors of the spectrum. Here the fine-structure is indicated by the numerical value of this phrase, 1,096, which is equal to 8 times 137.

"Justice" (צֶדֶק) is also an abbreviation of the expression, "a fine image" (צֶלֶם דַק), referring to the fine-structure that is apparent in these two types of righteous individual who represent the refined image of man.

In addition, "Justice" (צֶדֶק) is an abbreviation for another phrase that appears only once in the Bible, "thin and yellow" (צָהֹב דַק)[34] relating to a hair that is observed to grow in an open sore of a leper. The numerical value of the full spelling of this phrase (צַדִיק הֵא בֵית דַלֶת קוֹף) is 1,242—6 times 207, the numerical value of "light" (אוֹר). This reflects the six possible combinations of the three letters of "light" (אוֹר).

Of the seven colors of the spectrum,[35] yellow is the fifth, dividing the seven colors into five and two, the secret of the number 7. This is illustrated in the word "gold" (זָהָב), which is closely related to "yellow" (צָהֹב) and often interchangeable with it. The first letter of "gold" (זָהָב), has a numerical value of 7 and the following two letters (ה and ב) have numerical values of 5 and 2, respectively.

In Chapter 9, we saw that 515 is the sum of 137 (the numerical value of Kabbalah) and 378 [the numerical value of the esoteric word of Ezekiel's vision, *chashmal* (חַשְׁמַל)]. This implies that in our prayer we must follow through the stages of submission, separation and sweetening (see Chapter 7) regarding the Kabbalistic unifications in prayer.

The numerical value of "*Havayah* is your shadow by your right hand" הוי׳ צִלְּךָ) עַל יַד יְמִינֶךָ) is 410. Adding the *kollel* unit brings the sum to 411, which equals 3 times 137 etc.

The Arizal explains that the *Amidah* is the highest rung of the ladder and corresponds to the highest of the four worlds, *Atzilut* (the World of Emanation), where our soul meets Divinity.

The *Amidah* was initiated by Abraham as the service of prayer that connects man and God. The numerical value of "prayer" (תְּפִלָּה) is 515, which is also the numerical value of song (שִׁירָה), relating prayer closely to music. In prayer, we should create overtones above us, and by doing so aspire to sense how, "*Havayah* is your shadow by your right hand."[37]

Returning to the phrase with which we began this meditation, "But, in the image walks man" (אַךְ בְּצֶלֶם יִתְהַלֶּךְ אִישׁ),[38] we see that man was created in the image of *Elokim* (אֱ-לֹהִים), the five lettered Name of God, which has a numerical value of 86 (the same as "nature", הַטֶּבַע). *Elokim* is associated with the Divine power of creation that is embedded within natural causation. In physics, natural causation is the complementary aspect of synchronicity (see Chapter 6), which corresponds to the Name *Havayah*, the Name that is related to Divine Providence. As mentioned, the phrase in Psalms, "*Havayah* is your shadow" (הוי׳ צִלְּךָ) relates the Name *Havayah* directly to a "shadow" (צֵל), which is the two-lettered gate of "tone" (צְלִיל).

The image of *Elokim* is the compound sound that we hear reverberating through nature. The overtones—the four letters of *Havayah*—are available for those of us who wish to access them. They correspond to the four Worlds mentioned previously, the highest of which is *Atzilut* (the World of Emanation). Every spiritual world represents an overtone that is above the basic tone of this world. A well-trained "ear" can hear these overtones even in this world.

Man walks in the physical world of causation in the image of *Elokim*, the Name that reflects nature and science. The ideal is to connect to overtones that guide us from the shadow of the four letters of *Havayah*, the Name that represents Divine Providence in everything we say and do. When we connect to these overtones, we can begin to perceive the point at which Torah and science converge in harmonious union.

The song that results from their marriage is a "new song" (שִׁיר חָדָשׁ).[39] The numerical value of this phrase is 822, which equals 6 times 137; i.e., 137 is the average value of each letter.

Letter of God's Name	Spiritual World of Consciousness	Musical Overtone	Harmonic	
tip of the *yud*	*Adam Kadmon* (Primordial Man); super-conscious crown	fifth overtone	⅙	minor third
Yud	*Atzilut* (the World of Emanation); wisdom	fourth overtone	⅕	major third
Hei	*Beriah* (the World of Creation); understanding	third overtone	¼	major fourth
Vav	*Yetzirah* (the World of Formation); beauty (includes five *sefirot*)	second overtone	⅓	major fifth	(second octave)
Hei	*Asiyah* (the World of Action); kingdom	first overtone – one octave higher than the audible tone	½	first octave
The five letters of the Name *Elokim*	the manifestation of God in the natural world	audible tone	compound sound; pitch and four overtones	normal pitch

REFERENCE NOTES FOR CHAPTER 15

1. Max Born, "The Mysterious Number 137," *Proceedings of the Indian Academy of Science*, Section A, p. 547.
2. Richard Feyman, QED: The Strange Theory of Light and Matter, p. 129.
3. James G. Gilson, *Strong Quantum Coupling and Relativity*, 2002.
4. Psalms 100:1.
5. Ibid 29:2.
6. *Zohar Vayikra* 29a.
7. See our book in Hebrew, *Esa Einai*, Chapter 1.
8. Genesis 2:14.
9. Ibid 21:12.
10. See our book *Kabbalah and Meditation for the Nations*, p. 71 ff.
11. Genesis 2:4.
12. Isaiah 9:1.
13. Genesis 1:3.
14. *Bereishit Rabah* 2.
15. *Rashi* on Joshua 14:15.
16. Ezekiel 7:7.
17. *Sanhedrin* 98b; see also, *Eichah Rabah* 1:51.
18. II Chronicles 30:21; Rashi ad loc.
19. Psalms 62:12. See also Chapters 7 and 12.
20. Psalms 29:11.
21. *Devarim Rabah* 5:15..
22. Judges 7:13; see commentaries.
23. Genesis 1:27.
24. Ibid 9:6:
25. Psalms 39:7.
26. Genesis 1:27.
27. Job 31:26.
28. *Zohar Bereishit* 47a.
29. Deuteronomy 16:20.
30. Genesis 41:23.
31. Leviticus 13:30.
32. See our book, *Kabbalah and Meditation for the Nations*, pp. 71 ff.
33. Numbers 11:25.
34. Exodus 24:11.
35. Genesis 1:26.
36. *Zohar Bereishit* 150a.
37. Psalms 121:5.
38. Ibid 39:7.
39. Ibid 33:3, etc.

CONCLUSION

Since the discovery of quantum mechanics, science has been intrigued by the complementarity in nature. How can we explain the dual nature of reality, which simultaneously incorporates waves and particles (Chapter 1), chaos that turns into order (Chapter 2) and undeniable synchronicity alongside obvious causality (Chapter 6)? What explanation can be found for particles that apparently exist in superposition until they are measured, or for entangled particles that communicate instantaneously, defying the limitations of the speed of light? Scientists endeavor to develop a theory of everything that will tie all the different forces of nature into one whole. Attempting to unravel the mysteries of string theory, they have inadvertently met discoveries that they are unable to resolve. One of these mysteries is the paradoxical $R = 1/R$ diameter of the universe (Chapter 10).

Is there a way in which we can keep both poles of a complementary pair in view simultaneously? Niels Bohr suggested that complementarity can be resolved by finding the common source from which both extremities of the complementary pair originate (Chapter 2). This hypothesis is reflected in the multidimensional space assumed by M Theory. It is also an echo of the Kabbalistic perspective on marriage (Chapter 11).

Duality in the universe is emphasized in particular in the apparently bipolar philosophical realms of Torah and science. As with all complementary phenomena, the more we focus on one of these fields, the more the other seems to disappear from view. The goal of this book has been to explore the cutting edge of both realms in order to find their common origin and make the match between them.

As the book developed, various patterns emerged that link scientific findings to Kabbalistic concepts, especially through the common language of mathematics. One important key to this link is the number

137, the fine-structure constant and the electromagnetic coupling constant. 137 essentially marks the quantum leap from energy—a photon, which moves at the speed of light—to matter, in the form of an electron—the smallest matter particle—which moves at $\frac{1}{137}$ times the speed of light. 137 is also the numerical value of Kabbalah (קַבָּלָה), the field of Torah study that couples Torah with the mundane world at their origin.

Any coupling constant must contain within it aspects of both entities that it bonds. Indeed, as mentioned in the Introduction, the equation for the fine-structure constant includes in it both relative infinity (*c*, the speed of light) and relative infinitessimality (*h*, Planck's constant, which defines the smallest possible natural units of measurement).

In Chapter 8 we saw how in Kabbalah, 137 is related to feminine consciousness (Chapter 4). In Chapter 9 we saw how 137 is linked to communication and to the letter *vav*, which represents Joseph—the masculine force. Thus, 137, having both feminine and masculine properties, features in marriage (Chapter 11). In the marriage between Torah and science, Torah is the masculine element and science is relatively female.

God appears to us in the Torah with two general Divine Names. The essential Divine Name, *Havayah*, is related to Divine Providence (Chapter 15). It expresses God's absolute unity (the tip of the *yud*), which develops into two (the first two letters of *Havayah*, which spell the Divine Name, *Kah*) and then four (the four letters of *Havayah*), following the formula of 2^n (Chapter 5). ◄

The Divine Name *Elokim* is associated with the Divine power of creation that is embedded within natural causation. This Name is in the plural, relating to the duality apparent in nature (Chapter 2).

Revealing the inherent unity of the Giver of the Torah to the Jewish People (*Havayah*) and the Creator of nature (*Elokim*) will catalyze the discovery of scientific treasures that currently remain veiled from our knowledge. The combination of these two Divine Names is called "the Complete Name." As mentioned in Chapter 2, the numerical

▶ In the Book of Job, Elihu ben Berachel tells Job, "For God answers in one way and in two, to one who does not see it" (כִּי בְאַחַת יְדַבֶּר אֵ־ל וּבִשְׁתַּיִם לֹא יְשׁוּרֶנָּה).[1] The numerical value of the entire verse is 2,048, which equals 2^{11}. This is the secret of the "void" (בֹּהוּ). The numerical value of the first letter of "void" (ב) is 2 and the numerical value of the last two letters is 11, alluding to 2^{11}.

The numerical value of "void" (בֹּהוּ) is 13, which is the numerical value of "one" (אֶחָד).

The numerical value of "in one way" (בְּאַחַת) is 411, which is the numerical value of "chaos" (תֹּהוּ). Here we see a clear indication of chaos and void as a pair, in which the void is the initial stage of rectifying chaos (see Chapter 2).

value of this idiom, "the Complete Name" (שֵׁם מָלֵא) is 411, which equals 3 times 137. The classic phrase that combines these two Divine Names is "*Havayah* is *Elokim*!" (הוי' הוּא הָאֱלֹהִים).[2] In this phrase, apart from the two Names, there are four additional letters (הוּא הָ), which are a permutation of the four letters of the Goodly Name, *Akvah* (אי־הוה),▶ mentioned in Chapter 1 with reference to *da'at* (the *sefirah* of knowledge).▶▶ This Name is the key to simultaneously focusing on the spiritual aspects of reality and mundane reality. When the letters of this Name are spelled in full (אלף הי ואו הי), the numerical value of the filling is 137 (Chapters 1 and 2).▶▶▶

God created the earth in order that we inhabit it. This includes the *mitzvah* of procreation. The sages learn this principle from one verse in the Prophets that connects these two Names *Havayah* and *Elokim*, "For so said *Havayah*, Creator of heaven, Who is *Elokim*, Who formed the earth and made it, He established it; He did not create it for chaos, He formed it to be inhabited, 'I am *Havayah* and there is no other'" (Isaiah 45:18).▼▼▼▼

The Artist's Signature

In the Introduction to this book, we began with a metaphor of a unique jigsaw puzzle that, when put together, simultaneously reveals the properties of a still-life, a landscape and a portrait of the puzzle's assembler. The enigmatic picture that we see emerging on the puzzle is the reverse side that reveals the duality inherent in the Divine Name *Elokim*. The true picture that appears on the front fits together perfectly without any paradoxicality. This is the picture formed by studying the Torah. Becoming acquainted with this new picture is

---

▶ The Goodly Name *Akvah* (אי־הוה), does not appear explicitly in the Torah. It is derived in particular from the initial letters of the phrase, "The heavens and the earth" (אֵת הַשָּׁמַיִם וְאֵת הָאָרֶץ) at the beginning of creation.

▶▶ More specifically, the Name *Akvah* is associated with the five *chassadim* of *da'at*. The numerical value of the Name *Ekvi* (אהוי), associated with the five *gevurot* of *da'at*, is 22, the number of letters in the *alef-bet*.

▶▶▶ This spelling corresponds to the Divine Name *Sag* (סג), which has a numerical value of 63 (see p. 32). Beginning with 63 and adding another 137 each time [$f(n) = 137n + 63$] creates the following linear series:

63 200 337 474 611 748 885

The fourth number in this series is 474, the numerical value of *da'at* (דַּעַת). The next number in the series is 611, which is the numerical value of Torah (תּוֹרָה). The average value of each of the seven numbers is also 474, the numerical value of *da'at* (דַּעַת).

▶▶▶▶ The numerical value of the introduction to God's statement, "For so said *Havayah*, Creator of heaven, Who is *Elokim*, Who formed the earth and made it, He established it; He did not create it for chaos, He formed it to be inhabited" כִּי כֹה אָמַר הוי' בּוֹרֵא הַשָּׁמַיִם הוּא הָאֱ־לֹהִים יֹצֵר הָאָרֶץ וְעֹשָׂהּ הוּא כוֹנְנָהּ לֹא תֹהוּ בְרָאָהּ לָשֶׁבֶת יְצָרָהּ) is 3,836, which equals 28 [the numerical value of "strength" (כֹּחַ)] times 137. In this phrase the numerical value of "chaos" (תֹהוּ) is 411, which equals 3 times 137.

The numerical value of the statement with which God concludes this verse, "I am *Havayah* and there is no other" (אֲנִי הוי' וְאֵין עוֹד) is 234, which equals 9 times 26 (the numerical value of the Name *Havayah*). The numerical value of the entire verse is 4,070, which equals 110 times 37.

the key to putting together the pieces of the puzzle, which can then be viewed from either side.

The Divine Name, *Havayah*, is the signature of the Artist who created the jigsaw puzzle. It is encoded in every facet of both sides of the picture (as mentioned throughout our teachings).

The key that unlocks the ability to overturn the puzzle and see it from either side is Kabbalah and its numerical value, the number 137.

<hr>

REFERENCE NOTES FOR CHAPTER 16

1. Job 33:14.
2. I Kings 18:39.

Psalm 137

It is the custom to precede the blessing after eating bread with one of two chapters from Psalms. On weekdays when the *Tachanun*▸ prayer is said, we say psalm 137 and on Shabbat and holidays and certain special occasions, we say psalm 126.▸▸

The reading of these psalms after the eating of bread relates to the verse discussed in Chapter 4, "Not by bread alone does man live, but by all that originates from the mouth of *Havayah* does man live."[1]

These two psalms describe the travels of the Jewish People from their origin in Zion to exile in Babylonia (as described in Psalm 137), and their return from exile to rejoice in the redemption of Zion (Psalm 126). This also relates to the verse, "Moses wrote the origins of the journeys … and these are the journeys to their origins."[4] Thus, both these psalms relate to "origin" (מוֹצָא), which has a numerical value of 137.

The division of the Book of Psalms into 150 separate chapters is the most accurate division.▸▸▸ The ordinal number of Psalm 137 is self-referenced in the Psalm.

▸ *Tachanun* is a prayer of supplication in which we confess our sins and ask for God's forgiveness. It is said on regular weekdays.

▸▸ There were some great Chassidim, such as Rabbi Hillel of Paritsch, who always recited Psalm 126 before the blessing after meals.

The numbers 137 and 126 are two of the fillings relating to the Name *Akvah*, as mentioned in Chapter 8.

▸▸▸ Originally the Book of Psalms was divided into 147 chapters/psalms, corresponding to the lifespan of Jacob.[2] Later they were reorganized into the current division of 150 chapters. [3]

There, by the rivers of Babylon, we sat down and wept when we remembered Zion.	עַל נַהֲרוֹת בָּבֶל שָׁם יָשַׁבְנוּ גַּם בָּכִינוּ בְּזָכְרֵנוּ אֶת צִיּוֹן:
Upon the willows we hung our harps, for there our captors demanded of us words of song and our tormentors mocked us, "Sing for us a song of Zion."	עַל עֲרָבִים בְּתוֹכָהּ תָּלִינוּ כִּנּרוֹתֵינוּ כִּי שָׁם שְׁאֵלוּנוּ שׁוֹבֵינוּ דִּבְרֵי שִׁיר וְתוֹלָלֵינוּ שִׂמְחָה שִׁירוּ לָנוּ מִשִּׁיר צִיּוֹן:
[But] how can we sing a song of God on foreign soil?	אֵיךְ נָשִׁיר אֶת שִׁיר הוי' עַל אַדְמַת נֵכָר:

If I forget you, O' Jerusalem, may my right hand forget [to play music], may my tongue cling to my palate, if I will not remember you, if I do not hold Jerusalem as my highest joy.

אִם אֶשְׁכָּחֵךְ יְרוּשָׁלָ͏ִם תִּשְׁכַּח יְמִינִי תִּדְבַּק לְשׁוֹנִי לְחִכִּי אִם לֹא אֶזְכְּרֵכִי אִם לֹא אַעֲלֶה אֶת יְרוּשָׁלַ͏ִם עַל רֹאשׁ שִׂמְחָתִי:

Remember, God, the day Jerusalem [fell], and the words of the sons of Edom: "Raze it, raze it, down to its foundation!"

זְכֹר הוי' לִבְנֵי אֱדוֹם אֵת יוֹם יְרוּשָׁלָ͏ִם הָאֹמְרִים עָרוּ עָרוּ עַד הַיְסוֹד בָּהּ:

O Daughter of Babylon, it is you who will be annihilated. Praiseworthy is he who repays you in kind for what you have done to us.

בַּת בָּבֶל הַשְּׁדוּדָה אַשְׁרֵי שֶׁיְשַׁלֶּם לָךְ אֶת גְּמוּלֵךְ שֶׁגָּמַלְתְּ לָנוּ:

Praiseworthy is he who seizes your infants and dashes them against a rock.

אַשְׁרֵי שֶׁיֹּאחֵז וְנִפֵּץ אֶת עֹלָלַיִךְ אֶל הַסָּלַע:

▶ Psalm 137 is unique in that it begins with the letter *ayin*, which seldom begins a verse. It also concludes with the letter *ayin*. In addition, the letter *ayin* appears another seven times in this psalm. The name of the letter *ayin* (עַיִן) means "eye." The *ayin* at the beginning and the *ayin* at the end allude to the verse in Isaiah, "For they shall see eye to eye when God returns to Zion" (כִּי עַיִן בְּעַיִן יִרְאוּ בְּשׁוּב הוי' צִיּוֹן).[5] Like a pair of doves who gaze constantly into each others' eyes, this symbolizes the ultimate love and compatibility between God and His nation. This will be actualized at the ultimate redemption when God's Divine Providence will be envisioned and experienced by the Jewish People. The seven letters *ayin* in the middle of this psalm refer to God's careful providence over us throughout the exile and at every point on earth, as the Prophet Zachariah states, "upon one stone, seven eyes."[6]

The final verse of this psalm is one of seven verses in the Bible that has a numerical value of 1,820, which equals 70 (ע) times God's Name, *Havayah* (26). It is also the exact number of times that God's Name appears in the entire Torah.

In this psalm, the experience of a Jew in exile is depicted in a profound and poignant manner. One of the most famous and moving verses in the chapter is, "If I forget you, O' Jerusalem, may my right hand forget [to play music], may my tongue cling to my palate, if I will not remember you." This verse, quoted by the groom at the peak of the marriage ceremony, recalls national grief even at a time of great personal rejoicing. Every Jewish home is a miniature temple and the moment that it comes into being is a reminder that the Temple to God remains in ruins. It is also an expression of hope that just as this new Jewish home is being built, so too, the Temple will soon be reconstructed. ◀

There are exactly 84 words in Psalm 137, one half of which is 42, the number of journeys that the Jewish People traveled from their origins. The forty-second word in the chapter is "my tongue" (לְשׁוֹנִי), one of the five origins of the mouth. The next word is "to my palate"

(לְחֵכִי). The palate is another origin of the mouth. The division of the psalm into two halves between the "tongue" and the "palate" indicates that the tongue will never cling to the palate, for we will never forget Jerusalem. The total numerical value of all the words until the forty-second word is exactly 13,700, which equals 100 times 137, the number of this psalm.▸

Chapter 126

A Song of Ascents. When Havayah brings about [our] return to Zion, we shall be like dreamers.

Then our mouths will be filled with laughter and our tongues with songs of joy; then they will say among the nations, "Havayah has done great things for them."

Havayah has done great things for us; we will rejoice.

Return, Havayah, our exiles like springs in the desert.

[Then] those who sow in tears will reap with joyous song.

The one who walks along weeping, carrying a bag of seeds will return with joyous song, carrying his sheaves.

שִׁיר הַמַּעֲלוֹת בְּשׁוּב
הוי' אֶת שִׁיבַת צִיּוֹן
הָיִינוּ כְּחֹלְמִים:
אָז יִמָּלֵא שְׂחוֹק פִּינוּ
וּלְשׁוֹנֵנוּ רִנָּה אָז יֹאמְרוּ
בַגּוֹיִם הִגְדִּיל הוי'
לַעֲשׂוֹת עִם אֵלֶּה:
הִגְדִּיל הוי' לַעֲשׂוֹת
עִמָּנוּ הָיִינוּ שְׂמֵחִים:
שׁוּבָה הוי' אֶת שְׁבִיתֵנוּ
כַּאֲפִיקִים בַּנֶּגֶב:
הַזֹּרְעִים בְּדִמְעָה בְּרִנָּה
יִקְצֹרוּ:
הָלוֹךְ יֵלֵךְ וּבָכֹה נֹשֵׂא
מֶשֶׁךְ הַזָּרַע בֹּא יָבֹא
בְרִנָּה נֹשֵׂא אֲלֻמֹּתָיו:

The two complementary phrases in the final verse of this psalm have a special structure. Just as the two actions of coming and going are a complementary pair, so too, the two emotions of weeping and joyous song are also a complementary pair.

Translated literally, the last verse states, "[he who] walks will be walking" (הָלוֹךְ יֵלֵךְ) ... "[he who] comes will be coming" (בֹּא יָבֹא). In each

▸ Another number that appears in this psalm is 317, which is an integer permutation of 137. There are exactly 317 letters in the entire psalm. The sum of the numerical values of the eight places mentioned in the psalm—Babylon (בָּבֶל, 34, twice), Zion (156, צִיּוֹן, twice), Jerusalem (596, יְרוּשָׁלַיִם, three times) and Edom (51, אֱדוֹם, once)—is 2,219, which equals 7 times 317. This number is also the total numerical value of the first and final words of the chapter, together with the numerical value of God's Name, which appears twice in the chapter. The mathematical relationship between 137 and 317 is discussed in Chapter 3.

The numerical value of the entire chapter is 55 (△10, the numerical value of הַכֹּל, all) times 441 (21²), which is the numerical value of "truth" (אֱמֶת). The sum of these two numbers (55 and 441) is 496 (△31), the numerical value of "kingdom" (מלכות). Rabbi Abraham Abulafia taught that one of the intentions of the numerical value of malchut (the sefirah of kingdom) is "all is truth" (הַכֹּל אֱמֶת).

▶ The sum of the numerical values of the root forms "walks" (הָלוֹך) and "comes" (בֹּא), is 64, the numerical value of "prophecy" (נְבוּאָה). The sum of the numerical values of the two future verbs, "will be walking" (יֵלֵך) and "will be coming" (יָבֹא) is 73, the numerical value of "wisdom" (חָכְמָה). Together, their sum is 137.

Each of these expressions is followed by "carrying" (נֹשֵׂא). In the first instance, the sower carries seeds and in the second instance, he carries sheaves. The numerical value of "carrying" (נֹשֵׂא) is 351, which is the triangle of 26, the numerical value of the essential Name of God, Havayah. Together, the numerical value of the two appearances of this word is 702 (the diamond of 26), the numerical value of Shabbat (שַׁבָּת).

instance the root form is employed together with the future tense, to describe an ongoing action. The sum of these two expressions is 137 (121 ⊥ 16), the fine-structure constant that defines the splitting of spectral lines.◀ The use of the origin of the verb together with the future tense, describes the descent of the soul from its origin on high through its many travels on earth, as mentioned previously regarding the phrase, "the origins of their journeys" (Chapter 12).

The Sum of Two Squares

The numerical value of each of the two expressions—"[he who] walks will be walking" (הָלוֹך יֵלֵך) and "[he who] comes will be coming" (בֹּא יָבֹא)—is a perfect square. The numerical value of the first is 121, which equals 11^2, and the numerical value of the second is 16, which equals 4^2. This phenomenon reveals to us that the number 137 is the sum of two squares. It assists us in finding qualitative meaning to a quantitative expression.

Any number can be represented by no more than four squares. A perfect square number represents the inter-inclusion of the number within itself. Such a number is rare. In contrast, a number that is equal to the sum of four square numbers is not a novelty. A number that is the sum of two squares is a relatively rare phenomenon. In the case of 137, the difference between the two square roots, 4 and 11, is 7. The general formula for this function is $f(n) = n^2 \perp (n \perp 7)^2$ (see Chapter 11) that produces an infinite quadratic series, which is symmetrical between $n = -4$ and $n = -3$.

n = -5	n = -4	n = -3	n = -2	n = -1	n = 0	n = 1	n = 2	n = 3	n = 4
29	25	25	29	37	49	65	85	109	137

Beginning from $n = -3$, 137 is the eighth number. One mathematical property of any series with such feminine symmetry, is that when the sum of the first thirteen numbers of the sequence is divided by 13, the eighth number is produced. In the case of this series, the sum of the first thirteen numbers of the series is 1,781, which equals 13 times 137. Thus, 137 is a significant and unique member of this series.

As mentioned in Chapter 11, the qualitative significance of this mathematical function alludes to two consecutive Shabbatot. The Talmud teaches that if the Jewish People correctly observe two consecutive Shabbatot with all their laws, they will immediately be redeemed.[7] This function also relates to the concept of circumcision, which is carried out on the eighth day after birth, so that the day of the circumcision is the same day of the week as the day of the birth. If the baby is born on a Tuesday, for example, the circumcision will usually be carried out on the following Tuesday.

When a baby is born on Shabbat (whether it is male or female), it's life takes precedence over the sanctity of the Shabbat. If necessary during the birthing process, Shabbat is desecrated. This is generally true of any situation in which human life is endangered on Shabbat. The sages' logic is that is preferable to desecrate one Shabbat in order that the individual live and observe many more Shabbatot.

On the Shabbat following the birth of a male child the Shabbat may be desecrated once more in preference of circumcision. This is a unique exception to the rule that Shabbat may only be desecrated in favor of an individual's life. The exception to a rule proves the innate principle that defines it. The obligation to circumcise a Jewish child on Shabbat highlights the essential union between Shabbat and the Jewish People. Circumcision on the Shabbat is the epitome of keeping Shabbat. Thus, the redemption of two Shabbatot depends on a state of marital fidelity, represented by the act of circumcision, which is even higher than Shabbat.

As mentioned, Psalm 126 is recited before the blessing after meals, on Shabbat in particular. The two Shabbatot are alluded to in this psalm in an unusual way that augments the significance of the above mathematical series. The phrase, "Return, *Havayah*, our exiles" (שׁוּבָה הוי' אֶת שְׁבִיתֵנוּ) is spelled in one way (שבותנו), but read in another (שְׁבִיתֵנוּ). An alternate reading of the spelling changes the meaning from "our

▶ The maximal Jewish lifespan in the Torah is that of Isaac (see Chapter 7) who lived to the age of 180. The most common lifespan in the Torah is 137. Thus, reciting this psalm with the correct intentions after eating bread, the word of God's mouth, is good advice for living either 137 or 180 years (or both!)

exiles" to "our Shabbatot" (שְׁבוּתֵנוּ).[8] This alteration changes the numerical value of the word, the verse and the entire psalm. The numerical value of the entire psalm with this one word as it is written together with the numerical value of the entire psalm with the word as it is read is 24,660, which equals 180 times 137. ◀

REFERENCE NOTES FOR APPENDIX A

1. Deuteronomy 8:3.

2. See *Yerushalmi Shabbat* 16:1, *Y. Sofrim* 16:5, *Midrash Tehilim* 22.

3. See *Likutei Sichot, Toldot* (1) p. 75-76, note 31, esp. in the later addition to the note on p. 76.

4. Numbers 33:2.

5. Isaiah 52:8.

6. Zachariah 3:9.

7. *Shabbat* 118b.

8. See *Midrash Aggadah* ch. 28.

3, 4 AND 137[1]

Robert Fludd was a leading physician in Britain in the first half of the 17th century. He believed that an idea is objective only if it is related to alchemy or occult mysteries. His beliefs brought him into conflict with his contemporary, Johannes Kepler. For Kepler, objective science must be proven quantitavely or mathematically.

Wolfgang Pauli, one of the pioneers of quantum theory, referred to the argument between Kepler and Fludd to illustrate different approaches to science. Surprisingly for a physicist, Pauli not only harbored sympathy for the geometric symmetry with which Johannes Kepler described the cosmos, but also took great interest in the mystical philosophies of Kepler's opponent, Robert Fludd.

One of the points of disagreement between the two was whether the most perfect number was 3, as Kepler saw it, or 4, as perceived by Fludd.

Pauli had been torn between these two numbers in his own work. He suggested that the lack of a fourth dimension in Kepler's philosophy was due to his disregard of the time dimension. In contrast, Fludd defended the "quaternary." This, Pauli equated with the addition of the observer in the formulas of modern science. Pauli himself derived the exclusion principle by allowing for four, rather than three quantum numbers. This was a break with convention which his contemporary, Bohr, father of quantum parity, admiringly described as "complete insanity." Similarly, we can identify a transition from three to four in the addition of the time dimension to the three dimensions of space in modern physics.

Pauli was also intrigued by the fine-structure constant and believed that its deepest property was that it unites the two simple principles of three and four.▸

Mathematically speaking, 3 is the first significant triangular number

▸ The first integer in which the digits 3 and 4 appear together is 34, the prime root of 137. There are 11 possible pairs of numbers that meet the equation $3x + 4y = 137$.

$x =$	3	7	**11**	15	19	23	27	**31**	35	39	43
$y =$	32	29	**26**	23	20	17	14	**11**	8	5	2

Of the eleven numbers, two contain the number 11 itself ($x = 11$, or $y = 11$). The pair of 11 is 26 or 31, respectively. Both these numbers are numerical values of the Divine Names (the essential Name of God, *Havayah* = 26; *Kel* = 31). The sum of 11 and 26 is 37, the companion to 137, as we have seen throughout this book.

▶ The third, and highest, type of revelation is called inspiration (*hashra'ah*) and is represented by inspirational, or interface numbers, which are the sum of two consecutive square numbers. The first inspirational number (after $1 = 0^2 + 1^2$) is 5 ($1^2 + 2^2$).[2]

▶▶ The numerical value of, "There are three" (שְׁלֹשָׁה הֵמָּה) is 685, which equals 5 times 137.

▶▶▶ Despite this quantitative difference, the names of the three Patriarchs contain 13 letters ($13 = 4 + 4 + 5$; אַבְרָהָם, יִצְחָק, יַעֲקֹב) and the names of the four Matriarchs also contain 13 letters ($3 +$ שָׂרָה, רִבְקָה, רָחֵל, לֵאָה; $13 = 3 + 3 + 4$). The numerical value of "one" (אֶחָד) is 13, as is the numerical value of "love" (אַהֲבָה). This indicates an inherent connection of unity and love between the two numbers, three and four, and the Patriarchs and Matriarchs, as expressed in the verse, "One to one will approach, and [no] spirit shall come between them."[4]

▶▶▶▶ The numerical value of the entire verse in its written form is 1,836 which is the proton-electron mass ratio. When added to the pronounced form of spelling, with the additional *hei* (וְאַרְבָּעָה לֹא יְדַעְתִּים), the sum is 1,841. The sum of these two numbers is 3,677, the mid-point of which is 1,839, the neutron-electron mass ratio, which equals 3 times 613, the number of commandments in the Torah.

and 4 is the first significant square number. In Kabbalah, a triangular number represents the revelation of Divine light through a gradual evolutionary process (*hishtalshelut*). A square number represents the revelation of Divine light through "enclothement" (*hitlabshut*), by which all elements in the set are inter-included within themselves.◀

The inherent connection between three and four is stated in Proverbs, "There are three that are wondrous to me and four that I do not know" (שְׁלֹשָׁה הֵמָּה נִפְלְאוּ מִמֶּנִּי וְאַרְבָּע לֹא יְדַעְתִּים).[3] The context indicates that this verse refers to the reality of nature. "Three" is "wondrous," which indicates that it corresponds to the super-conscious crown.◀◀ "Four I do not know" corresponds to the unknowable head at the apex of the super-conscious. Four is an expression of absolute uncertainty, such as the Heisenberg uncertainty principle.

We would expect both words in this verse to be of the same gender. However, "three" (שְׁלֹשָׁה) appears in the masculine form, and "four" (אַרְבָּע) is written in the feminine form (אַרְבָּע). Indeed, there are three Patriarchs and four Matriarchs.◀◀◀ Yet, in this verse "four" is vowelized, and therefore read, as the masculine version (אַרְבָּעָה). Thus, "four" is seen in the feminine form, but heard in the masculine form.

Seeing is a faculty related to wisdom, the father figure of the mind, while hearing is related to understanding, the mother figure of the mind. This manifests the quality of inter-inclusion in the two numbers. Inter-inclusion is also reflected in the fact that the phrase referring to "three" contains four words, while the phrase referring to "four" contains three words.

The final letters of the first four words of this verse (שְׁלֹשָׁה הֵמָּה נִפְלְאוּ מִמֶּנִּי) are a permutation of God's Name, *Havayah*, and their numerical value totals 26. This phenomenon merits attention in its own right. The sum of the numerical values of the final letters of the second phrase (וְאַרְבָּע לֹא יְדַעְתִּים) is 111, which is the numerical value of "wonder" (פֶּלֶא), a root that appears explicitly in the first phrase. Together, the numerical values of the seven ($4 + 3$) final letters—the seal of this verse—total 137.◀◀◀◀

More About Three and Four

The sum of the numerical values of "three" (שְׁלֹשָׁה) and "four" (אַרְבָּעָה) is 913, which is the numerical value of *Bereishit* (בְּרֵאשִׁית), the first word of Genesis. Another phrase, "three things together" (שְׁלֹשָׁה דְּבָרִים יַחַד) also has a numerical value of 913.▶ It is mentioned in the works of medieval Kabbalist, Rabbi Abraham Abulafia. The first word of both phrases, "three" (שְׁלֹשָׁה), is identical, therefore the numerical value of the second half of each phrase is also identical. This equality indicates the inter-inclusion of two (the minimum required to make a plural) "things" that become "four." The basic example in Kabbalah of this principle is in the male and female aspects of reality, which, when inter-included within one another produce the ultimate whole—the female within the male (2) and the male within the female (2), four in all.

▶ The initial letters of this phrase spell out the Divine Name *Shakai* (שדי).

The first verse in the Torah contains 28 letters. The method for finding the middle letters of a verse that has an even number of letters is to divide the number of letters by 2 until we reach an odd number that has a middle. In the case of this verse, $\frac{28}{2}$ is 14 and $\frac{14}{2}$ is 7. The middle of 7 is the number 4. The four middle letters, the fourth of each group, are the middle letters of the entire verse. In this unique case in the Torah, the four letters spell out a meaningful word, "three" (שְׁלֹשָׁה):

ב	ת	י	שׁ	א	ר	בּ
ם	י	ה	ל	א	א	ר
ם	י	שׁ	ה	ת	א	א
ץ	ר	א	ה	ת	א	ו

In the table below, we see how the very first verse of Genesis relates to the argument between Kepler and Fludd.

word in verse	numerical value	divisors	sum of divisors
בְּרֵאשִׁית	913	1, 11, 83, 913	1,008
בָּרָא	203	1, 7, 29, 203	240
אֱ־לֹהִים	86	1, 2, 43, 86	132
אֵת	401	1, 401	402
הַשָּׁמַיִם	395	1, 5, 79, 395	480
וְאֵת	407	1, 11, 37, 407	456
הָאָרֶץ	296	1, 2, 4, 8, 37, 74, 148, 296	570
		Total:	**3,288** (= 24 times 137)

▶ The sum of the divisors of *Bereishit* (בְּרֵאשִׁית, 913) alone is 1,008, which equals 42 times 24. The sum of all the other numbers is 2,280, which equals 95 times 24. This divides the number 137 into 42 and 95, as discussed in detail in Chapter 12.

The sum of 3, 4 and 137 is 144, which equals 12^2, where 12 is the product of 3 and 4.

666 and 137

Let's take the three numbers 3, 4 and 137 as the beginning of a quadratic series. We can discover the next numbers in the series by taking finite differences between the three numbers 3, 4 and 137:

$$\begin{array}{ccccccc} \mathbf{3} & \mathbf{4} & \mathbf{137} & 402 & 799 & 1328 & 1989 \\ & \mathbf{1} & \mathbf{133} & 265 & 397 & 529 & 661 \\ & & \mathbf{132} & 132 & 132 & 132 & 132 \end{array}$$

The base of this series is 132, which equals 3·4·11.

Since the sum of 3 ⊥ 4 is 7, we have developed the series to the seventh place.

According to a basic mathematical rule, the sum of all the numbers to the seventh place must be a multiple of 7. Indeed, the numbers total 4,662, which equals 7 times 666, which is the average of all seven numbers. The number 666 is equal to 18 times 37 (the companion number of 137). It is also equal to the triangle of 36.

666 has no apparent significance in science. However, because of its ostensibly sinister spiritual significance in certain non-Jewish religious texts, it is marked as a "cult-number" and some attempt to decipher it. Perhaps the most sinister thing about this number in Jewish teachings is that it is the numerical value of Shechem ben Chamor (שְׁכֶם בֶּן חֲמוֹר), who raped Leah's daughter, Dinah, and subsequently wished to marry her.[5]

On a more optimistic note, the numerical value of "advantage" (יִתְרוֹן) is 666. Thus, in general, this number has positive significance. "Advantage" (יִתְרוֹן) appears ten times in the Bible, each time in Ecclesiastes. One verse includes this word twice, "And I saw that wisdom has an advantage over folly, as light has an advantage over darkness."[6] (In Kabbalah, this is interpreted to mean that light has a greater advantage when it appears out of darkness.)▶

The above series is our first "witness" to a connection between 666 and 137. However, in order to set this relationship as fact, we need a second witness. The numerical value of "Let there be luminaries" (יְהִי מְאֹרֹת), the continuation of the creation of light, is 666.

The first property described by the inverse of the fine-structure constant relates to spectroscopy. When an atom makes a transition from one energy state to another, it emits light as spectral lines. These lines may be split into two or more finer lines, called the fine structure. The lines arise from the interaction of the orbital motion of an electron with its quantum mechanical spin. The amount of splitting is characterized by the fine-structure constant, which is nearly equal to $\frac{1}{137}$. This means that one quality of 137 is the splitting of one spectral line into two.

The verse in Psalms states, "God spoke one and I heard two, for God has might."[7] This is the classic reference to one unit that splits into two. It refers in particular to the ability to bear paradoxes. In the

▶ The numerical value of "wisdom has an advantage over folly, as light has an advantage over darkness" (יִתְרוֹן לַחָכְמָה מִן הַסִּכְלוּת כִּיתְרוֹן הָאוֹר מִן הַחֹשֶׁךְ) is 2,701, which equals 73 [the numerical value of "wisdom" (חָכְמָה)] times 37. The numerical value of the two words "advantage" (יִתְרוֹן) is 36 times 37 and the remaining value is exactly 37^2.

Zohar, the development from 1 to 2, then from 3 to 4 is the secret of the emanation of the ten *sefirot*: $1 \perp 2 \perp 3 \perp 4 = 10$, as seen in Chapter 3.

We can illustrate this mathematically with a series that begins with 1 and 2, then continues with 3, 4 and 137. Since there are five given numbers in this series (1, 2, 3, 4 and 137), we must follow it through with a four-dimensional formula, in which the highest value of n is n^4:

1	2	3	4	137	666
	1	1	1	133	529
		0	0	132	397
			0	132	264
				132	132

As in the previous series, the base of this series is also 132. But, here we see that 666 appears explicitly as the sixth number in the series, directly following 137. ◄ So here we have two different "witnesses" that identify the relationship between 137 and 666 through their connection to 3 and 4. ◄◄

▶ The sum of the first five numbers of this series is 147 (the lifespan of Jacob). Adding 666, the next number in the series, totals 813, which is the numerical value of three significant phrases in the account of creation: "And God said, 'Let there be light' and there was light'" (וַיֹּאמֶר אֱ-לֹהִים יְהִי אוֹר וַיְהִי אוֹר); "and God divided between the light and the darkness" (וַיַּבְדֵּל אֱ-לֹהִים בֵּין הָאוֹר וּבֵין הַחֹשֶׁךְ); "And God said, 'Let us make man'" (וַיֹּאמֶר אֱ-לֹהִים נַעֲשֶׂה אָדָם). The next (seventh) number in the series is 1,987, bringing the total to 2,800, which equals 7 times 400 (the average value of the seven numbers), which is the numerical value of the letter *tav* (ת), the final letter of the *alef-bet*.

▶▶ The difference between 666 and 137 is 529, which equals 23^2. 529 is the numerical value of "pleasure" (תַּעֲנוּג). See Chapter 12 for an explanation of the relationship between 23, 37 and 137.

REFERENCE NOTES FOR APPENDIX B

1. See also our book *913: The Secret Wisdom of Genesis*, Chapter 7.

2. See our book 913, *The Secret Wisdom of Genesis*. p.xiv; also our book in Hebrew, *Einayich Bereichot Becheshbon*, pp. 27-28.

3. Proverbs 30:18.

4. Job 41:8..

5. Genesis 34:1-4.

6. Ecclesiastes 2:13.

7. Psalms 62:12. See Chapter 15.

THE STORY OF JOB

Just as in every Biblical text, there are infinite secrets hidden in every nuance of the Book of Job. By contemplating it through the lens of our faith and through the teachings of the sages, we can discover profound Torah mysteries.

The Book of Job addresses the contradiction that is sometimes apparent between the acts of the righteous and their fate in this world. The sages teach us that the later chapters allude to even more esoteric secrets regarding the science of creation and the psychology of man. For this reason, it is fascinating to reveal that the number 137, the fundamental coupling constant of the physical world, appears in many places in the book.

In some communities, and in Yemenite communities in particular, the Book of Job is read on the Ninth of Av, the national Jewish day of mourning, commemorating the date on which both Temples were destroyed. This custom is appropriate because Job's suffering reflects the pain and sorrow of our national grief. And, in the Book of Job, we find answers to some of mankind's most poignant queries regarding the fate and destiny of man.

There are varying rabbinical opinions in the Talmud[1] regarding in which era Job lived. One opinion states that Job lived during the time of the Patriarchs. Another opinion, according to the Midrash, is that before he enslaved the Jewish People, Pharaoh consulted with his three advisors, Job, Jethro and Balaam. Jethro, who later became Moses' father-in-law and a convert to Judaism, advised Pharaoh not to enslave the Jewish People. The evil Balaam advised him to enslave them. Job abstained from offering his advice and remained silent. A third opinion states that Job lived during the Babylonian exile. Yet another opinion states that the story of Job is merely a parable and that he never existed at all.

The book begins with a description of Job, a righteous man who prospered in various ways. He had seven sons and three daughters, a huge herd of sheep, cattle, camels and donkeys, and abundant crops. At this point, the story tunes in to a unique Biblical dialogue between Satan and God. Satan has returned from wandering around the earth to see all the bad things that people are doing and he reports back to God. God asks him if he has noticed Job's righteousness. Satan counters God's question with a question of his own, claiming that Job is so comfortable in life that it's not surprising that he is faithful to God. Satan, who is powerless to do anything without God's permission, requests license to challenge Job's righteousness by confiscating all his prosperity. God agrees and, in a series of terrible calamities, all Job's possessions are taken from him in one fell swoop and his house collapses upon all of his ten children, who are killed at once.

Seeing that Job has not succumbed to cursing God for what happened, but actually blesses Him, Satan approaches God once more. He asks Him for permission to cause Job even more distress to see if he will remain faithful to God. Once again, God agrees. Then, Job becomes physically afflicted. He still does not curse God, but this time he does not bless Him either, but remains silent. His wife turns to Job and says, "Bless God and die!" The word "bless" in this context can imply the opposite of blessing. Indeed, many commentaries explain that Job's wife suggested that he curse God and by doing so, incur the death penalty and finally be released from his sufferings. However, Malbim, an important Biblical commentary, makes the unique observation that Job's wife meant that he should truly bless God once again in his agony, but she meant it sarcastically. She was troubled by the idea that if he would bless God again, as he did after the first four catastrophes, he would surely die this time for doing so. The sages teach that although he never says anything against God with his lips, he does sin in his heart.[2]

Job has three friends◄ who sense his terrible anguish from afar, and they come to visit and console him. Each of the three chastises him for

▶ The sages explain that Job's three friends experienced telepathic communication. They felt Job's affliction from a distance of 300 *parsaot* (approximately 1,200 km.). They were considered true friends, so much so that at the end of the passage discussing this matter, Rava declares, "This is what people [mean when they] say, 'Either friends like those of Job or death.'"[3] Elsewhere in the Talmud we find the phrase, "Either a partner or death,"[3] meaning that death is preferable to living life alone. Here, the phrase is much more potent, insinuating that death is preferable to a life without friends with whom one has a telepathic connection!

Telepathy is a classic example of the entanglement of souls. Entanglement of two elementary particles produces non-local instantaneous communication (telepathy) between them, even when they are at opposite ends of the universe.[5]

complaining against God's providence. After Job's three friends has each spoken, they are silent. A fourth friend, Elihu ben Berachel,▶ who has respectfully sat in silence all the while, then speaks. His words are a long discourse explaining to Job the real secret of apparently unwarranted suffering. Nachmanides interprets this to refer to the secret of reincarnation. Even if we are aware of everything we have done in our current lifetime, we can never know what our deeds were in previous lifetimes, or the effects of our primordial soul-source. Since this is so, in general, we cannot truly understand the effects of Divine Providence in current own lifetime. We can only strengthen our faith that everything that befalls us in this life is a rectification for some aspect of our soul, whether or not it is apparent.

Once Elihu has completed his discourse, God Himself appears directly to Job and poses fifty rhetorical questions about the secrets of creation.▶▶ The fifty questions are clues to the fifty gates of understanding. After hearing these fifty questions, Job admits that he really knows nothing and that he has no more questions.

God then rewards Job for his exceptional faith, and He tells Job's three older friends that they must offer sacrifices to atone for the pain that they caused Job with their chastisement of him; He also asks Job to pray for them. Finally, God doubles the prosperity that Job was granted before the catastrophes and blesses him once more with seven sons and three most beautiful daughters. At the end, we learn that Job loved his daughters so much that he granted them an inheritance together with his sons, and he himself lived contentedly for another 140 years.

The Speakers in the Book of Job

More than any other book in the Torah, the Book of Job is a book of conversations between various speakers. There are a number of human speakers, Job, his wife, his three friends and finally, Elihu,▶▶▶ the fourth friend, a mysterious messianic figure▶▶▶▶ who appears towards the end of the book.

▶ The numerical value of Elihu ben Berachel (אֱלִיהוּא בֶן בְּרַכְאֵל) is 358, the numerical value of Mashiach (מָשִׁיחַ). Ultimately, Mashiach is the only one who will offer an acceptable answer to why we suffer in exile and why the redemption is taking so long.

The Midrash[6] teaches that Mashiach will be born on the Ninth of Av. This relates to the reading of the Book of Job on that day and the unannounced arrival of the messianic figure of Elihu.

▶▶ In his introduction to *Sefer Yetzirah*, Ra'avad teaches that these questions are most profound questions concerning the Workings of Creation and the Workings of the Chariot, the most esoteric parts of the Torah.

▶▶▶ The sages interpret that Elihu's full name, "Elihu ben Berachel the Buzi from the family of Ram" (אֱלִיהוּא בֶן בְּרַכְאֵל הַבּוּזִי מִמִּשְׁפַּחַת רָם), refers to the family of Abraham. The initial letters of his names are the letters of Abraham (אַבְרָהָם).

▶▶▶▶ On two of the seven occasions that Elihu's name is mentioned in the Book of Job, it is spelled without the final *alef* (אֱלִיהוּ). This spelling of his first name alludes to the Prophet Elijah (אֵלִיָּהוּ), who will ultimately usher in the messianic era.

Of these human speakers, the only one whose name is not mentioned explicitly in the script is Job's wife. The Aramaic translation states that her name was Dinah.[7]

In Chapter 11 we saw that the numerical values of Job (אִיּוֹב) and Dinah (דִינָה) are the midpoints of 37 and 137, respectively. If we add together the numerical values of all the human speakers in this book, Job (אִיּוֹב; 19) and Dinah (דִינָה; 69), Job's three friends Elifaz (אֱלִיפַז; 128), Bildad (בִּלְדַּד; 40) and Tzofar (צוֹפַר; 376) and finally, Elihu (אֱלִיהוּא; 53), the total is 685, which is 5 times 137. Since Job and his wife are a couple, we will consider them here as one unit, in accordance with the Talmudic dictum that, "a man's wife is like his own body" (אִשְׁתּוֹ כְּגוּפוֹ).[8]◀ The total number of human speakers is thus five. In this case, the exact average of all five names is 137.

In addition to the humans who speak in the Book of Job, there are two non-human speakers—Satan and God Himself. God appears in two different guises. At the beginning of the book, God projects "harsh judgment," allowing Satan to afflict Job. At the end of the book, because of all his suffering, Job merits that God appears to him directly; this time, God projects "soft judgment." God then reveals to Job the science of creation[9] and the psychology of man[10] by means of posing fifty rhetorical questions.

The sum of the numerical values of Satan (שָׂטָן; 359) plus God (Havayah) in two guises (2 times 26) is 411, which equals 3 times 137. In a most remarkable way, the average numerical value of these three names, like the average number of the five human speakers, is also 137.

Job and the Fiftieth Gate of Understanding

The Talmud[11] states that Moses was the author of the Book of Job. The fact that this is the only book authored by Moses other than the Torah indicates its particularly holy status. Job's name appears in the book exactly 50 times without a prefix.◀◀ This corresponds to the fifty questions at the end of the book, which in turn, correspond to the fifty gates of understanding.

▶ The numerical value of "a man's wife is like his own body" (אִשְׁתּוֹ כְּגוּפוֹ) is 822, which equals 6 times 137.

▶▶ As mentioned, the numerical value of Job (אִיּוֹב) is 19. The sum of 19 ⊥ 50 (the number of times his name is mentioned) equals 69, the numerical value of the name of his wife, Dinah (דִינָה).

Job's name appears another 6 times with a prefix.

Moses, too, is closely connected to the number fifty. When he received the Torah on Mt. Sinai he merited all fifty gates of understanding. After the revelation at Sinai, when the Jewish People sinned with the Golden Calf, the fiftieth gate of understanding was removed from him and returned only at the moment of his passing.[12] The Zohar[13] explains that when the letter *nun* (נ), which has a numerical value of 50, representing the fiftieth gate, entered Moses' name *Moshe* (מֹשֶׁה), it was transformed into *Mishnah* (מִשְׁנָה), the index and foundation of the Oral Torah. The transition between the Written Torah and the Oral Torah occurred at the very last moment of Moses' life.

Job's greatest trial in this world was living on the verge of passing away from this world while still alive. Thus, the most appropriate moment for Moses to write this book was at the moment of his passing, when he merited the revelation of all fifty gates of understanding.

This corresponds to the opinion that Job never existed on our physical plane, except as a parable. The best time to perceive a true parable is at a moment when the consciousness is in a state of annulment, wandering through no-man's land, either before sleep, before wakening, or at the moment of death.

Chaotic Healing Energy

At the beginning of the Book of Job, his children die as the result of a windstorm.[14] At the end of the book, God appears out of a windstorm[15] and reveals to Job the fifty secrets of creation. As Rashi points out, God "smote him (Job) with a storm and with a storm He healed him."[16] This is an allusion to the two levels of God's Name, mentioned above. Through the first/lower level, indicating "harsh judgment," God permitted Satan to smite Job. Through the second level, God displayed His power to transform the same windstorm into a positive phenomenon.

A windstorm is a classic symbol of chaotic energies that break the vessels. The messianic vision is that the great lights of chaos enter into rectified vessels. This can be compared to the use of atomic energy for productive purposes rather than for warfare.

The final appearance of "chaos" (תֹהוּ) in the Bible is in the Book of Job, in the verse, "He stretched out the north over chaos, and suspended the earth over emptiness" (נֹטֶה צָפוֹן עַל תֹּהוּ תֹּלֶה אֶרֶץ עַל בְּלִי מָה).[17] The initial phrase, "He stretched out the north over chaos," relates to chaos, as is clear. The numerical value of the second phrase of the verse, "and suspended the earth over emptiness" (תֹּלֶה אֶרֶץ עַל בְּלִי מָה) is 913, the numerical value of the first word of the Torah, *Bereishit* (בְּרֵאשִׁית), "In the beginning." This is an indication that chaos preceded the initial act of creation.

The Temple was destroyed when the chaotic energies of God's Name (at the first/lower level) were unleashed via Nebuchadnezzar as punishment for the unrectified state of the Jewish People. It is our duty to harness that same chaotic energy to reconstruct the Temple. This is the special quality of the month of Av, the month of the lion, as the sages taught, "There rose a lion (Nebuchadnezzar) in the Month of the Lion (the month of Av) and destroyed the lion [Ariel (אֲרִיאֵל), a name for the Temple, from the word "lion" (אַרְיֵה) in Hebrew)." But in the future the rectification will be that, "there will come a Lion (God), in the Month of the Lion and will rebuild the lion (Ariel)."[18]◄ This Midrash captures the two levels of the Name of God. The first level works through the negative windstorm energy at the beginning of the Book of Job. The second level converts that same energy into infinitely positive energy as it becomes clothed in rectified vessels.

▶ The phrase quoted in the Midrash that describes God as a lion is, "A Lion roared, who does not fear Him" (אַרְיֵה שָׁאָג מִי לֹא יִירָא).[19] The numerical value of this phrase is 822, which equals 2 times "chaos" (תֹהוּ), or 6 times 137.

REFERENCE NOTES FOR APPENDIX C

1. *Baba Batra* 14a-16b.
2. *Rashi,* Job 2:11.
3. *Baba Batra* 16b.
4. *Ta'anit* 23a.
5. For more on Telepathy, see our article "Kabbalah and Telepathy," http://www.torahscience.org/communicat/kabbalah_and_telepathy.htm.
6. *Eichah Rabati* 1:51.
7. *Targum,* Job 2:9.
8. *Berachot* 24a.
9. The Kabbalistic subject of the Workings of Creation, referring to the study of the mysteries of science.
10. The Kabbalistic subject of the Workings of the Chariot, referring to the study of the mysteries of human psychology.
11. *Baba Batra* 14b.
12. *Sefer Shlah, Parashat Va'etchanan.* See also, *Zohar Shemot* 115a.
13. See *Tikunei Zohar* 14b.
14. Job 1:19; 9:17.
15. Ibid 38:1.
16. Rashi on Job, 38:1.
17. Job 26:7.
18. *Psikta Derabi Kahana* 13.
19. Amos 3:8.

GEMATRIA, FILLING AND SUBSTITUTION SYSTEMS

Gematria

Gematria (גִּימַטְרִיָא, numerical value) is the numerical value of Hebrew letters, words, or phrases. Comparing numerical values can reveal correlation between different concepts. The assumption is that numerical equivalence is not coincidental. God created the world with His "speech," thus, each letter represents a unique creative force. The numerical equivalence of two words reveals that they have a common origin.[1]

There are three *gematria* systems for individual letters:

1. **Absolute value (מִסְפָּר הֶכְרֵחִי)**, also known as **normative value**. Each letter is given the value of its accepted numerical equivalent. *Alef* (א, the first letter) equals 1, *bet* (ב, the second) equals 2, and so on. The tenth letter, *yud* (י), is numerically equivalent to 10 and successive letters equal 20, 30, 40, and so on. The letter *kuf* (ק), near the end of the alphabet, equals 100; and the last letter, *tav* (ת) equals 400.

   The "final" forms of the letters *chaf, mem, nun, pei* and *tzadik* (ך ,ם ,ן, ף ,ץ), are used when these letters conclude a word. Their numerical value is generally the same as the standard form of the letter. However, sometimes the numerical value of the final *chaf* is considered as 500, the final *mem*, 600, etc. Following this alternative system, the Hebrew alphabet becomes a complete cycle. The final *tzadik* equals 900 and the numerical value of the *alef* in the next cycle is 1,000. Indeed, the same spelling is used for *alef* (אָלֶף), the name of the letter, and *elef* (אֶלֶף), meaning "one thousand."

2. **Ordinal value (מִסְפָּר סְדוּרִי)**. Each of the 22 letters is given an equivalent from one to twenty-two.

3. **Reduced value** (מִסְפָּר קָטָן, *modulus* 9 in mathematical terminology). Each letter is reduced to one digit. For example, א (=1), י (=10), and ק (=100) all have a numerical value of 1; ב (=2), כ (=20), and ר (=200), all reduce to a numerical value of 2, and so on. Thus, the letters only have nine numerical values.

Letter	Normative value	Ordinal value	Reduced value
א	1	1	1
ב	2	2	2
ג	3	3	3
ד	4	4	4
ה	5	5	5
ו	6	6	6
ז	7	7	7
ח	8	8	8
ט	9	9	9
י	10	10	1
כ	20	11	2
ל	30	12	3
מ	40	13	4
נ	50	14	5
ס	60	15	6
ע	70	16	7
פ	80	17	8
צ	90	18	9
ק	100	19	1
ר	200	20	2
ש	300	21	3
ת	400	22	4

In the ordinal and the reduced systems, the numerical value of the five final letters are generally the same as the same letters in their regular form. However, they are sometimes given an independent value. For example, the final *nun* (ן) is considered 14 or 25. Similarly, its reduced value is 5 or 7.

Final Letter	Normative value	Ordinal value	Reduced value
ך	500	23	5
ם	600	24	6
ן	700	25	7
ף	800	26	8
ץ	900	27	9

4. The **Integral Reduced value (מִסְפָּר קָטָן מִסְפָּרִי)** is a fourth gematria system. Here, the total numerical value of a word is reduced to one digit.

If the value of the word exceeds 9, the integer values of the total are repeatedly added to each other to produce a single-digit figure. No matter which of the other three systems is used to calculate the numerical value of the word, the integral reduced value remains the same.

For example, "loving-kindness" (חֶסֶד) has three letters. The absolute or normative value of *chet* (ח) is 8, *samech* (ס) is 60, *dalet* (ד) is 4. The sum of these three figures is 72. Thus, the integral reduced value is 9.

The ordinal value of *chet* (ח) is 8, *samech* (ס) is 15, *dalet* (ד) is 4. The sum of these three values is 27. Thus, the integral reduced value is again 9.

Finally, the reduced value of *chet* (ח) is 8, *samekh* (ס) is 6 and *dalet* (ד) is 4. The sum is 18. Again, the integral reduced value is 9.

Another example, "grace" (חֵן) is often used an appellation for Kabbalah. Its absolute value is 58, or 708, when the final *nun* is 700.

Its ordinal value is 22 according to the normal reckoning, or 33 when the final *nun* is assigned a value of 25.

Its reduced value is 13, or 15 if the final *nun* is assigned the value of 7. In all three cases, the integral reduced value is either 4 or 6.

In the *Tikunei Zohar*, reduced value relates to *Yetzirah* (the World of Formation). Similarly, we can assign each of the four gematria systems to one of the four Worlds, which further correspond to the letters of God's Name, *Havayah*:

Letter of *Havayah*	Calculation Form	World
Yud	Absolute value	*Atzilut*
Hei	Ordinal value	*Beriah*
Vav	Reduced value	*Yetzirah*
hei	Integral reduced value	*Asiyah*

Milui

When the *milui* (מִלוּי, "filling"; pl. מִלוּיִם, *miluim*) is used, the numerical values of the names of the letters are considered. In such cases, the letters used to spell the main letter are "pregnant" letters.

For example, the name of the letter *chet* (ח) is spelled *chet-yud-tav* (חֵית). The letters *yud* and *tav* are pregnant in the *chet*. Similarly, the name of the letter *nun* is spelled *nun-vav-nun* (נון). The *vav* and the final *nun* are pregnant in the opening *nun*. The numerical value of "grace" (חֵן) when these two letters are filled is thus 418 (the numerical value of חֵית) plus 106 (the numerical value of נון). Thus, the value of "grace" (חֵן) with its *milui* is 524. The value of the *milui* alone is 466.

A second level of filling can be added by filling each of the letters of the filled spelling with their "pregnant" letters. This process of filling letters can theoretically be continued indefinitely. The second filling of the letter suffices for most purposes.

The numerical value of "grace" when filled with a second filling (חית יוד תו נון וו נון) is 1,068.

Certain letters (e.g., *hei*, *vav* and *pei*) have more than one possible *milui*. For example, the name of the letter *hei* (ה) can be spelled *hei-alef* (הֵא), *hei-hei* (הֵה) or *hei-yud* (הֵי). This results in a numerical value of 6, 10, or 15 respectively. Similarly, the name of the letter *vav* (ו) can be spelled *vav-vav* (וו), *vav-alef-vav* (וָאו), or *vav-yud-vav* (וִיו), resulting in a numerical value of 12, 13, or 22, respectively. Each different numerical equivalent represents a different aspect of Divine influence.

For example, when the *vav* in the second filling of "grace" is spelled with an *alef* (ואו), or a *yud* (ויו), the numerical value is 1,069, or 1,078, respectively.

Mispar Kidmi [Primordial (Accumulative, or Progressive) Value]

The numerical value of a letter is sometimes calculated using *mispar kidmi*. This value is the sum of the absolute values of the letters up to and including the letter under observation.

Letter	Mispar Kidmi	Letter	Mispar Kidmi	Letter	Mispar Kidmi
א	1	ט	45	פ	405
ב	3	י	55	צ	495
ג	6	כ	75	ק	595
ד	10	ל	105	ר	795
ה	15	מ	145	ש	1,095
ו	21	נ	195	ת	1,495
ז	28	ס	255		
ח	36	ע	325		

For example, the primordial value of "grace" (חֵן) is 36 (ח) plus 195 (ו), which equals 231.

Substitution Systems

An additional technique is the systematic substitution of one letter for another.

Three of the most common substitution methods are *Atbash*, *Albam* and *Atbach*.

1. *Atbash:* In this method, the last letter of the *alef-bet* (*tav*) is

substituted for the first (*alef*), the second-to-last (*shin*) is substituted for the second (*bet*), and so on. The name **atbash** is formed by the first two pairs: אַ"תְּבַּ"שׁ.

כ	י	ט	ח	ז	ו	ה	ד	ג	ב	א
ל	מ	נ	ס	ע	פ	צ	ק	ר	ש	ת

2. *Albam*: In this substitution method, the letters are divided into two groups and aligned in order opposite one another. The name **albam** is formed from the first two pairs: אַ"לְבַּ"ם.

כ	י	ט	ח	ז	ו	ה	ד	ג	ב	א
ת	ש	ר	ק	צ	פ	ע	ס	נ	מ	ל

3. *Atbach*: Here, a similar substitution to the **Atbash** transformation is performed, but the *alef-bet* is divided into three groups (corresponding to the numerical values of the letters as integers, tens, and hundreds) and arranged in reverse order. Again, the name **atbach** is formed by the first two pairs: אַ"טְבַּ"ח.

ת	ש	ר	ק		צ	פ	ע	ס	נ	מ	ל	כ	י		ט	ח	ז	ו	ה	ד	ג	ב	א
ק	ר	ש	ת		ת	כ	ל	מ	נ	ס	ע	פ	צ		א	ב	ג	ד	ה	ו	ז	ח	ט

When the final letters complete the hundreds-series, the last group is augmented to:

ץ	ף	ו	ם	ר	ת	ש	ר	ק
ק	ר	ש	ת	ר	ם	ו	ף	ץ

REFERENCE NOTES FOR APPENDIX D

1. See *Tanya, Sha'ar Hayichud Ve'ha'emunah*, chapters 1 and 12.

GLOSSARY

Note: all foreign terms are Hebrew unless otherwise indicated. Terms preceded by an asterisk (*) have their own entries. For further reading, see our book *What You Need to Know About Kabbalah*.

Abba (אַבָּא [Aramaic], "father"): the *partzuf* of *chochmah*.

Adam Kadmon (אָדָם קַדְמוֹן, "Primordial Man"): the spiritual World above *atzilut. Adam Kadmon* corresponds to the *sefirah* of *keter*.

Adar (אֲדָר): the twelfth month from *Nisan*.

Adni (אֲדֹנָי): The pronunciation of God's holy Name that relates to the *sefirah* of *malchut*. The correct pronunciation is used only when reciting a complete Scriptural verse or in liturgy, particularly as the alternative pronunciation of the Name *Havayah* outside of the Temple.

Arich Anpin (אֲרִיךְ־אַנְפִּין [Aramaic], "the long face" or "the infinitely patient one"): the external *partzuf* of *keter*.

Arizal (אֲרִיזַ"ל) Rabbi Isaac Luria [acronym for "the Divine Rabbi Isaac (Luria) of blessed memory" (הָאֱ-לֹקִי רַבִּינוּ יִצְחָק זִכְרוֹנוֹ לִבְרָכָה)] (1534-1572); a renowned Torah scholar and Kabbalist.

Asiyah (עֲשִׂיָּה, "Action"): the lowest of the four spiritual *Worlds, below *Yetzirah. Asiyah* corresponds to the *sefirah* of *malchut*.

Atbash: a letter substitution system by which the last letter of the *alef-bet* is substituted for the first, the second-to-the-last for the second, and so on (see Appendix D). The name *atbash* is formed from the first two letter pairs: א"ת ב"ש.

Atik (עֲתִיק) shortened form of *Atik Yomin*.

Atik Yomin (עֲתִיק יוֹמִין [Aramaic], "the ancient of days"): the inner *partzuf* of *keter*,.

Atika Kadisha (עֲתִיקָא קַדִישָׁא [Aramaic], "the holy ancient One"): in some contexts, this term is a synonym for *Atik Yomin,* in others, for *keter* in general.

Atzilut (אֲצִילוּת, "Emanation"): The fourth spiritual World, above *Beriah. Atzilut* corresponds to the *sefirah* of *chochmah*.

Av (אָב, "father"): the fifth month from Nisan.

Ba'al Shem Tov (בַּעַל שֵׁם טוֹב, "Master of the Goodly Name [of God]"): Title of Rabbi Israel ben Eliezer (1698-1760), founder of the Chassidic movement (see *Chassidut*).

Ba'al teshuvah (בַּעַל תְּשׁוּבָה, "one who returns"): 1. One who does *teshuvah* for a sin. 2. Generic term for someone who returns to adhere to Jewish law after a period of estrangement from tradition. Often used in contrast to a *tzadik*, who has not undergone such a period.

Beriah (בְּרִיאָה, "Creation"): The third spiritual World, above *Yetzirah* and below *Atzilut*. Beriah corresponds to the *sefirah* of *binah*.

Binah (בִּינָה, "understanding"): the third of the ten *sefirot*.

Brit (בְּרִית, "covenant") 1. *Milah*, 2. The ceremony at which *milah* is performed. 3. Euphemism for the male reproductive organ.

Chabad (חַבַּ"ד) acronym for *chochmah*, *binah*, *da'at*: 1. the three intellectual *sefirot*; 2. the branch of *Chassidut* founded by Rabbi Shneur Zalman of Liadi (1745-1812), emphasizing the role of the intellect and meditation in the service of God.

Chanukah (חֲנֻכָּה, "Dedication"): the eight-day holiday celebrating the victory over Hellenism and the restoration of the *Temple* after its defilement by the Greeks.

Chashmal (חַשְׁמַל, "glow," "fire"): a key, mystical word in the vision of Ezekiel interpreted to mean "silent speech" or "silence-circumcision-speech."

Chassidut (חֲסִידוּת, "piety" or "loving-kindness"): 1. Behavior that goes beyond the letter of the law. 2. The movement within Judaism founded by the *Ba'al Shem Tov*. 3. The teachings of this movement.

Chayah (חַיָּה, "living one"): the second highest of the five levels of the soul, above *neshamah* and below *yechidah*.

Chesed (חֶסֶד, "loving-kindness"; pl. *chasadim*; חֲסָדִים): 1. the fourth of the ten *sefirot*. 2. a manifestation of this attribute, specifically in *da'at*.

Cheshvan (חֶשְׁוָן): The eighth month from *Nisan*.

Chochmah (חָכְמָה, "wisdom"): the second of the ten *sefirot*.

Da'at (דַּעַת, "knowledge"): the third *sefirah* of the intellect, counted as one of the ten *sefirot* when *keter* is not enumerated. It is the conscious manifestation of the *Keter*.

Elokim (אֱלֹהִים): The pronunciation of God's holy Name that relates to the *sefirah of *gevurah. The correct pronunciation is used only when reciting a complete Scriptural verse or in liturgy.

Elul (אֱלוּל): The sixth Hebrew month from *Nisan.

Gematria (גִּימַטְרִיָא, numerical value [Aramaic]): the technique of comparing Hebrew words and phrases based on their numerical values.

Gevurah (גְּבוּרָה, "might"): the fifth of the ten *sefirot.

Havayah (הוי'): the pronunciation of a permutation of the four letters of the Tetragrammaton ("four-letter Name"). The Divine Name of God associated with the *sefirah of *tiferet. This Name may only be pronounced in the Holy Temple. When reciting *Torah verses or liturgy, it is read as if it were the Name *Adni.

Also referred to as "the essential Name" (שֵׁם הָעֶצֶם), "the unique Name" (שֵׁם הַמְיֻחָד), and "the explicit Name" (שֵׁם הַמְפֹרָשׁ).

Hod (הוֹד, "acknowledgment," "thanksgiving"): the eighth of the ten *sefirot.

Ima (אִמָּא [Aramaic], "mother"): the *partzuf of *binah.

Iyar (אִיָּר): The second month from *Nisan.

Kabbalah (קַבָּלָה, "receiving" or "tradition"): the esoteric dimension of the Torah.

Keter (כֶּתֶר, "crown"): the first of the ten *sefirot. *Keter is the super-conscious power of the soul, *da'at is the conscious dimension.

Kislev (כִּסְלֵו): The ninth month from *Nisan.

Kohen (כֹּהֵן, "priest"; pl. kohanim, כֹּהֲנִים): a descendant of Aaron (the brother of Moses). The kohanim officiate in the Temple rites.

Levi (לֵוִי, "levite"; pl. leviim, לְוִים): a descendant of Levi who is not a *kohen. The leviim perform assistive functions in the Temple rites.

Lubavitch (לִיוּבַּאוִויטְשׁ, "City of Love" [Russian]): 1. the location of the *Chabad movement from 1812 to 1915. 2. the *Chabad movement.

Magid of Mezeritch, (הַמַּגִּיד מִמֶּזְרִיץ', "The Preacher of Mezeritch"): Rabbi Dov Ber (1704-1773), the second leader of the *Chassidic movement after the *Ba'al Shem Tov.

Maimonides: see *Rambam.

Malchut (מַלְכוּת, "kingdom"): the last of the ten *sefirot.

Mashiach (מָשִׁיחַ, "anointed one," "messiah") also, **Mashiach ben David**

379

(מָשִׁיחַ בֶּן דָּוִד, "*Mashiach* the son of David"): the descendant of King David who will reinstate Torah-ordained monarchy, rebuild the Holy *Temple, and gather the exiled Jewish people to their homeland.

Mashiach ben Yosef (מָשִׁיחַ בֶּן יוֹסֵף, "*Mashiach* the son of Joseph") the predecessor to *Mashiach*, who will rectify certain aspects of reality in preparation for *Mashiach (ben David)*.

Mazal (מַזָּל, "sign"; pl. מַזָּלוֹת, *mazalot*): 1. a spiritual channel of Divine benevolence (from the Hebrew root "to flow" [נ-ז-ל]). 2. specifically, the thirteen tufts of the "beard" of *Arich Anpin*. 3. a physical embodiment of such a spiritual conduit, such as a star, planet, constellation, etc. 4. specifically, the twelve constellations of the zodiac.

Menorah (מְנוֹרָה): 1. the seven-branched candelabra in the sanctuary of the Holy *Temple in Jerusalem. 2. The eight or nine-branched candelabra used to light candles on *Chanukah (in Modern Hebrew, *Chanukiah*, חֲנוּכִּיָּה).

Midot (מִדּוֹת, "measurements"): a collective term for the six *sefirot* from *chesed to *yesod.

Midrash (מִדְרָשׁ, "seeking"; pl. מִדְרָשִׁים, *Midrashim*): the second major body of the oral Torah (after the *Talmud*), consisting of linguistic analyses of the Biblical text, especially in the form of narrative material.

Milah (מִילָה, "circumcision") or **brit milah** (בְּרִית מִילָה, "covenant of circumcision"): 1. the rite of circumcision, performed on a Jewish male, usually on the eighth day after his birth. 2. specifically, the first phase of this rite in which the foreskin is removed.

Mishnah (מִשְׁנָה, "teaching by oral repetition"): The *Mishnah* is the basic compendium of laws (each known as a *mishnah*) comprising the Oral Torah. It was elaborated on and redacted by Rabbi Judah the Prince in the second century CE.

Mitzvah (מִצְוָה, "commandment"; pl. מִצְוֹת, *mitzvot*): one of 613 commandments given by God to the Jewish people, or seven commandments given by God to the nations of the world, at Mt. Sinai. 2. one of the seven commandments instituted by the sages. 3. idiomatically, any good deed.

Mitzvot: plural of *mitzvah*.

Nefesh (נֶפֶשׁ, "creature," "psyche"): 1. the life-force of any living being; the lowest of the five levels of the soul.

Neshamah (נְשָׁמָה, "soul"): the third level of the five levels of the soul, above *ruach and below *chayah.

Netzach (נֶצַח, "victory," "eternity"): the seventh of the ten *sefirot.

Nisan (נִיסָן): The first Hebrew month.

Omer (עמר); 1. A measure of grain, 2. The 49 day period between the second day of *Pesach and *Shavuot.

Partzuf (פַּרְצוּף, "persona," "profile"; pl. פַּרְצוּפִים, partzufim): the rectified state of an individual *sefirah, in which it transforms into a complete set of intellectual and emotional powers. Unlike single *sefirot, partzufim can interact with one another. In their completely formed state, each partzuf contains a male and a female aspect.

Sefirah	Partzuf	
*keter	עַתִּיק יוֹמִין *Atik Yomin	"The Ancient of Days"
	אֲרִיךְ אַנְפִּין *Arich Anpin	"The Long Face"
*chochmah	אַבָּא *Abba	"Father"
*binah	אִמָּא Ima	"Mother"
Six emotive *sefirot, *midot	זְעֵיר אַנְפִּין Z'eir Anpin	"The Small Face"
*malchut	נוּקְבֵיה דִזְעֵיר אַנְפִּין Nukvei d'Z'eir Anpin	"The Female of the Small Face"

Pesach (פֶּסַח: Passover): 1. The sacrifice offered on the 14th of *Nisan, 2. The seven day festival between 15th-21st Nisan, during which it is forbidden to eat leaven bread.

Radla: see *Reisha d'lo Ityada*.

Rambam [רמבּ"ם, acronym of Rabbi Moses ben Maimon (רַבִּי משֶׁה בֶּן מַימוֹן)]: Maimonides (1135-1204), medieval scholar, physician and philosopher.

Rebbe (רַבִּי , "teacher" or "mentor"): 1. In the Talmud, Rebbe refers to a sage from the Land of Israel, 2. a term used to describe or address a teacher of Torah. 3. leader of a branch of the Chassidic movement.

Reisha d'lo Ityada (רֵישָׁא דְּלָא אִתְיְדַע) [Aramaic], "the unknowable head"): the highest of the three heads of *keter. Superconscious faith.

Rosh Chodesh (רֹאשׁ חֹדֶשׁ, "Head of the Month") : the first day of the Jewish month, a day of celebration.

Rosh Hashanah (רֹאשׁ הַשָּׁנָה, "Head of the Year") : the Jewish New Year, celebrated on the first day of *Tishrei.

Ruach (רוּחַ, "spirit"): The second level of the soul above *nefesh and below *neshamah.

Sages: see *Torah.

Sefirah (סְפִירָה, pl. סְפִירוֹת, *sefirot*): One of the channels of Divine energy or life force. Each *sefirah* also possesses an inner experience. The order of the *sefirot* and their inner qualities is depicted in the following chart:

Name				Inner experience
*keter	כֶּתֶר	"crown"	אֱמוּנָה תַּעֲנוּג רָצוֹן	1. "faith" 2. "pleasure" 3. "will"
*chochmah	חָכְמָה	"wisdom," "insight"	בִּטּוּל	"selflessness"
*binah	בִּינָה	"understanding"	שִׂמְחָה	"joy"
*da'at	דַּעַת	"knowledge"	יִחוּד	"union"
*chesed	חֶסֶד	"loving-kindness"	אַהֲבָה	"love"
*gevurah	גְּבוּרָה	"strength," "might"	יִרְאָה	"fear"
*tiferet	תִּפְאֶרֶת	"beauty"	רַחֲמִים	"mercy"
*netzach	נֶצַח	"victory," "eternity"	בִּטָּחוֹן	"confidence"
*hod	הוֹד	"splendor," "thanksgiving"	תְּמִימוּת	"sincerity," "earnestness"
*yesod	יְסוֹד	"foundation"	אֱמֶת	"truth"
*malchut	מַלְכוּת	"kingdom"	שִׁפְלוּת	"lowliness"

Shabbat (שַׁבָּת, "Sabbath"): the day of rest beginning sunset on Friday and ending at nightfall on Saturday.

Shavuot (שָׁבֻעוֹת, "weeks"): the holiday celebrating the wheat harvest and the giving of the Torah (cf. *Omer).

Shechinah (שְׁכִינָה, "Indwelling"): the Divine Presence in the universe.

Shekel (שֶׁקֶל, "weight"): a unit of weight of silver (and therefore of money), equivalent to approximately 12.3 gram.

Shema (שְׁמַע, "hear): a compilation of three Biblical passages (*Deuteronomy* 6:4-9, 11:13-21, *Numbers* 15:37-41) recited twice daily, beginning with this word, or sometimes, the first verse alone. The first verse is the fundamental profession of monotheism, "Hear O Israel, *Havayah* is our God, *Havayah* is one."

Shevat (שְׁבָט): The eleventh month from *Nisan.

Shofar (שׁוֹפָר, "ram's horn"): blown on *Rosh Hashanah*, and to be blown when the *Mashiach* arrives.

Sivan (סִיוָן): The third month from *Nisan.

Sukah (סֻכָּה, "hut"): hut resided in on the holiday of *Sukot*.

Sukot (סֻכּוֹת, "Huts"): the seven-day festival celebrated from 15th-21st Tishrei, during which we are commanded to reside in a temporary hut, whose roof is made of plant-matter.

Talmud: (תַּלְמוּד, "learning"): the written version of the greater part of the Oral *Torah, comprising mostly legal material, including much narrative and some mystical teachings.

The *Talmud* includes the *Mishnah* and its elaboration in the *Gemara* (גְּמָרָא, "completion").

Talmud Bavli ("The Babylonian *Talmud*"): the elaboration of the *Talmud* completed in Babylonia in the sixth century.

Talmud Yerushalmi ("The Jerusalem *Talmud*"): the elaboration of the *Talmud* completed during the third century in the Land of Israel.

Tanya (תַּנְיָא, "It has been taught"): the fundamental work of *Chabad* *Chassidut*, written by Rabbi Shneur Zalman of Liadi (1745-1812).

Tamuz (תַּמּוּז): The fourth month from *Nisan.

Tefilin (תְּפִילִין, "phylacteries"): two small, black, leather cubes containing parchment inscribed with certain *Torah* verses. *Tefilin* are attached to the head and arm by adult men during weekday morning prayer.

Temple (or "Holy Temple"; Hebrew: בֵּית הַמִּקְדָּשׁ, "house of the sanctuary"): The central sanctuary in Jerusalem. The first Temple was built by King Solomon (833 BCE) and destroyed by the Babylonians (423 BCE); the second Temple was built by Zerubabel (synonymous, according to some opinions, with Nehemiah, 353 BCE), remodeled by Herod and destroyed by the Romans (68 CE); the third, eternal Temple will be built by *Mashiach*.

Teshuvah (תְּשׁוּבָה, "return"): repentance. See also **Ba'al Teshuvah*.

Tetragrammaton: see **Havayah*.

Tevet (טֵבֵת): tenth month from **Nisan*.

Tevunah (תְּבוּנָה, "comprehension"): the lower of the two secondary **partzufim* which develop from the partzuf of **Ima*.

Tiferet (תִּפְאֶרֶת, "beauty"): the sixth of the ten **sefirot*.

Tishrei (תִּשְׁרֵי): the seventh month from **Nisan*.

Torah (תּוֹרָה, "teaching"): 1. The Bible, especially the Five Books of Moses (also called "the Written Torah"), 2. The entire corpus of Jewish teachings derived from the Bible, including the "Oral Torah" (**Mishnah, *Talmud, *Midrash* and the **Zohar*).

Tzadik (צַדִּיק, "righteous individual"; pl. צַדִּיקִים, *tzadikim*): an individual who has overcome the evil inclination and has converted its potential into good.

Tzadikim: plural of **tzadik*.

Worlds: The four spiritual levels of consciousness **Atzilut, *Beriah, *Yetzirah* and **Asiyah* and the fifth level of **Adam Kadmon*.

Yechidah (יְחִידָה, "single one"): the highest of the five levels of the soul, above **chayah*.

Yesod (יְסוֹד, "foundation"): the ninth of the ten **sefirot*.

Yetzirah (יְצִירָה, "Formation"): the second lowest of the four spiritual **Worlds* above **Asiyah* and below **Beriah*. *Yetzirah* corresponds to the six **sefirot* of the **midot*.

Yom Kippur (יוֹם כִּפּוּר, "Day of Atonement"): the holiest day of the Jewish year, marked by fasting and **teshuvah*, particularly through confession of sin.

Z'eir Anpin (זְעֵיר אַנְפִּין [Aramaic], "the small face"): the **partzuf* of the **midot*, corresponding to the emotive faculties of the soul. In general, the concept of "finitude" or "finite power" is identified with *Z'eir Anpin*.

Zohar (זֹהַר, "Brilliance"): one of the basic Kabbalistic texts of the Oral **Torah*, recording the mystical teachings of Rabbi Shimon bar Yochai (2nd century). The Zoharic literature includes the *Zohar* proper, the *Tikunei Zohar*, and the *Zohar Chadash*

INDEX

rectification   32, 34-36, 44-45, 55, 58, 63, 64, 77, 85, 102, 108, 109, 111, 176, 189, 204, 261, 266, 274, 295, 365, 368

Rectification, World of   35, 36, 44, 58, 123

Redeemer, see *Mashiach*

redemption   7, 40, 78-79, 147, 183, 215, 268, 278, 280, 299, 313, 351, 352, 355, 365

relativity, see general relativity, special relativity

repentance, see *teshuvah*

reproductive system   326

righteous individual, see *tzadik*

Rishon Model   59

river of time   294-296, 298

Rogachover, see Ruzhin, Rabbi Yosef

run and return   227, 276, 287

Ruth   194, 267

Ruzhin, Rabbi Yosef   80

S

sabbatical year   297, 317

Samson   266-267

Samuel (prophet)   109, 125, 215

Sarah   27, 39, 188, 190-191, 193, 204, 213, 265, 278, 310

Satan   133, 168, 364, 366-367

Saul, King   127, 267-268, 287

scales   312-313, 321-323, 325, 327

scarab

Schneerson, Rabbi Menachem Mendel (the Lubavitcher Rebbe)   86, 101, 139, 160,

Schneerson, Rabbi Menachem Mendel (Tzemach Tzedek), see Tzemach Tzedek

Schneerson, Rabbi Yosef Yitzchak   181

Schneerson, Rebbe Dov Ber   211

School of Hillel   272

Schrödinger, Erwin   152-153, 197-198

Schrödinger's cat   197-199

science, integration of in education   105

scribe   15

seasons   40-41, 293, 297, 321

Seder night   79

Sefer Yetzirah   15, 35, 54, 157, 247, 263, 307, 312, 316, 318, 324, 365

sefirah, sefirot

   keter, crown (faith, pleasure, will)   26, 30-32, 41, 55, 56, 66, 95, 121, 123, 136, 160-161, 173-174, 180, 184, 190-191, 209, 225, 301, 309

chochmah, wisdom (selflessness)   26, 30-31, 38-39, 41, 55-56, 95, 121, 123, 136, 147, 159, 174, 190-191, 209, 212, 229, 309, 320

binah, understanding (joy)   26, 30-31, 39, 41, 55-56, 78, 95, 121, 147, 158-159, 174, 190-192, 208, 229, 283, 320

da'at, knowledge (unification)   30-32, 41, 116, 123, 190-192, 204, 208, 229-230, 275, 301

chesed, loving-kindness (love)   30-31, 55-56, 93, 95, 161, 174, 189, 191, 225, 247, 249, 252, 284-285

gevurah, might (fear)   30-31, 55-56, 93, 95, 161, 174, 189, 225, 284-285

tiferet, beauty (compassion)   30-31, 55-56, 93, 95, 174, 189, 192, 225, 239 283, 285, 320

netzach, victory (confidence)   30-31, 55-56, 70, 93, 95, 174, 189, 273, 284-285

hod, acknowledgment (thanksgiving)   30-31, 55-56, 70, 95, 174, 189, 191-192 273, 284-285, 324

MORE BOOKS BY RABBI GINSBURGH

GENERAL

What You Need to Know About Kabbalah

190 pages

The Hebrew Letters

Channels of Creative Consciousness
502 pages

The Wondering Jew

Mystical Musings & Inspirational Insights
284 pages

Kabbalah and Meditation for the Nations

216 pages

SCIENCE AND MATHEMATICS

913: The Secret Wisdom of Genesis

160 pages

137: The Riddle of Creation

400 pages

Wisdom: Integrating Torah and Science

216 pages

The Breath of Life

Torah, Intelligent Design and Evolution
186 pages

Lectures on Torah and Modern Physics

184 pages

The Torah of Life

Nutrition + Nervous System
44 pages

RAISING A JEWISH FAMILY

Consciousness and Choice

Finding Your Soulmate
284 pages

The Art of Education

Internalizing Ever-New Horizons
302 pages

The Mystery of Marriage

How to Find True Love and
Happiness in Married Life
500 pages

Gal Einai · www.inner.org

Psychology and Meditation

Anatomy of the Soul

144 pages

Transforming Darkness into Light

Kabbalah and Psychology
192 pages

Living in Divine Space

Kabbalah and Meditation
288 pages

A Sense of the Supernatural

Interpretation of Dreams and
Paranormal Experiences
208 pages

Frames of Mind

Motivation According to Kabbalah
256 pages

Health and Youthfulness

Body, Mind and Soul

Kabbalah on Human Physiology,
Disease and Healing
342 pages

The Twinkle in Your Eye

Kabbalistic Remedies for Preserving
Youth and Memory
202 pages

Leadership

Awakening the Spark Within

Five Dynamics of Leadership
that can Change the World
200 pages

Rectifying the State of Israel

230 pages

לזכות

ר׳ מיכאל בן רות

ר׳ יצחק בן לאה

ר׳ אליהו בן אביגיל

ומשפחתם שיחיו

ולעילוי נשמת

ר׳ משה ז״ל

בן ר׳ גבריאל רפאל ותמר ז״ל

Dedicated to
the everlasting memory
of

Miriam and Jerome Katzin

(מרים ושמואל יוסף הכהן קצין)

www.ingramcontent.com/pod-product-compliance
Lightning Source LLC
Chambersburg PA
CBHW082145150426
42812CB00076B/1911